CONTENTS

ROYAL GEOGRAPHICAL SOCIETY
WITH THE INSTITUTE OF BRITISH GEOGRAPHERS

EXPEDITION HANDBOOK

EDITED BY
SHANE WINSER

NEW EDITION

P
PROFILE BOOKS

First published in Great Britain in 2004 by
Profile Books Ltd
58A Hatton Garden
London ECIN 8LX
www.profilebooks.co.uk

Typset in Minion by MacGuru
info@macguru.org.uk

Printed in Great Britain by Biddles Ltd, www.biddles.co.uk

A CIP catalogue record for this book is available from the British Library.

ISBN 1 86197 044 7

INTRODUCTION

Shane Winser

Acquiring new knowledge about the world in which we live and sharing it with others for the benefit of us all is at the heart of the geographical exploration and research supported by the Royal Geographical Society (with the Institute of British Geographers) since its foundation in 1830.

Many of the great names of geographical exploration from the Victorian era are associated with the RGS–IBG and are part of its heritage: Livingstone, Burton and Speke in Africa; Scott and Shackleton in Antarctica; and more recently the Everest expeditions, to name but a few. This process of encouraging geographical research in partnership with teams from the host countries not only continues to the present day, but is also perhaps more extensive than at any time in the past.

The Society's Expedition Advisory Centre (EAC), funded by Shell International Limited, provides information, training and advice for anyone embarking on a scientific or adventurous expedition. It is the leading such centre in the world. Each year it assists more than 500 teams, ranging from modest research, conservation and community activities organised by school, youth and university groups, to more adventurous projects.

In addition to this advice, the RGS–IBG also provides financial support to geographical research projects. The Society awards in excess of £100,000 in grants, each year, to some 80 expedition teams and field research projects involving more than 400 individuals. In 2003 these teams will be visiting 47 different countries worldwide to explore the vast human, natural and physical dynamics of geography.

This new and fully revised second edition of the *Expedition Handbook* is designed to help anyone involved in an expedition or fieldwork project either as an organiser or member, whatever the purpose: exploration, research, education or discovery.

In particular the EAC aims to raise the safety, quality and effectiveness of all expeditions overseas. Through the *Expedition Handbook* we hope particularly to encourage expeditions to:

- undertake well-designed and appropriate fieldwork projects
- be safe and responsible to both people and the environment
- make a useful educational contribution to the participants and the host country
- share expedition results widely

The *Expedition Handbook* is divided into seven sections.

Section 1: Planning and organisation gives advice on how to turn your dream into reality, and the leadership styles you might want to adopt in putting together your team. There are greater opportunities than ever before for people of all ages and abilities to take part in expeditions and fieldwork, and a new chapter offers approaches to providing an inclusive experience for those with disabilities. The RGS–IBG particularly encourages host country participation on overseas projects; suggestions are provided on how this can be achieved. Practical issues relating to maps, navigation and catering are also addressed.

Section 2: Finance and fund-raising consists of three chapters suggesting a variety of ways to finance your chosen projects, starting with the budget and techniques for financial management, through to approaching grant-giving organisations and commercial companies for money and sponsorship. Charity fund-raising expeditions that try to raise money for worthy causes are now a regular occurrence and advice on these is also given.

Section 3: Safe and responsible expeditions is essential reading. Preparing a formal risk assessment is a prerequisite for any expedition and fieldwork, particularly when involving young people. There are chapters on the legal context to this, other aspects of expeditions, in particular copyright, and the recent UK government guidance on school visits, as well as vital information on insurance and expedition medicine. A new chapter on minimising your impact on both the environment and the people among whom you work reflects the Society's commitment to encouraging responsible expeditions.

Section 4: Expedition logistics provides practical tips from a team of world-class explorers, field scientists and travellers on living and working in mountains, hot deserts, tropical forests and polar environments, or undertaking caving or underwater expeditions.

Section 5: Field research is intended to suggest possible topics for field research projects in a variety of different environments. It begins with a step-by-step guide to designing a research project and gives specific advice on research in tropical forests and wetlands, arid lands and savannahs, the tundra and periglacial regions of the world.

Section 6: Expedition transport is about how to get the best out of 4 × 4 vehicles, bicycles, camels, horses and canoes to travel in remote and challenging environments.

Section 7: Recording your expedition provides advice on how to communicate your findings and experiences through photography, sound recording, radio broadcasts and film-making, and gives tips on lecturing and writing up your expedition report and scientific papers so that there is a lasting record of your achievements.

The diversity of the authors' approaches reflects the many different approaches to organising expeditions and fieldwork. However, we all share a common belief that the best way to learn about the world and all of its component parts – its geography – is to go out and see it for yourself. There is no better way to understand the complexity of the world than to learn from original experience as part of an international team. We hope this edition of the *Expedition Handbook* will help you achieve this.

Shane Winser

SECTION 1
PLANNING AND ORGANISATION

1 PLANNING AN EXPEDITION

Nigel Winser

> Projects, hopes and resolutions jostled in my brain clamouring for attention. I could not wander from day to day. I had to plan.
>
> PETER BOARDMAN, MOUNTAINEER

The start point of planning an expedition is simple. Initially you need three things: a piece of paper, a pencil and a quiet corner where no one will disturb you. Pause – and let your mind mull over the kind of project you want to undertake. This may be linked to your own interests and training or to your own passion for outdoor activities. Be it a school field trip, a gap year project or a university research expedition, think about the activities you want to undertake and the fieldwork you really would like to do. Is it to be an exploration or an adventurous journey? Solo or in a team? Is it to the nearby hills – or does a remote and challenging environment beckon? Or is it to be a geographical study that contributes to a better understanding of our world? In the A to Z of fieldwork today, the potential research topics are infinite.

While mulling over, begin to jot down a list of things that you would like to achieve, both personally and as part of a team. List the kind of people you would like to involve and work with. List what you think the tangible end products might be that would help you measure it as a success. New knowledge, new skills, a new language, new international friends, new cultures, new adventures? A scientific paper, an expedition report, a portfolio of stunning images, a published book, a television film? The longer the lists, the greater the challenge. Nevertheless, do make that first list. It's a cornerstone from which to build.

Then take a deep breath to assess honestly whether you have the skills to embark on such a commitment. Perhaps you might like to trim the list or tackle a lesser challenge. Give it careful thought. In your heart you will know when you are ready to accept the responsibility of organising a project of your own and whether your ideas are worth following. Share your plans with your family, your friends and other potential members of the team. Depending on their comments, review your list again!

As part of this analysis, you then need to address your responsibilities. Begin by listing who might want a share or have a stake in this great new plan of yours, because it is they to whom you will need to be accountable. They might include some or all of the following:

- your hosts and the local community with whom you will work, travel or stay
- your team, their partners and their families
- your organisation, school, university or research institution
- your sponsors and other supporters who will make the project possible
- your discipline – and any representation you might be making.

If this list of stakeholders becomes too heavy a burden for you to shoulder, pull out now before you start raising the hopes of others. If not, return to your list and review the objectives again. From here on, there are a number of ways to convert your list into a realistic plan to organise an expedition regardless of purpose, size, destination and cost.

AN AIM WITH SUPPORTING OBJECTIVES (Figure 1.1)

This is what you do next.

1. On your list of priorities, highlight those aspects of your project that are clearly the most important to you. Consult and discuss these with your team – or at least with potential members of the project. Those priorities that cause uncertainty relegate for the time being. From this jumble of ideas, try to form some semblance of order so that you can begin to focus on what you hope will be a single clear aim of what you want to do – your project.
2. After consultation, write down the aim of your expedition or project as a single clear statement – **the aim**. *The aim of my expedition next year is to ...* Then share this short statement with those currently involved with the project to see whether your initial idea can withstand some scrutiny. You will soon know when you have sufficient consensus to proceed!
3. Then from your list take those other objectives that you would like to achieve, time and funds permitting. These will be your **supporting objectives**. You may need to decide here which of these you may need to sacrifice to achieve the aim. Differentiating between the aim and the supporting objectives is a vital stage in your planning. The model in Figure 1.1 might help.

To help understand this critical point, look at the two following projects: one a mountain science project and the other a climbing expedition.

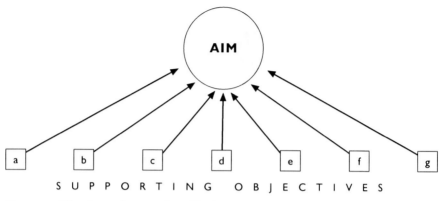

Figure 1.1 *The aim and supporting objectives*

1. A **mountain science** project might aim to take sediment cores in high-altitude lakes to determine long-term climate change patterns. Supporting objectives might be:
(a) to map the geology of the region;
(b) to make botanical collections for the local herbarium;
(c) to undertake physiological experiments for altitude sickness;
(d) to assess the impact of tourism;
(e) to make a photographic record for lecturing and website use;
(f) to make a video for educational purposes; and
(g) to climb a nearby mountain or two.
2. The aim of a **climbing expedition** may be to put at least two members on the top of a particular mountain and return safely. Supporting objectives might be:
(a) to try a new approach to the mountain;
(b) to climb some smaller peaks on the route in for training and to get fit;
(c) to undertake some geological/botanical collections;
(d) to undertake a medical survey:
(e) to study crafts in a local village;
(f) to write some popular articles; and
(g) to identify other mountain peaks in the area for future expeditions.

In both cases, if all goes well, all the objectives can be met. However, should there be a shortage of funds and time constraints, the supporting objectives can be altered slightly without jeopardising the aim.

To be able to write down your aim and the supporting objectives on one sheet of paper is a good position to be in early on in your planning, especially if you have a consensus!

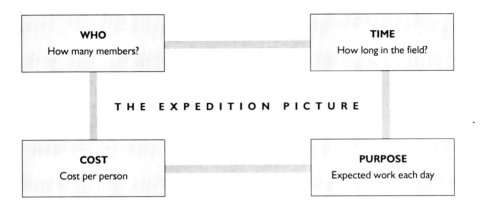

Figure 1.2 *The expedition picture*

THE EXPEDITION PICTURE (Figure 1.2)

Once you and the team agree your aim, you can start to develop your more detailed plans. In sharing your plans with others, it is helpful to have a clear mental picture of the kind of project that you have in mind, particularly concerning the four parameters of purpose, size, length and costs.

These four factors are the cornerstones of your expedition and it is helpful to agree these as early as possible, to avoid misunderstandings at a later date. For example, do you see your project as an international team of researchers taking sediment cores in a high mountain lake, or do you see yourself as a group of climbers summiting a new mountain peak? Both are laudable. In short, you are agreeing a brand image. This will help you with the planning and the fund-raising. More importantly it helps confirm that all members share the same clear vision of the type of expedition they are joining.

CRITICAL PATH PLANNING

Having tied down your expedition into a manageable form, you can now do your critical path planning. There are five steps here.

Appreciating the challenge

An early appreciation of the overall expedition plan is important even if the details may not be finalised. There are a large number of headings to consider and a glance at the planning checklists may help you to start grouping the issues you need to consider, so you can identify your priorities (see Appendix 1 for a sample planning checklist and Appendix 2 for a sample reconnaissance checklist).

Assessing the time spans involved

Make a list of tasks that need to be addressed and make an intelligent guess on how long it may take to execute each one, for example:

Appointing the team: numbers, skills, roles, responsibilities, host country members	2 months
Science plan: deciding on your field methodology	4 months
Logistics plan: working out how much water, food, fuel you need – where and when	1 month
Insurance: to research and then decide the level of insurance cover you need	2 months
Equipment: design, procurement, testing, trials, packing	6 months

Most tasks will take longer than you think. Unless you begin to delegate tasks, you will almost certainly run out of time, e.g. for your fund-raising plan, to identify the application dates for key grant-giving organisations. This will give you the deadline for when you will need your project plans in a presentable form ready for evaluation. The deadline for Royal Geographical Society–Institute of British Geographers (RGS–IBG) Expedition Research Grants is the end of January (for expeditions going into the field during the summer months), and the end of June for those going into the field over the winter months. So you will need to have tied down your detailed research plans, host country involvement and budget by one of these dates if you want to get RGS–IBG approval and funding. Each topic needs to be individually assessed.

Freighting of scientific equipment is another task that often catches people out because it requires a long lead-in time (I am assuming that you cannot afford air cargo on this occasion).

Equipment procurement	8 weeks
Packing	4 weeks
Delivery to docks	1 week
Sailing time	6 weeks
Clearance through customs	5 weeks
Transport to project site	1 week
Contingency (delays)	4 weeks
Total	**29 weeks (or 6.7 months!)**

Making up a flow chart

It can now be quite fun to fit together the various headings into a flow chart, with a list of the main headings going down the left side of the chart and the expected time-line for each set out across the page. This can start on the back of an envelope, and be upgraded to a spreadsheet.

Critical path timeline

If you are confident that the overall plan fits together, now prepare your own 'critical path' timeline to help you clarify key decisions and actions that need to be made by when, and which ones are governed by what deadlines (permission, funds, flights, etc.). Some are movable; some are not. Pulling all that together as a business plan will require some intelligent guessing. There are many models that you can develop, depending on the complexity and size of your project. An example of a proven critical path plan for a school expedition to Iceland can be seen in Appendix 3.

The detailed priorities list

In the hurly-burly of an expedition it is helpful to know what your current priorities are and which of the outstanding tasks require most of your time. A final checklist of tasks to achieve each week/month is extremely useful. An example of such a countdown, which can be used as a sample checklist, is in Appendix 4. This represents a typical school or university expedition covering 24 months with 12 months planning, 3 months in the field and 9 months writing up. It assumes that you already have a good idea of your aim, supporting objectives, team size and an outline budget, and that you have done some initial research to test the feasibility of your plan with your colleagues and other advisers. You will need to adapt it to your own timetable should you be planning a winter or Easter project. Do NOT follow the plan without considering your own criteria. Simply try to use it as a starting point for your own project.

Good luck with your planning. All geographical field projects, whatever their purpose, are very hard work indeed. So convince yourself that spending the next two years organising and running an overseas project will be worthwhile. They are not for the faint-hearted. It is your commitment and enthusiasm that will make your project a success and this will be driven by your motivation to achieve the aim you set yourself at the beginning.

FURTHER READING

Blashford-Snell, J. and Ballantine, A. (1977) *Expeditions the Experts' Way.* London: Faber & Faber.

Deegan, P. (2002) *The Mountain Traveller's Handbook: Your companion from city to summit.* Manchester: British Mountaineering Council.

Edwards, D. (2000) *Exploring New Frontiers: A guide to planning expeditions and team research projects in the field.* Glasgow: Royal Scottish Geographical Society.

Keat, W. (ed.) (2000) *Expedition Guide.* Windsor: The Duke of Edinburgh's Award Scheme.

Land, T. (ed.) (1978) *The Expedition Handbook.* London: Butterworths.

Lorie, J. (ed.) (2000) *The Traveller's Handbook.* London: Wexas.

Putnam, R. (2002) *Safe and Responsible Expeditions.* Newark, Notts: Young Explorers' Trust.

Young Explorers' Trust (2002) *YET Expedition Manual.* A web-only publication (www.theyet.org).

For a full list of RGS–IBG Expedition Advisory Centre publications call +44 20 7591 3030 or see www.rgs.org/eacpubs

2 LEADERSHIP AND TEAMWORK

Chris Loynes

A leader is best
when people barely know that he exists,
not so good when people obey and proclaim him
worst when they despise him.
Fail to honour people,
they fail to honour you.
But of a good leader who talks little,
when his work is done, his aim fulfilled,
they will all say, we did this ourselves.

LAO TZU (C. 450 BC) in HEIDER (1985)

… leaders are neither born nor made – they grow.

MARY COX

Expeditions are demanding experiences full of opportunities and challenges. These challenges are part of the reason why people choose to face unfamiliar situations such as difficult weather conditions or terrain, changes to travel plans or using new equipment in a strange place – all in the name of adventurous pursuit, scientific study or a community project. The many problems to be faced on an expedition can, of course, never be fully predicted. This uncertainty is part of the fun of expeditions, but it can also be the source of conflict and danger. It is the management of these diverse and changing factors that is the concern of the leader, a complex and dynamic role. Leadership is about working with people: to maximise the experience for everyone; to resolve problems that arise; and to get the job done safely and happily! This chapter briefly outlines some of the factors that contribute to effective leadership.

LEADERSHIP

Leaders are people who take on a particular role in relation to other people and 'leadership' is the set of behaviours that they use. The 'qualities' approach to leadership suggests that leaders possess certain attributes or qualities that make them suitable for leadership. It is probable that everyone possesses latent qualities that lend themselves to being a leader, which may or may not emerge depending on circumstance and inclination. When they do emerge, their effectiveness will depend on the confidence, experience and maturity of the individual.

Leaders, however, also require a range of specific skills to undertake the job. These can be described under three headings:

1. **Hard skills**, e.g.
   knowledge of human physiology
   knowledge of the environment
   technical skills related to the activity
   safety methods, planning and administration.
2. **Soft skills**, e.g.
   understanding of individual and group psychology
   communication skills.
3. **Meta-skills**, e.g.
   judgement, perception
   creativity
   problem-solving.

The way in which leaders blend their personal qualities with the various skills that they possess to respond to certain situations results in the leader's style. Style can mean the distinctive way in which someone always does things (rigid/set) or it can be seen as the way chosen at a particular time to cope appropriately with a particular situation (flexible). As expeditions involve such a diverse range of circumstances, a leader will be required to adopt a flexible approach involving a range of styles.

The key skill, then, for all expedition leaders is choosing what style of behaviour to use to respond to problems or resolve conflicts, i.e. deciding how to decide.

The issues

- The aim or focus of the expedition: although all leaders are concerned with both the accomplishment of a task and the development (needs) of the group members, it is important to be clear which of these is the main purpose of the expedition. This will affect the way decisions are made as well as their outcome.
- The stage of development of the group: as an expedition progresses, so the capability of the group develops and this will affect the style of decision-making and the role of the leader as time passes.

9

• The nature of the situation: a planning decision about logistics requires a different style of decision-making to that in an emergency. The situation and the skills and resources available to deal with it will be balanced by the leader, who can then select an appropriate style.

The focus of the expedition

The motivation to undertake an expedition will stem from one of two focuses:

1 to achieve a certain task
2 to provide an experience for a certain group of people.

This determines the focus of the leader. Although it is probable that both motives are involved in most expeditions, it is vital to be clear to yourself and your members which is the ultimate aim.

John Adair's action-centred leadership model (1988) helps to explore this (Figure 2.1). This model demonstrates the three main areas to which a leader must attend if a task is to be achieved with maximum satisfaction to all. Often, an inexperienced leader focuses his or her attention on getting the task done and is unaware that effective leadership requires him or her to consider the group and the individual as well. Without this attention, the task may never be completed because of such problems as a breakdown in communications (team) or low morale (individual). This model was developed for application in industry and for this purpose the sphere on top is the task area so, despite the need to attend to individuals and to group interaction, the task is the main focus. The job of industry is to make a profit. Without this, a firm cannot exist and the secondary outcomes of providing fulfilling work for the individual and achieving group cooperation will be impossible.

This orientation for the model applies to the expedition with the ultimate aim of achieving a task such as the ascent of a peak, or the undertaking of a particular piece of field research (i.e. task-focused expeditions). Expedition literature is full of examples of expeditions that that have failed to complete their task because individuals have put their personal ambitions before those of the team. There are also many examples of effective leadership, one of the most famous being Shackleton's commitment to saving his men shipwrecked in the Antarctic.

Many expeditions, particularly youth expeditions, have a different focus. For these it is appropriate to turn the model round so that the individual focus is at the top, i.e. the aim of the expedition is personal development (Figure 2.2).

In this case, working as a team to achieve a task is a means of completing the ultimate aim of personal development. The task and team will be chosen to suit the development of the individuals best. This may mean that the group members choose their own tasks and the leaders primarily act in support of these decisions. The leader must certainly be prepared to allow variation in the job being undertaken to suit the

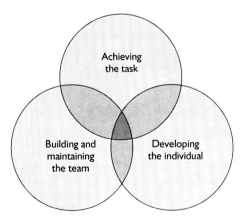

Figure 2.1 *John Adair's action-centred leadership model (Adair, 1988)*

needs of the personalities. In this case, it may be better to state the aim of the expedition as providing opportunities for mountaineering experience rather than the ascent of a certain peak. Perhaps the first decision a leader should make is whether the expedition focus is a specific task or personal development.

Example
A serious river crossing lies between the base camp and the expedition area.

- **Task focus:** the important thing is to gain access to the expedition area. The leader may decide to build a bridge to overcome the obstacle.

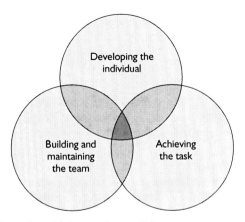

Figure 2.2 *Adaptation of model for youth expeditions*

- **Personal development focus:** the main need is to teach the members how to undertake their own river crossing so that independent decisions can be made safely later in the trip. A training session is organised and crossings supervised until the leader is satisfied that they are safe.

DECISIONS AND GROUP DEVELOPMENT

It is unusual in the field for people to be told to lead an expedition. Generally, they don't get given the job to do, but choose to undertake it because it was their idea. Ownership of the idea can be hard to let go because it often requires the commitment of one person to provide the energy to make the trip happen. This will inevitably be the person whose idea it was. However, if the trip is to be a success, the group needs to be involved and share in the leader's enthusiasm and commitment. He or she will need to share knowledge, skills and the control of decision-making.

Decision-making

Tannenbaum and Schmidt's (1968) model (Figure 2.3) is a simple way to look at control in the group. Although there are six steps described, these are intended to function as a continuum.

Tells

The leader assesses the situation, the group and their resources, selects a course of action and tells the group what to do. He or she may or may not consider what the group think and may or may not give reasons. Members do not participate in the decision-making. "This is the way we are going to cross the river."

Sells

As before, the leader makes the decision without consulting the group, but communicates it to the group giving reasons and explanations. "I want you to cross the river like this because …"

Tests

The leader identifies a problem and proposes a solution to the group, giving his reasons. He or she invites questions and discussions before finalising the decision. "How are you feeling about this river crossing?"

Consults

The leader presents the problem and relevant background information then asks members for their ideas on how to solve it. The leader then decides on the course of action. "Do you need any help to do this?"

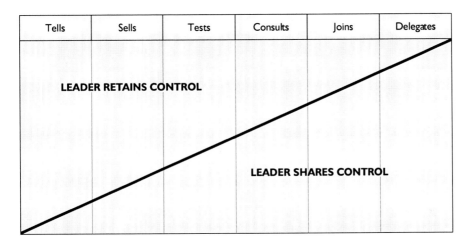

Tells	Sells	Tests	Consults	Joins	Delegates

LEADER RETAINS CONTROL

LEADER SHARES CONTROL

Figure 2.3 *Tannenbaum and Schmidt's (1968) decision-making continuum*

Joins
The leader identifies the problem and then joins in the discussion. The decision will be a joint one. "How shall we go about this?"

Delegates
The leader or a group member identifies the problem. The group are asked to make their own decision as to the best action to take. The leader may participate in the discussion and will support the decision once taken. "You sort out how to get across."

Sharing control leads to an effective team, well motivated and efficient, but the decision about which style to use also depends on the skills and resources available and the effectiveness of the group in using these and in working together.

Group development
Tuckman and Jensen (1977) identify four stages through which a group progress when they are brought together to complete a task: forming – storming – norming – performing. As they become acquainted with the problem and aware of each other's strengths and weaknesses, they become more effective.

Forming
The forming phase occurs as group members make initial contact with each other and the leader. It is characterised by a feeling of uncertainty. There is a high aware-ness of the leader as the group seek direction and endeavour to come to terms with the task and each other.

Storming

As the group gain in confidence, and control of decision-making begins to shift from the leader to the group, relationships go through a more searching and aggressive phase. The transition can result in conflict but the shift in control is essential if a functional group is to materialise. This is the "shake down".

Norming

The result of the storming phase is that the group learn to "rub along". Roles are defined, strengths and weaknesses recognised, and compromises renegotiated. The group have arrived at a shared goal and an acceptable culture to live by.

Performing

All the group's energy is directed at achieving the goal. Everyone is playing their full part and problems are tackled openly. This phase is often described as synergetic, i.e. the value of the whole is greater than the sum of its parts. Morale and achievement are high.

SITUATIONAL LEADERSHIP: A DYNAMIC APPROACH

The Hersey and Blanchard model, called situational leadership (Hersey and Blanchard, 1969), is a way to look at leadership style in relation to the development of a group (Figure 2.4). This includes their skill level as well as their ability to work together. The appropriate style adopted by the leader will help to ensure this development.

When a group forms, the leader's main concern is to familiarise them with the skills needed to undertake the task. In this situation, the focus will be on the task and "telling" is the most effective leadership style. This might equate with recruitment and the defining of roles in an expedition.

A "selling" style is adopted when the group start wanting to share control of the venture and the leader needs to explain the reasons for making certain decisions. The focus will now be on relationships as well as the task, a most demanding time. There is often conflict ("storming") but this is a necessary stage and an effective leader will tackle this positively. With some members of the team involved in the planning, this may well occur at home. With others it may well be in the field.

Once the team are familiar with their roles and have the skills and resources to undertake them, it is important for the leader to share the decision-making with the group. Participation involves testing, consulting and joining as styles and the leader's focus will move from the task to concentrate on relationships. On some expeditions leadership never progresses beyond this stage to ensure that the leader's experience is fully used.

An effective group will, after some experience, be able to perform and will gain the most satisfaction when decision-making is delegated and sharing control is at its greatest.

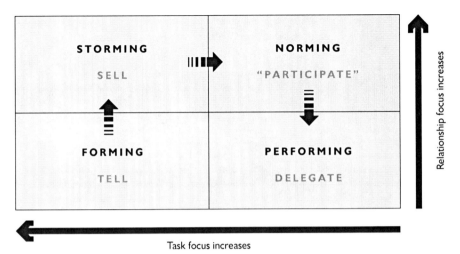

Figure 2.4 *Hersey and Blanchard's (1969) situational leadership model superimposed on Tuckman and Jensen's (1977) group model*

Notice the relationship between this model and Adair's spheres of task, team and individual. Here "individual" and "team" are combined in "relationship" and the model suggests a time sequence for which spheres, or group of spheres, the leader needs to attend to when taking account of the group development process.

Example
Tackling river crossings on an expedition in Iceland: on the first occasion the group were instructed (told) how to cross. At the next few crossings the leader explained (sold) the reasons for selecting certain sites and using certain methods. Groups then took increasing responsibility for these decisions (participated) until the leader allowed independent trips out (delegated).

Of course this is a simplistic model. An expedition is not one but many tasks and individuals vary in the skills that they possess and the rate at which they integrate into a team. Certain expedition tasks, e.g. safety, will always need to remain in the leader's control. On the other hand, the leader may never have control over areas of specialism in the team. In this case the specialist is the "leader" in this area. New and unexpected situations may throw an experienced team back to square one for a time and extended chains of communication may mean that only the leader has all the information to make decisions, despite the experience of the team. Nevertheless, it is a useful model for increasing a leader's awareness of a range of styles open to him or her and identifying some of the situations in which they might be used.

FURTHER INFORMATION

References

Adair, J. (1988) *Effective Leadership: a modern guide to developing leadership skills.* London: Pan Books.
Heider, J. (1985) *The Tao of Leadership.* Aldershot: Wildwood House Ltd.
Hersey, P. and Blanchard, K.H. (1969) *Management of Organisational Behaviour.* New York: Prentice-Hall.
Tannenbaum, R. and Schmidt, W. (1968) *How to Choose a Leadership Pattern.* Harvard Business Review.
Tuckman, B.W. and Jensen, M.A. (1977). Stages of small group development revisited. *Group and Organisation Studies* 2(4), 419-27.

Further reading
Barnes, P. (2002) *Leadership with Young People.* Lyme Regis: Russell House.
Langmuir, E. (1995) *Mountaincraft and Leadership.* Manchester: MLTB.
Loynes, C. (ed.) (1988) *Adventure Education and Outdoor Leadership* 5(1). Special edition devoted to outdoor leadership.
Morrel, M. and Capparell, S. (2002) *Shackleton's Way: Leadership Lessons from the Great Antarctic Explorer.* London: Penguin Putnam.
Ogilvie, K. (1993) *Leading and Managing Groups in the Outdoors.* Penrith: IOL.
Priest, S. and Gass, M. (1997) *Effective Leadership in Adventure Programming.* Leeds: Human Kinetics.
Ringer, T.M. (2002) *Group Action.* London: Jessica Kingsley Publishers.

Useful addresses
Basic Expedition Leadership Award, British Sports Trust, Clyde House, 10 Milburn Avenue, Oldbrook, Milton Keynes MK6 2WA. Tel: 01908 689180, email: admin@bst.org.uk, website: www.bst.org.uk/bel.html
Institute for Outdoor Learning, The Barn, Plumpton Old Hall, Plumpton, Penrith, Cumbria CA11 9NP. Tel: 01768 885800, fax: 01768 885801, email: institute@outdoor-learning.org, website: www.outdoor-learning.org
National Outdoor Leadership School (NOLS), 284 Lincoln Street, Lander, WY 82520-2848, USA. Website: www.nols.edu

Expedition and field leader's logbook
For those who want to develop their qualifications to lead an expedition, the RGS (with IBG) and the Young Explorers' Trust have published a logbook in the belief that leaders require varied kinds of experience in the field as well as formal training in technical and people (soft) skills. This logbook aims to:

- provide a framework for leaders wishing to develop their expedition skills with training and experience
- record this training and experience
- act as a reference for leaders and organisations wishing to assess the training and experience of the holder.

This logbook is intended to complement existing logbooks that expeditions might also use (e.g. *Mountain Leader Logbook*). The *Expedition Leader's Logbook* is 30 loose-leaf, punched colour pages. Download the A4 logbook or A5 logbook to suit your needs from www.rgs.org/je

Register of personnel available for expeditions

A register for those with skills to offer expeditions and fieldwork projects (e.g. doctors, nurses, mechanics, scientists) is maintained by the RGS–IBG Expedition Advisory Centre (EAC). Those wishing to join this register should ask for the appropriate form. Send us an email with your CV (email: eac@rgs.org).

Leaders looking for members for their own expeditions are welcome to consult this register at the EAC. If you are unable to visit London, please supply us with a detailed "job description". There may be a small charge to cover photocopying. Contact us with your requirements.

Bulletin of expedition vacancies

The EAC also publishes a bulletin of expeditions and fieldwork projects seeking members, often with specialist skills. If you wish to recruit members for your expedition, please send details of your requirements including dates, costs and skills needed. The bulletin is updated every two months.

3 INCLUSIVE EXPEDITION PRACTICE

Suresh Paul and Karen Darke

The standard image of the explorer or field scientist suggests that venturing into remote and challenging parts of the world is an activity only for those who are physically strong. However, this is far from the case and there are many individuals with a wide range of disabilities who are actively involved in expeditions and field-work. Examples include circumnavigation of the UK by sea kayak with a visual

Figure 3.1 *Iceland field trial (© Suresh Paul)*

Figure 3.2 *Up among the rigging on a Raleigh International expedition (© Paul Harris)*

impairment, journeying by electric wheelchair across Iceland, hand-cycling through central Asia, the Himalayas, and Iceland by paraplegic and amputee explorers, a range of Himalayan projects that have included people with learning disabilities in both planning and implementation stages, scientific fieldwork in the Andes – to mention just a few.

Building on these scattered inspirational examples, the first Disabled Explorers' Conference was held by the Royal Geographical Society (with the Institute of British Geographers or IBG) in 1995 and was ground-breaking in setting the tone and standard for inclusive practice in the expedition world. Also in 1995, the Disability Discrimination Act came into force in the UK, ensuring that the rights of disabled people are protected by law, and helping to remove both the physical and attitudinal barriers faced by disabled people.

Since then, there has been a noticeable growth in the number of expedition and fieldwork projects being undertaken in an inclusive format (involving both disabled and non-disabled members). There have been several successful expeditions with inclusive teams that have attracted media attention, and so raised the profile of what disabled explorers can achieve. In addition, expedition providers are increasingly open to the meaningful involvement of disabled people in their activities.

One noticeable area of development in inclusive outdoor opportunities in the UK has been through the "Adventure for All" group of outdoor centres, which have developed a range of opportunities for disabled people to participate in outdoor activities. There are also a number of initiatives that have focused their efforts on providing opportunities for specific disability groups: BackUp, focused on people with spinal cord injury, and Adventure Guide Dogs for the Blind, to name but a few.

National governing bodies of individual sports within the UK have also been working towards an inclusive mainstreaming agenda for a number of years. The concept of mainstreaming is simply making it possible for disabled people and other minority groups within society to be able to access sport as part of a mainstream club or at a standard facility, be it a club house, or outdoor or leisure centre.

The RGS–IBG now include a strong input from disabled explorers in their annual expedition planning conferences, supporting the inclusion of people with disabilities in all aspects of exploration. An "inclusive expedition" project being run at the RGS–IBG is challenging the barriers faced by disabled people when seeking full inclusion in expedition opportunities and offers a range of practical resources and support networks for disabled explorers and expedition planners.

This all sounds very impressive, and there have certainly been big changes in the last decade regarding inclusive opportunities, but there is still much that can be done.

KEY PRINCIPLE: A PEOPLE-CENTRED APPROACH

People are central to any project; being able to understand the needs of the team is arguably the key to the success of any team. Our understanding of people and their needs starts from our understanding of our past experiences and ourselves. It can therefore be daunting when considering the needs of people from a different cultural, racial or physical perspective or background. When organising an expedition or fieldwork project, each one of us faces specific barriers, which need to be overcome to make the project happen. To facilitate this, there needs to be a common language.

Language and terminology
Language is a powerful tool. The need is to ensure that the language used by the team is inclusive, both in the way in which it is used within the team and in the way the image of the project is presented to the outside world. Current terminology falls into two main categories: "people with disabilities" and "disabled people". Neither is wrong; intention and context are everything. Some of the models of disability summarised below may help in developing understanding and thus appropriate use of language.

Medical
In the past, disabled people were all too often considered to be a problem

requiring a medical solution, with individual personalities and achievements hidden, and the person being defined purely by the medical nature of their health state.

Social
In the social model of disability, the whole person is considered, with the emphasis on the removal of barriers and the view that it is the environment that creates the disability, not the individuals themselves.

Functional
The functional model is the favoured and current model used by the World Health Organisation. This leads on from the social model and accepts that barriers to participation are created by society. The aim is to understand the nature of a person's needs in a practical manner to ensure that positive actions can be taken to create new and progressive opportunities for all. There is no blame on an individual, environment or society. It encourages a partnership approach by providing a structure that promotes an understanding of an individual's impairment. Table 3.1 shows a comparison of the social and medical models of disability.

TABLE 3.1 COMPARISON OF THE SOCIAL AND MEDICAL MODELS OF DISABILITY

Social	Medical
Owned by society as a whole	Owned by an individual
Not preventable	Preventable
Solution is to eliminate discrimination	Solution is to find a medical cure
Disabled person is valued	Disability is a problem to be solved

Development of an understanding of a person's needs can be partly achieved by training and reading, but there is no substitute for working with the person concerned. An open, positive and direct approach is often most effective in achieving partnership, understanding and workable solutions to any challenges faced by the team.

Definitions of disability
Disability means the loss or limitation of social opportunities to take part in the normal life of the community on an equal level with others resulting from physical or social barriers (Barnes, 1992).

The word "impairment" is used where there is a functional limitation within the individual, the cause of which may be physical, mental or sensory. In broad terms, disability can be broken down into the following areas:

- learning disability
- educational and emotional disabilities
- sensory impairments – visual or hearing
- communication disabilities
- physical impairments.

BARRIERS TO PARTICIPATION

We all face problems or require additional support in particular areas to make it possible for us to participate in a particular project. The barriers that a disabled person will face when seeking full inclusion need to be considered at the earliest stages of planning and organisation – not just before you start the field phase. Most of the barriers to participation fall under following headings:

- **Environmental:** physical access to the urban, rural or wilderness environment (see access model below).
- **Attitudinal:** managing the attitudes of team members, any linked institutions such as schools or "controlling" bodies, and also external societal attitudes.
- **Legal:** check that any insurer is aware of the nature and make-up of the team. Be specific to avoid complications in the future should the worst happen.

Figure 3.3 *Model of access and inclusive participation (© Suresh Paul)*

A useful model of access and inclusive participation is illustrated in Figure 3.3. It involves three key factors:

1. **Presentation**: marketing of the project should include positive language and images.
2. **Preparation**: there should be open discussion with the individuals to ensure that needs and personal objectives are understood.
3. **Organisation and planning**: balance the needs of the project and the needs of all the individuals concerned.

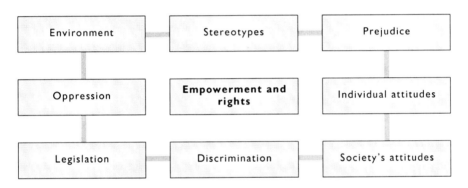

Figure 3.4 *The cycle of oppression*

Working with the "cycle of oppression" (Figure 3.4) is a helpful way to overcome barriers and create a positive inclusive experience for all concerned. Barriers can become cumulative, creating an ongoing negative experience for the individual and team unless the cycle is challenged in as many ways as possible.

PLANNING A SUCCESSFUL INCLUSIVE EXPEDITION

Expedition planning can be viewed as a matter of balancing the needs of the team, task and individuals, but often this is done without considering the social implications of the project or the background to the individuals concerned. Project planning is often seen as just a "straight-line process" either task led or team led.

Successful inclusive planning is indeed a matter of balancing the needs of the team, task and individuals. Consider planning as a circular process, with problem-solving, negotiating and researching, and then reassessing and redefining as the project evolves with the team. The need as ever is for good communication from the outset. This will lead to good planning and therefore a successful expedition.

Figure 3.5a *Karen Darke on the Interventure Canada to Alaska Sea Kayak Expedition (© Suresh Paul)*

Figure 3.5b *Alistair Hodgson on the Coppermine River Expedition, Canada (© Suresh Paul)*

The team

The need is always to treat the individual as an individual, but there is a requirement to consider how the needs of each of the team members interrelate. It can be useful to consider and discuss ways in which each of the team members is able to support the others. Consider also the balance of the team and the ability of the individuals to work in partnership (e.g. buddying) to ensure that both the project objectives and the individual and team needs are met.

Communication

The aim of communication is to ensure that the intentions of all are understood. This can be challenging enough with a group of individuals with differing life experiences, dialects and backgrounds. It can be even more daunting when there is an added dimension such as a group member with a disability. Prioritise the information that you wish to share and set up the appropriate environment. A range of approaches that allow the team members to focus, understand, absorb, reflect and react in a positive way should be used. To assist in this communication process, you may want to consider the following:

- the environment for the session and its accessibility to all of the team
- the use of practical sessions to demonstrate techniques (e.g. safe moving and handling)
- background noise, especially for team members with hearing impairments
- the use of colour coding or tactile marking (various textures or sensations) for

Figure 3.6 *Interventure Canada to Alaska Sea Kayak Expedition*
(© *Suresh Paul*)

team members with visual impairments
- if appropriate, learn Basic Sign Language
- structure and content of the session
- timing and duration.

Talking about physical issues

Physical impairments vary enormously, and so it is important to understand some of the specific ways in which an individual's disability affects them physically. It can be helpful to structure your communication about needs and the possible challenges that an individual and the team may have to face in a particular environment, and this also ensures that partnership and openness evolve with the project.

These conversations should consider factors such as muscle strength, endurance, levels of physical control, sensory function and the likely effects of the environment being visited.

Explore the project goals and ways of working that will ensure that all of the team are able to contribute and participate to their fullest. One of the hardest things is to decide that a particular option or activity is not suitable for all the team. However, if the options have been explored in practical terms, it can be much easier to say no without disappointment.

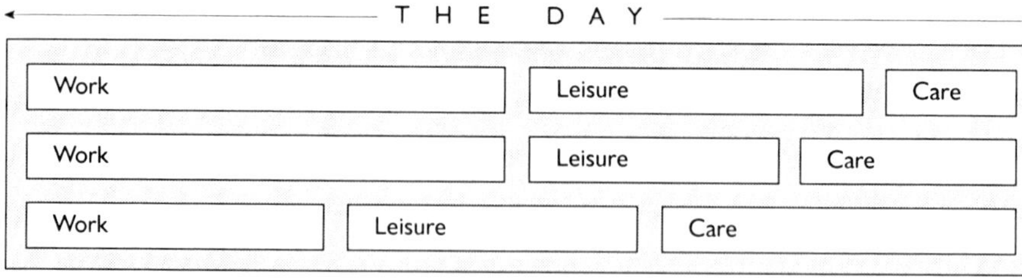

Figure 3.7 *Diagram of "The Expedition Day" (© Suresh Paul and Graham Kay)*

Time management
It is worth taking some time to consider the daily needs of the individual members to ensure that work and personal needs are balanced. Team members with a disability may need more time for personal care, so schedule activities so that all team members have time and space to cater for their own needs. This does not mean doing jobs for people. The need is for greater efficiency and flexibility by all to ensure that the team continues to function as a whole and is not split into subgroups. An understanding of everybody's needs within the team is important so that no one feels that they are wasting their time and not achieving. It can be tempting to force the pace or just get jobs done, but this achieves little. The team will develop only if it is able to adapt to ensure that all of the team is able to contribute.

Safe handling (Table 3.2)
In the field it can often be difficult for people with mobility impairments to move around independently. It can be tempting to address the problem by using the team to provide assistance, but this is not an optimum solution – the individual may feel dependent and it may also put the team at risk. Therefore, try to choose a field site that reduces the need for assistance and promotes independence.

This cannot always be achieved, and if the environment dictates a large amount of moving and handling take time to consider all the aspects of the process. In such a highly changeable environment it is difficult to offer absolute rules, but consider training the team in safe handling techniques and acquire any equipment that may be useful before you depart. When on location, always check, "Is lifting really required?", "Can I get some help?", "Share the load?" The following points may be useful to help ensure that your team works safely together.

TABLE 3.2 **SAFE HANDLING**

Plan – SAFE	Consider – LITE	Execute
Stop	**L**oad	**S**pine in line
Assess	**I**ndividual	**S**lide
Formulate	**T**ask	**B**alance the load
Execute	**E**nvironment	**R**isks

© Suresh Paul 2002

Equipment

The packing, storage and transportation of equipment should be made as simple as possible, to help team members as well as border guards and customs officials. Consider the following:

- The size and shape of individual packages to make them easy and safe to handle by as many team members as possible.
- Pre-expedition familiarisation and briefing with regard to more complex pieces of equipment.
- Colour coding and differentiation (between the background and object), numbering and texture to help identification of individual packages by any team members with visual or hearing impairments.

Note that colour and texture can be useful for people with hearing impairments because the appropriate use of colour can reduce the number of signs or simplify the language used by the team when talking about the equipment or organising packages.

Legal issues: consent and risk

The legal issues surrounding safe and responsible expedition practice are no different whether or not your project includes disabled people. However, some fear and confusion remain about the issue of consent. As with any expedition project, the need is for all team members to maintain control over the risks that they take and the hazards to which they are prepared to be exposed:

- Ensure that your research and preparation for the project identify all the potential hazards to the team members and the project. Assumptions and not confronting issues during the planning stage can only lead to trouble while in the field.
- When in the field it is important to ensure that the hazards within a given

operational environment are minimised and the risks to the team explained in such a manner that all concerned are able to understand.
- If, as a leader, you are in a position where you hold a higher duty of care, it is important that you explain your actions and hold discussions to ensure that the team is able to work with you to best effect.

If there is a need to simplify the situation or to ensure that it is possible to keep a situation under control, it can be useful to categorise risk. One model used in Outdoor Education is the "traffic light model", which can be explained and reinforced during the planning and preparation by the team:

- **Green:** very low or little risk, which allows all those involved in an activity to relax and accept a balanced but open challenge; this in turn allows the team the possibility for exploration.
- **Yellow:** medium risk activities; close to appropriate help if required.
- **Red:** areas or activities that, if not conducted in an appropriate manner, may cause harm and therefore require direct monitoring.

This model can be used as a basis of an exercise that helps the team develop a greater level of understanding of the nature of risk and what can be considered as successful risk assessment.

BENEFITS OF INCLUSIVE EXPEDITIONS

This chapter is written as a start point to help you prepare and implement an inclusive expedition. It has focused on possible tools to dissolve barriers and lead to a successful experience for all involved. It is worth remembering the many benefits of an inclusive expedition. A successful inclusive expedition is a powerful way of challenging social barriers, encouraging participation, promoting access to science and adventure, and removing stereotypes of disability.

The nature of the challenges encountered by an inclusive team often require creative thinking, innovative problem-solving and flexible team working which all contribute to the personal development of those involved.

MEDICAL ISSUES AND HEALTH CARE IN THE FIELD

Regardless of disability, the key principles of health care and first aid remain the same. Casualties in the outdoors will still suffer from the effects of the environment. The need is still for prevention. Good nutrition, equipment, appropriate clothing, good hygiene and sanitation are still the priorities. Do not replace your thinking and experience in this area, but build on first-aid principles and common sense, working with individuals during the preparation phase of the project to ensure that they are able to develop their own daily living strategies:

- Try not to make everything a medical matter.
- Work to ensure that a team member is able to maintain independence as far as possible.
- Ensure that you have the correct permissions and paperwork to carry any non-standard medication, dressings and medical supplies, and include letters from your consulate or country representative in the area in which you are staying.
- If an individual team member requires regular medication or equipment, try not to package all the supplies into one bag, to help ensure that an emergency is not created if the pack is lost.
- Package medical supplies appropriately.
- Consider the need for a grab-and-go bag containing the vital supplies for an individual team member should the worst happen and an evacuation is required.
- Test any specialist medical equipment under "simulated" environmental conditions before departure (e.g. for a polar expedition, do catheters freeze in very cold conditions?).

There are an increasing number of people in society who have a wide range of conditions that are not obvious at first glance. The need is to ensure that you know about the medical conditions of all of your team members. Use a well-worded and confidential pre-expedition medical questionnaire backed up by a one-to-one interview. This should allow you to explore issues such as coping with hidden disabilities like diabetes and epilepsy when in the field.

CONCLUSION

The key to successful inclusive expedition practice is understanding the needs of the team, task and individuals, and fostering open communication from the outset.
In the planning stage, remember to consider:

- sharing information and understanding team member requirements
- time management
- safe moving and handling
- grouping and buddying
- appropriate use of equipment
- thoughtful packing
- consent and risk assessment tools
- insurance.

Inclusion is a process and not necessarily an end result. Challenging barriers is a natural part of exploration. Vision is the key to good planning and inclusive practice is the key to visionary team work.

FURTHER INFORMATION

RGS–IBG Inclusive Expedition and Fieldwork Practice website

Since the first RGS–IBG Disabled Explorer's workshop in 1995 much has been achieved to make field science and exploration accessible to disabled people. A partnership between the RGS–IBG and Shell helped provide practical support and encouragement for inclusive fieldwork practice, under the guidance of Shell secondee Dr Karen Darke. This project was driven by legislative requirements surrounding access to education for disabled people. Further details and information sheets on this topic can be found at www.rgs.org/inclusive

Suggested reading

Barnes, Colin (1992) *Disabled people in Britain: a case for anti-discrimination legislation*. London: C. Hurst and Co.

Duke of Edinburgh's Award (2003) *Special Needs: Over to You*. The Award Scheme Ltd. www.theaward.org.

Gregory, W. (1996) *The Informability Manual*. London: HMSO.

Access

Centre for Accessible Environments (1999) *Designing for Accessibility Environments*. London: Centre for Accessible Environments.

Fieldfare Trust (2001) *BT Countryside for All*. Sheffield: British Telecom/Fieldfare Trust

Royal National Institute for the Blind (1995) *Building Sight*. London: HMSO.

Fieldwork

Geography Discipline Network: www.glos.ac.uk/gdn/disabil/
   Issues in Providing Learning Support for Disabled Students Undertaking Fieldwork and Related Activities.
   Providing Learning Support for Students with Mobility Impairments.
   Providing Learning Support for Blind or Visually Impaired Students.
   Providing Learning Support for Deaf or Hearing Impaired Students.
   Providing Learning Support for Students with Mental Health Difficulties.
   Providing Learning Support for Students with Hidden Disabilities and Dyslexia.

Inclusive canoeing

Smedley, G. (1995) *Canoeing for Disabled People*. West Bridgford, Notts: British Canoe Union.

Ripley, K. and Scandrett, S. (date of publication not known) *Signs for Canoeists*. Avon Deaf Children's Society, 8 Fairlawn Road, Montpellier, Bristol BSA6 5JR.

Wortham, A. and Zeller, J. (1990) *Canoeing and Kayaking for Persons with Physical Disabilities*. American Canoe Association, 7422 Alban Station Blvd., Suite B-232, Springfield, VA 22150, USA. Tel: +1 703 451 0141, website: www. acanet.org

Safety

Bailey, H. (1994) *Leisure Activities – Safety Guidelines*. SCOPE.

Putnam, R. (1994) *Safe and Responsible Expeditions*. Newark, Notts: Young Explorers' Trust.

Sports and activity coaching

Bremner, A. (1992) *Coaching Deaf Athletes*. Australian Sports Commission.
Goodman, S. (1995) *Coaching Athletes with Disabilities*. Australian Sports Commission.
Goodman, S. (1996) *Coaching Wheelchair Athletes*. Australian Sports Commission. Australian Sports
Commission, PO Box 176, Belconnen, ACT 2616. Tel: +61 262141111, email: asc@ausport.gov.au,
website: www.ausport.gov.au
Goodman, S. (1998) *Coaching Athletes with Cerebral Palsy*. Australian Sports Commission.
Hokey, K. and Goodman, S. (1992) Coaching *Athletes with Vision Impairments*. Australian Sports
Commission.
Nunn, C.J. (1994) *Coaching Amputee and Les Autres Athletes*. Australian Sports Commission.

Useful addresses

Adventure for All (AfA). Website: www.adventureforall.org.uk
    AfA is a group of leading residential outdoor activity centres that are primarily for people with
    disabilities or special needs.
Equal Adventure Developments Ltd, Glenmore Lodge, Aviemore PH22 1QU. Tel: +44 1479 861372,
    email: suresh@equaladventure.co.uk, website: www.equaladventure.co.uk
The Centre for Accessible Environments, Nutmeg House, 60 Gainsford Street, London SE1 2NY.
    Tel: +44 20 7357 8182, fax: +44 20 7357 8183, email: cae@globnet.co.uk
    The Centre for Accessible Environments is a charity that provides architectural information to make
    the built environment easier to use for people with a wide range of impairments – the team has
    produced a range of information sheets and publications for architects, planners and project
    managers.
English Federation of Disability Sport, Manchester Metropolitan University, Hassall Road,
    Alsager ST7 2HL. Tel: +44 161 247 5294, fax: +44 161 247 6895, email: federation@efds.co.uk,
    website: www.efds.co.uk
    The umbrella organisation that develops and coordinates sport for disabled people in England.
The Fieldfare Trust, 67a The Wicker, Sheffield S3 8HT. Tel: +44 114 270 1668, email: info@fieldfare.org.uk,
    website: www.fieldfare.org.uk
    The Fieldfare Trust has campaigned for the development of access to the countryside for disabled
    people. The charity has run a number of national schemes in the UK that have challenged barriers; the
    Trust's team possess a wide range of resources for those wishing to make a field centre or non-urban
    site accessible to all.
Royal Association for Disability and Rehabilitation, 12 City Forum, 250 City Road, London EC1V 8AF.
    Tel: +44 20 7250 3222, fax: +44 20 7250 0212, email: radar@radar.org.uk, website: www.radar.org.uk
    The umbrella body that represents disability issues nationally in the UK.
SCOPE, PO Box 833, Milton Keynes, MK12 5NY, England. Tel: +44 808 800 3333,
    email: cphelpline@scope.org.uk, website: www.scope. org.uk

4 WORKING WITH THE HOST COUNTRY

Nigel Winser

This chapter reminds the planner of the importance of making early contact with your hosts, wherever in the world they may be, and describes how to get permission to undertake your expedition or project.

WHY WORK WITH THE HOST COUNTRY?

There are several good reasons why you need to work with your host country from the outset, apart from just simple good manners.

The first is that there may be a legal requirement to do so. Without such authority, permission to enter the country may be denied. Second, there is a moral obligation to respect the territory of others and to ask permission to enter. Third, there are added benefits to you, to your counterparts and to the wider academic community from working with those with local knowledge and expertise. Much research worldwide could not have been achieved without such involvement, especially when there is a need for local knowledge and expertise. Finally, the project as a whole has a much greater chance of succeeding if counterparts are involved in the planning stages, as so much can be achieved with help in-country. There may be funding agencies that specify the need to have a proven commitment to host nation cooperation, before certain grants are made available.

So, early on, take the host country's perspective. Think locally. Consider the views of those with whom you will be working, and in particular identify those people who are likely to have a stake in the project, as members, advisers or supporters. These stakeholders may operate at a national, regional or local level, but whoever you work with the principle will be the same: to build respect and understanding. All your efforts to establish good relations with your host country at every level will yield invaluable dividends for your expedition, its follow-up work, future projects and ultimately for international concord.

WHO TO WORK WITH

First, you need to identify the key organisations with which you should be working. This will depend on the nature of your project. If you are working in a protected area, the national parks authority will need to be approached. If you are doing any mapping work, you will probably want to work with the national survey and mapping department. If you are mountaineering in a popular mountain region, you are likely to need a climbing permit. If you are undertaking any kind of scientific research, you may need to clear this at ministry level, often through a specific body established to vet and approve scientists visiting the country. In all cases, having a sponsoring body, such as a research institute, a university, school, non-governmental organisation (NGO), mountaineering club or other body is helpful and often essential. Making early contact with these organisations will be a key element of your planning and is likely to need to be over a year before you go into the field. The British Council provides an invaluable service giving advice and helping with introductions and contacts in the host country. The British Council has staff in most countries, so, if your project has research, training and other educational objectives, making early contact will be a good investment.

HOW TO GET IN TOUCH

Try to establish a good rapport with those bodies that you want to work with early on. This will mean presenting enough of your own plans to show that you are doing your homework, while leaving enough room to involve and integrate ideas that might be suggested locally. This is both common sense and good manners. In developing a link, often by email now, keep the line open for ways of working together:

- What are the local priorities that might be integrated with your own interests?
- Are there any suggested topics, locations or on-going projects to which you could contribute?
- Who could join you in the field (e.g. students, researchers, local climbers)?

When you make your first approach, a clear well-written letter that is 100 per cent accurately typed on one side of headed paper, with any supporting information attached, is still the most effective way of making a good first impression. Research thoroughly to whom to write and check their position and correct title. This is getting easier and easier using the Internet. It makes all the difference.

If you can afford it, a planning visit is an excellent investment. There is nothing like a face-to-face meeting to help build mutual understanding. If you are hoping to recruit expedition members or employ translators and field guides, this is also an excellent opportunity to meet them before a commitment is made on either side.

IN THE FIELD

Continue to 'think local' and be sensitive to cultural and religious differences. Learn as much as you can of the national and local language(s). Even elementary language skills will raise levels of trust dramatically. This in turn should give you confidence and make you a more relaxed and natural person to do business with, and improve your chances of having a harmonious team working together, with the minimum of misunderstanding. I know that this may appear a tall order if this is one of your first projects, but it is important.

FOLLOW-UP RESEARCH AND REPORTS

Your expedition will be collecting data and possibly samples. The intellectual property rights and ownership need to be discussed beforehand – and included in any agreements about collections and type of specimens, deposit of raw data, report feedback, and ownership of photographs and artwork.

Access to all the information collected, including photographs, should be made available to government representatives. Some will want copies of the original field notebooks. Some will be happy with the final report. Be clear about the level at which you are operating, especially if you need to take soil, rock, plant or animal samples. Usually this is possible only if you are working through the national museum or herbarium and there can be stiff penalties if you try to smuggle even a pebble out – without discussing the implications. Try to keep a high level of trust and communications throughout – and, if in doubt, ask.

Sharing the results of a project with local members is a courtesy often forgotten in the rush to get back to the UK. Think carefully how the results of the project can be shared locally, and how all future publications should have some kind of counterpart involvement. Joint authored publications in research and for education purposes are a good goal to aim for. Agreeing the distribution list and making sure that those who can benefit from the project get copies of the report is essential. Assuming that you have established good host country links, it won't be difficult to agree a plan that benefits all members. This might include the need to produce an executive summary that is translated into the local language(s). Looking ahead, perhaps you will want to create opportunities for new-found friends to visit the UK, to meet others interested in the research or project undertaken. This may include opportunities for further training that might not be available locally.

GAINING OFFICIAL PERMISSION TO UNDERTAKE YOUR RESEARCH

Rules and regulations for conducting research are constantly changing, and permission may be required at several levels, which can include national, regional and local permits, and perhaps endorsement from key government departments or organisations concerned with your particular area of study or activity. Some countries have set procedures and have appointed representatives in specific ministries to deal with applications. Some procedures are more complex than others, but it is not unusual for permits to take over 12 months to obtain. Make it your responsibility to check current rules, and do not rely on the guidelines set out in a past report, because this may well be out of date.

As mentioned earlier, even if you are not carrying out research you may still need permission. Mountaineers visiting the Himalayan ranges will find permits, peak fees, and even codes of conduct and environmental levies are now par for the course. School and other youth projects may find themselves directed to ministries of sport, education, youth or tourism. The permutations are endless, but the key is to establish early and good contact with those who can advise on the correct procedures and right people to contact.

For those who have not yet established these contacts or are unsure whether high-level permission is required, the following 10-point plan of action, although lengthy, is usually successful, provided you start early enough:

1. Send a neatly typed letter on headed paper to the Ambassador or High Commissioner of the country that you intend to visit. Most countries have a representative in London and addresses can be found in the London telephone directory and also on their website. Ask for details on how to apply for permission to undertake the research/mountaineering/community project that you describe, what visas and/or permits are required and to whom such an application should be submitted. Your letter should include a short statement summarising your overall plan and location in which you are working, in case the plans require special permissions or you are working in a sensitive region or protected area.

2. Send a copy of this letter to the British Ambassador or High Commissioner in the capital city of your intended country with a polite letter of notification. He or she will normally pass it to the Second Secretary to answer. Similarly, copies of these may be sent to the appropriate desk officer at the Foreign and Commonwealth Office, King Charles Street, London, for information.

3. Approximately two weeks later, follow up the first letter with a telephone call, asking to make an appointment with a representative in the London Embassy. This will ascertain the progress of your letter and establish a personal contact. If things go wrong later, this courtesy visit will have been an investment. It goes without saying that you must be respectfully dressed when you visit the Embassy and be courteous at all times, even if things are not going your way. There may

be delays but usually you will have a positive response within a few weeks. On some occasions, an Embassy representative will say that there is no need for a visit, and they then advise the correct procedure accordingly. On some occasions they may simply say that you don't need a research permit and that you may travel on a tourist visa. Please always double-check this, because this seemingly easy option now may cause difficulties once you are in the field.

4. Upon receipt of advice from the Embassy in London, complete the application forms neatly and carefully, making sure to answer all sections and provide any photographs, fees and evidence of authority. The advantage of getting these papers early is self-evident. There may be certain requirements, such as a sponsoring letter from a local research institution, that is required before the application can be accepted.

5. The forms are usually sent either to the London representative (Embassy or High Commission) or direct to the appropriate ministry in the host country. Make sure that you keep photocopies of everything you send, and again send copies of the application to both the British Ambassador and the Foreign Office. In some cases these can be sent electronically. Await replies.

6. If nothing has been heard after a month, write again (even more politely) with copies of the previous correspondence, asking if there is any other information that is required, and state that you would be happy to visit the host country to sort out any further applications if required. (You of course will have to decide whether this is something you can afford.)

7. If nothing has been heard a month later you may like to enlist the help of your new-found diplomatic friends here or over there, who by now have quite a file on your project. Ask if there is any way they can help. Emphasise the support you have for your project within the host country. The amount of effort they will give to processing the permissions for your project will depend on how easy you have made it to process your application, and the importance (relevance to the host country) that they attach to your project. This will be enhanced by any endorsements that you have received from major national (host country) and international institutes.

8. If none of the above has produced any response whatsoever, I would suggest you need to reassess the situation. But this is unlikely and I think you will be surprised at the cooperation that you will receive from all those you contact. However, do make contact with those who know the country well to check if there are any specific reasons why there might be a hold-up.

9. Once you have your permission, make sure that you get in touch and keep the government ministry with whom you will be liaising informed. Send regular updates on your progress. Write to ask for an appointment as soon as you arrive. If you need to make contact with other ministries, ask for letters of introduction. This may be important if you are expecting temporarily to import any freight without having to pay customs duty.

10. Finally make sure that you fulfil any agreements that you signed while applying for permission. This may include presenting a copy of your preliminary report with key ministries before leaving the host country and depositing a duplicate set of biological specimens with the appropriate natural history museum. Failure to do this can cause serious repercussions to you and to others who follow in the future.

Don't forget that it will be your positive international attitude with impeccable manners that makes you the field diplomat, a prerequisite for all expedition members, at all times. Good luck.

USEFUL WEBSITES

British Council: www.britcoun.org
Commonwealth Youth Exchange Council: www.cyec.org.uk
Foreign and Commonwealth Office: www.fco.gov.uk/travel
The Commonwealth Secretariat: www.thecommonwealth.org
Windows on the World – schools linking: www.wotw.org.uk

5 CATERING FOR EXPEDITIONS

Nigel Gifford

The job of catering officer on an expedition is usually the least popular – because "we do it every day", very little time is spent on planning the food and yet it is a most vital element. The feeding of an expedition is like the fuel that drives an engine. No fuel, no drive. The wrong fuel will reduce performance.

FORESEEING SOME OF THE PROBLEMS AND OVERCOMING THEM

During an expedition, food takes on an exaggerated but vital importance and without proper consideration the whole venture could become a mediocre experience. It is dangerous to consider food only as a fuel, because it also has a most important effect on morale.

The problems

At worst, an expedition can fail if food consumption is higher than estimated and supplies run out. Poor storage or contamination can produce the same results, as can expeditions mislocating food dumps.

There is the possibility of having plenty of food but of the wrong sort. Eating the same food every day is not only boring but can cause revulsion and even nausea in the most stoic expedition trenchermen! Today, more than ever before, we "suffer" from well-educated palates and are used to having variety in our meals. Lack of variety brings underlying problems in its wake, the most dangerous being the effect on morale. Petty jealousies can occur over the more delicious and select items in an otherwise plain diet. Someone is guaranteed to take more than his or her fair share, and it is not unknown for mature people to fight over food.

The time taken in preparing food can also be a cause of aggravation, especially when you are hungry, or if dehydrated foods are being used and there is insufficient water. Poor cooking equipment, such as inefficient stoves, and cooking pots that are too small or hard to clean, will cause frustration and lead to wastage, and a meal that looks unappetising is undoubtedly hard to swallow.

Once the meal is finished, washing up can be difficult. Think about how this will be tackled. Will you take tin foil? The amount of washing up can be reduced if all dishes are wiped clean with soft paper before they are washed in hot water.

The solution is in the planning

Many problems are encountered only after it is too late, when it requires considerable time and finance to correct the situation. However, with careful planning most can be avoided.

The nature of the expedition and the volume of food needed must be the first considerations. From this basis, more elaborate ideas can be researched:

- How many meals are required each day?
- Is there sufficient time/daylight to prepare these meals?
- Will food consumption be increased as the group is living away from their normal environment?

These are not difficult questions to answer and usually result in two meals a day with a midday snack. This saves cooking time and allows flexibility in day-to-day organisation.

Next, decide the total meal requirement for the trip. One person requires 1 × breakfast, 1 × snack, 1 × main meal each day. Simple multiplication gives:

Number in the party × Number of days in the field = Total person/days food

The importance of variety has already been mentioned and it is very difficult for one person to decide what to take. Ask the expedition members for ideas. Question them on their likes and dislikes. This will involve them in the feeding programme and makes them feel that the catering officer cares about their needs. A simple questionnaire may help.

Having gathered together these suggestions, work out a few sample menus. The "skeleton method" is often the easiest way of doing this. Start with the main item of the meal and then what is needed to go with it. Begin with the main meal of the day, and work through the snack meal and breakfast. Do not forget the drinks to accompany each meal. The ideal expedition feeding programme is as near to a normal daily diet as possible with nothing unusual about it. Do not get carried away and try to make the meals exotic. If you are stuck for ideas, browse around the local supermarket or look through simple cookery books.

NUTRITION

Nutrition is usually the last consideration for the catering manager. This is because it is important to provide foods that your team will eat and will enjoy eating. Food has a psychological importance as well as a physiological role. A perfectly balanced meal that is not eaten, for whatever reason, has no nutritional value and all your efforts will have been wasted. And, remember, it is very unlikely that any nutritional deficiency will manifest itself if the food is good quality and balanced to some degree. The time to look closely at this aspect is in extreme environments. In Arctic winter conditions, for example, you will need a very high fat content to help keep the body warm.

PROCURING RATIONS

Where to buy the food: the cost implications

From the outset you must be aware of the cost implications of the food that you will eat. This will include not only its purchase, but transportation and the fuel and equipment needed to prepare it. Don't get carried away with the idea that food given to the expedition in the UK is free. For example:

Cost of food bought in UK + Freight to host country + Transport from point of arrival to base camp = Cost of food eaten in host country.

Do not take the availability of food items in remote places for granted; some villages and small towns grow or buy enough only to support themselves and have none to spare for visitors. Shops in remote areas, on the other hand, may surprise you with the range of their stock. If they have good communications they may be able to order food for you. This provides an income for the local people and is a demonstrable way of providing benefit to a remote area.

Try to research exactly what foods are available in the host country with the up-to-date costs. A reconnaissance is invaluable for assessing food prices and availability in remoter locations.

Specialised rations usually have to be imported. The requirements of a mountaineering expedition, for example, may well be such that the high-altitude elements of the ration programme need to be selected and packaged in the UK where they will have been assessed for palatability, texture, weight and durability of packaging. One must accept that such specialised rations and their transportation are expensive.

If you don't buy your food in the host country, and have to obtain it in the UK, first find out what you can obtain through sponsorship by either donation or discount. The balance can be bought from a supermarket or cash and carry.

Remember, at the cash and carry you can usually buy only by the packet or container, not individual items. Specialist foods usually come direct from the manufacturer. Please note that civilians cannot always buy military composite rations from official sources and that their sale to civilians may be restricted.

Types of ration
The relative merits of the various types of ration are outlined below.

Fresh local food
There is no substitute for fresh food for both taste and nutrition.

Meat
All local meat, whether flesh, fowl or fish, should be freshly killed. In all but the coldest climates, it is unwise to purchase any quantity that cannot be consumed within a couple of days. If you have a permanent base camp, consider keeping a few chickens, a small sheep or goat, providing that you are willing to slaughter it yourself. Fishing can sometimes supplement the expedition's diet. Some meats are more dangerous than others. If you eat partly cooked or raw beef, at the worst you may get beef tapeworm that stays in your gut, but, if you eat half-cooked or uncooked pig, you may get pig tapeworm which migrates into your muscles and into your brain. For this reason Muslims, Orthodox Jews and Coptic Christians do not eat pork. Store meat in as cool a place as possible, raised to allow free circulation of air and protected, with muslin, from flies. If beef or lamb acquires a sickly smell and becomes slimy, you may be able to save it in time if you wash it in a strong brine solution. Certain spices do help tenderise meat and have a redeeming effect.

Vegetables and fruit
Although the variety may be limited, vegetables and fruit in season will be cheap and plentiful. Choose those that are fresh and ripe, and not bruised or blemished. Dates, grapes, etc. have fragile skins and are easily infected. Fruit from trees may be contaminated by pesticide. Correct storage will reduce deterioration and wastage. Keep them in a cool place where the air can circulate. They should be regularly sorted, those that are badly bruised are discarded and the less damaged eaten immediately. Green vegetables do not keep well and should be eaten as soon as possible.

If you eat raw fruit or vegetables, they must be washed in sterilised or boiled water, wiped and peeled. For salads choose cucumbers or tomatoes, and avoid lettuce unless you soak it properly in a sterilising fluid (e.g. Milton 2) for at least 30 minutes and then dry it. Weight for weight, lettuce has 300–400 times greater surface area than a cucumber or tomato. The acid in your stomach will kill a certain number of nasty germs but if you overdo it some will get through and make you ill. Most vegetables in less developed countries are reared with the aid of human faeces ("night soil"), and are likely to have a lot of these germs on their surface.

TABLE 5.1 **EXAMPLES OF FRESH AND TINNED FOODS***

Basic item	Weight (oz)	(g)	Alternatives	Weight (oz)	(g)
Meat (beef)	8	240	Lamb or pork or offal	8	240
			or (c) corned beef	8	240
			or (c) stewed steak	8	240
			or (c) steak and kidney pudding	10.5	315
			or (c) meat and vegetables	18	540
			or chicken	8	240
			or rabbit	8	240
Bacon	1.5	45	Luncheon meat	1.5	45
			or eggs (large)	1 egg	
			or eggs (small)	2 eggs	
Sausages (fresh)	1.75	52.5	(c) sausages	1.75	52.5
			or (c) luncheon meat	1.5	45
			or eggs (large)	1 egg	
			or eggs (small)	2 eggs	
			or whole fish	7	210
			or fish – headed and gutted	5.25	157.5
			or fish fillets	3.5	105
			or boned kippers	3.5	105
			or smoked haddock fillets	3.5	105
			or (c) beans	5.75	172.5
Fish fillets	1.5	45	Fish, whole	3	90
			or fish – headed and gutted	2.25	67.5
			or boned kippers	1.5	45
			or smoked haddock fillets	1.5	45
			or (c) sardines	0.5	15
			or (c) salmon	0.75	22.5
			or (c) herrings	1	30
			or (c) beans	2.5	75
			or fresh potatoes	21	630
Large eggs	1 egg		or small eggs	2 eggs	
Canned milk (fl. oz)	5	150	Fresh milk	12.5	375
			or powdered milk (mixed)	12.5	375
Cheese (Cheddar)	0.5	15	Processed cheese	0.5	15
			or (c) cheese	0.5	15
			or chocolate	1	30
Butter (fresh)	0.5	15	Butter concentrate	0.5	15
			or margarine	0.5	15
Margarine	1.5	45	(c) margarine	1.5	45
Cooking fat	0.5	15	Margarine	0.5	15
			or local cooking oil	0.5	15
Bread	12	360	Flour	9	270
			or biscuits	9	270
			or potatoes, fresh in lieu of each oz of bread	3	90
			or vegetables, fresh in lieu of each oz of bread	1	30
Flour	2.5	75	Bread	3.33	99.9
			or rice	2.5	75
			or potatoes, fresh in lieu of each oz of flour	4	120
			or vegetables, fresh in lieu of each oz of flour	1	30

Basic item	Weight (oz)	(g)	Alternatives	Weight (oz)	(g)
Rice	0.25	7.5	Macaroni	0.25	7.5
			or semolina	0.25	7.5
			or cornflour	0.25	7.5
			or spaghetti	0.25	7.5
			or ice cream	1.25	37.5
Breakfast cereals	0.75	2.5	Rolled oats	1	30
			or rolled oats and sugar	1	30
			or breakfast cereals	0.25	7.5
			and (c) milk	0.75	2.5
Marmalade	3	90	Jam	1.5	45
Honey	1	30	Marmalade	1	30
			or syrup	1	30
			or sugar	2	60
Tea	0.5	15	Coffee	1	30
			or instant coffee powder	0.25	7.5
Dried fruit	0.5	15	Jam	0.5	15
			or marmalade	0.5	15
			or syrup	0.5	15
			or fresh fruit	3.75	112.5
			or apple solid pack	1.5	45
Fresh fruit	5	150	Melons, mangoes, papayas, pineapples or bananas	6	180
			or lemons, oranges and grapefruit	5	150
			or other fresh fruit	4	120
			or (c) fruit	2	60
			or dried fruit	0.5	15
Vegetables (fresh)	8	240	(c) vegetables	4	120
			or (c) beans	4	120
			or dried pulses	2	60
			or dehydrated vegetables	0.75	22.5
			or frozen vegetables	4	120
Onions (fresh)	1	30	Dehydrated onions	0.08	2.4
			or fresh vegetables	1	30
Potatoes (fresh)	20	600	(c) potatoes	13.25	397.5
			or bread	3.25	97.5
			or mashed potato powder	4	120
			or dehydrated potatoes	2.75	82.5
Lemon/orange powder	0.25	7.5	Tea	0.08	2.4
			and sugar	0.25	7.5
			and (c) milk	0.5	15
			or jelly powder	0.75	22.5
			or sugar	0.75	22.5
Salt (culinary)	0.25	7.5			
Salt (table)	0.175	5.25			
Custard powder	0.08	2.4			
Baking powder	0.08	2.4			
Tomato purée	0.25	7.5			
Pickles	0.08	2.4			
Pepper (pinch)					
Mustard (teaspoon)					
Vinegar (teaspoon)					

*Amounts suitable for one person per day, (c) = canned

Milk, ice cream and fruit juices
Local milk, ice cream and fruit juices should be avoided. Bottled drinks with metal caps (but not those with corks) should be safe, although the necks should be rubbed well after the tops have been removed and before the drinks are consumed. When cooling a drink, place the ice outside and not inside the container. If you are sampling locally cooked foods, choose well-cooked hot spicy meals, and avoid salads and ice cream.

Supermarket food bought in the UK

It is unlikely that you would want to feed a large expedition entirely on fresh local food. Very few people enjoy unfamiliar food for any length of time, despite what they say in the UK before they leave (this also includes experienced expeditioners who adamantly think to the contrary!). Familiar foods bought at a supermarket or cash and carry before you go will provide essential basics that cannot be obtained in the host country, as well as the luxury items that add interest and lift morale.

Generally, the cheaper the price, the poorer the quality and, in the case of tinned goods, the higher the water content. With meats, the fat content will be higher in cheaper products. Test the product range yourself before buying in bulk. Familiar products are often welcomed by members.

Expeditioners eat primarily with their eyes; in other words they will be immediately attracted to food that looks colourful, interesting and familiar. Having appraised the food with a quick glance, their next assessment will be by smell, and last, having selected the food of their choice and put it on their plate, it must pass the final test of taste. At any of these stages people may well decide that, because the food has not met their standard at any one of these levels, they are no longer hungry, the result being no food intake, wastage of time in preparation and fuel, and poor effect on morale.

Consequently, the catering officer should give continual thought to the attractiveness of meals to appease the expeditioners' feeding requirements and tastes, to the extent that selection of both tinned and food items should consider the presentation of the product by its wrapper or easily recognisable household brand name. For example, John West fish products are invariably presented in a cardboard wrapper that shows a cooked and garnished meal comprising the food item inside the can. Such effective marketing becomes the garnish of the canned ration!

Food in extreme temperatures

Remember that foods bought in the UK in a warm supermarket may not be in the same state as this when stored in the Arctic or tropics. Favourite foods such as cheese, butter, some biscuits and chocolate tend to freeze solid or melt, leaving them impossible or unappetising to eat. This can be a big disappointment at the time. Therefore, it is important to consider the physical state of familiar foods when exposed to extreme temperatures.

Specialist expedition foods

These types of foods are designed for a specific purpose, and should by no means be considered to be the only answer to expedition rations, even for expeditions working in remote harsh environments. They are nothing more than a convenience and, where fresh alternatives are readily available and cost-effective, they should not even be considered. They are in essence the food planner's final resort to allow the expedition to achieve the objectives and overall aim.

Three main types are currently being marketed for expeditions: dehydrated foods, accelerated freeze-dried food and boil-in-the-bag meals.

Dehydrated foods (Table 5.2)

The dehydration process is a harsh one involving high temperatures and prolonged heat treatments. Consequently, the resultant food does not retain its original texture. The exception is pre-cooked and dehydrated cereals, rice and some vegetables. Meat-based meals tend to provide the "meat" in the form of textured vegetable protein, which is processed from soya beans. The protein in soya beans is a food technologist's dream because it can be spun and woven to resemble different textures. It is also a neutral colour with a bland flavour, so many different colours and flavours may be added to produce a range of dishes. It has been used most extensively to produce textures that imitate meat – this is unfortunate because the consumer then compares it with meat, with less than favourable results. Soya protein granules are also available which are used to produce minced meat-based dishes. These products contain edible gums and starch to give the required thickness, but beware of eating too many of them because your diet could take on the consistency of wallpaper paste!

TABLE 5.2 DEHYDRATED FOODS

Advantages	Disadvantages
Lightweight, and cheaper than freeze-dried foods	Tends to shrivel and change colour during processing
Wide variety of flavours, and manufacturers	Must be soaked or cooked before eating
More available overseas	Requires to be fully cooked with much water
	Uses more fuel than freeze-dried, but probably less than fresh food

Another source of protein to compete with soya in dehydrated meals is a type of fungus known as Quorn. This might sound unappealing but food technologists have produced some very acceptable and elaborate dishes.

Accelerated freeze-dried food (Table 5.3)
This is a much less harsh preservation technique than dehydration, and as a result it tends to preserve the original flavour and texture of the food when rehydrated. The preservation process involves freezing the food and putting it in a vacuum. The water in the food is then driven off as a vapour by sublimation. This uses very little heat and therefore causes less damage to the food. After the food is freeze-dried it is sealed in moisture-proof packets containing nitrogen. Provided that these are not punctured, the food can remain preserved for a number of years, the lack of oxygen and water preventing deterioration. Most expedition foods available in America are freeze-dried. There is a large selection of exotic meals available, such as chicken and cashew nuts with wild rice.

TABLE 5.3 ACCELERATED FREEZE-DRIED FOOD

Advantages	Disadvantages
Low bulk and lightweight	Fragile and easily damaged
Immediate meal available in five minutes	Quickly digested/hungry again quickly
No preparation or expertise required, just add hot or cold water	Two-person portion of main meal = one expedition portion
Sterile with long storage life	Cannot be eaten where water is not available in sufficient quantities

Boil-in-the-bag meals (Table 5.4)
These have the advantage of not dirtying the pan and, as the water is not contaminated it can be used for making a hot drink or soup. In conditions where, for a limited period, no fuel is available the meal can be eaten cold or sucked frozen like a lollipop. These are heavier than dehydrated rations but have the advantage of not having to be reconstituted. The contents are unaffected by extremes of temperature, humidity and salinity.

TABLE 5.4 BOIL-IN-THE-BAG MEALS

Advantages	Disadvantages
Does not need reconstituting	Heavier than dehydrated or freeze-dried
Less bulky than tins or fresh food	Fragile and easily damaged
Superior flavour/texture	Most expensive of the specialist expedition foods
Can be eaten cold without water	

"Tinned equivalent"
This is a military term used to devise a suitable ration for an expedition or exercise that is based on an established fresh scale of rations, e.g. you may establish that each member of the expedition will be entitled to a daily fresh meat element of 8 oz (240 g). However, in the field, meat may not be reliably supplied; you may alter your feeding programme to mix fresh and tinned rations, or dried main meal elements to suit the environment, work load, availability, costing, transportation and other criteria as they arise. Examples are shown in Table 5.5.

"CILOR"
This is also a military term, being an abbreviation of "cash in lieu of rations". A useful tip for the expedition planner, who, having decided the ration scale per person for the expedition and evaluated its monetary value in local currency, may find that the issuing of the ration allowance in cash enables a person on the move to purchase his or her own cooked meal. If the person should wish to purchase outside of the ration value, this is of course his or her own financial responsibility and not that of the expedition as a whole. Similar guidelines could be applied in an emergency, where any over-expenditure should be charged against the contingency fund and claimed back later from insurance.

Lightweight rations
Producing a lightweight ration for 48 or 72 hours is seldom a problem. A few favourite items, whether savoury, chocolate or biscuit, combined with a small canned or dehydrated main meal if carrying a stove, and some teabags and coffee usually seem to suffice. In summer conditions, in a European environment, it is not unusual for parties to travel with fresh bread, cheese, garlic sausage, fruit juice drinks or even sweetened powder additives to dissolve with uncontaminated water. Perhaps distributed through the party will be a simple stove and some fuel as a safety measure in case the weather should turn. It is not always necessary to carry a stove – common sense dictates the time and conditions when one is not required. Remember, the main reason for cooking food is to break down sinews and starches, making items easily digestible and more palatable. There is little increase in energy value by cooking foods, whether fresh, tinned or dehydrated, but the boost to morale is beyond any argument. For the expeditioner, there are certain points to remember when designing this part of the ration programme. A lightweight ration should give the maximum calorific intake for minimum weight, be easy to prepare and as varied as possible. When under physical or mental strain, the appetite seems to decrease and, on occasions, the desire to eat is reduced by fatigue.

For short periods of around 7–10 days, before returning to a place where fresh food is available, a lightweight ration can be considered along the following lines. Remembering that the basal metabolic rate (the amount of energy used only for the

TABLE 5.5 **TINNED RATION SCALE FOR FOUR PEOPLE FOR ONE DAY**

	Basic item	Weight (oz)	(g)	Alternatives	Weight (oz)	(g)
1. Breakfast	Baked beans in tomato sauce	16	480			
	Oatmeal blocks	5 × 1				
	Sausages	16	480	or bacon grill	16	480
				or baconburger	15	450
2. Main meal	Cuppasoup (choice)	3				
	Steak and onion casserole	16	480	or stewed steak	16	480
				or vegetable goulash	16	480
				or chicken/vegetable		
				curry	16	480
				or corned beef	12	360
				or steak and kidney		
				pudding	16	480
				or chicken supreme	16	480
	Mashed potato powder	6	180	or pre-cooked rice	10	300
	Carrots	10	300	or processed peas	10	300
				or mixed vegetable	10	300
	Apple pudding	22	660	or canned pears	24	720
				or rice pudding	24	720
				or chocolate pudding	17	510
				or mixed fruit pudding		
				or fruit salad	24	720
3. Snack	Tinned cake	10	300	or luncheon meat	16	480
				or (c) fish	16	480
				or hamburgers	15	450
	Selection of jams	9	270			
	Tinned cheese	8	240			
	Margarine	8	240			
	Chocolate bars	4 × 2				
4. Drinks and	Teabags	2	60			
sundries	Instant coffee	0.5	15			
	Dried milk	3	90			
	Sugar	14	420			
	Salt	1	30			
	Mustard powder	0.125	3.75			
	Small tin opener	1	30			
	Toilet paper	25 sheets				
	Plastic reclosure	2	60			
	Paper towels	8	240			

process of being alive while at rest) has a representative value in a European-style climate of about 1600 kcal per day, any calorific intake beyond these limits in a 24-hour period is a bonus to the already existing resources in the body and the daily energy output.

From the reports of various expeditions working in Alaska, the Yukon, Africa, tropical forest areas and the Himalayas, the requirement of 5000 kcal per day has been proven unrealistic in practical terms, and it is better to assume that an expedition member will actually consume 3500–4200 kcal per day when in the field living on a lightweight ration, as a result of many factors from food repulsion to a stove failing to work correctly. Therefore, the multi-choice system of producing lots of small but varied items is the one now favoured; small tins of cheese, containers of jam, miniature chocolate bars and ring-pull cans of pâté are all worth considering. Some rations built on the multi-choice system of wide variety, but relying on small portions, have as many as 36 different items and thus allow for the widest change in taste, palatability and food texture on a day-to-day basis. Some expeditions spread this type of variety through four or six different ration packs by using the same type of items and varying the flavour, thus keeping weight, packaging and volume the same, while increasing the choices even more.

As we all know, it's hard enough to achieve 100 per cent success with every meal we eat at home, but on an expedition you have to try to achieve the same acceptance with restricted food items in an inhospitable environment! A lightweight ration, by its very nature, makes this type of presentation and appreciation very nearly impossible. One of the ways to make a ration look attractive is to purchase cans or dehydrated foods that have colourful wrappers, which show a photograph of a prepared and finished dish on the label, nicely served and garnished. Another way is to ensure a variety of food colours: green beans, sweetcorn, instant potato, noodles, beef stew. Sometimes ingenuity takes over; in fact on some expeditions cooking competitions have started up between individuals or partners in a tent. I remember a friend producing a multicoloured instant whip, garnished with crumbled Hardtac biscuit, revolting now, but quite delicious at the time! Food cravings become quite common too; only recently I heard of an expeditioner who gained a liking for blackberry-flavoured apple flakes, cooked with an overdose of onion Oxo cubes!

Although there will be complaints and faults with the rations, they should be based on the fact that, by its very nature, a lightweight ration is a restricted form of eating, and not on the fact that the food has been badly prepared. Consideration also has to be given to the main components of a lightweight ration: should it rely on tinned (wet) or dehydrated (dry) main meal elements? Only the planner can consider which will suit his or her needs best for the project being undertaken. The main advantage of a wet ration is that little liquid is required to make the food item edible, whereas with a dehydrated ration a reasonable quantity of water in some form must be available. Consider too that, although the ration may be lighter by using a dry lightweight programme, extra fuel (and therefore extra weight) may be necessary to melt liquid to make it edible; cooking times may also be longer and, at altitudes where water boils at a lower temperature, rehydration may be a problem (although this is usually only so over 21,000 feet). Therefore packing heavier canned weight may in the long run prove to be lighter for the overall work effort.

There may also be environmental factors that affect the lightweight feeding programme. In extremely cold climates, a high fat content may be required; in hot climates, light, soft-fibre foods may be best, allowing for a high liquid content; certain types of food and container may not keep in extremes of temperature – the fat in biscuits may go rancid; internal packaging may burst at altitude, as a result of the differences in air pressure inside and outside the plastic or synthetic wrapping; processed cheese may go stale over a long period and chocolate may deteriorate if not processed for the climate by the manufacturers. All these points should, where possible, be borne in mind before making the final decision on what items to include in the lightweight ration and, having made the final recommendation, it is then advisable to check calorific values, to ensure that each ration is around the 4000 a day mark. I have never really found it important to worry about calorific values, and those who do usually seem to end up taking a ration full of items that make up the exact daily energy requirement, which is stodgy and boring, and eventually nobody wants to eat any of it. In these instances, such ordinary items as cans of sardines, tubs of margarine and sweet biscuits turn into items to be coveted, even stolen, or hidden from others!

By giving people the type of items that they want to eat and forgetting the calorific recommendation, it makes more money available for selected items, reduces volume of packaging and weight, and keeps acceptability and morale high.

Emergency rations (Table 5.6)

These are essential and should not be overlooked. The food should be of high carbo-hydrate content and of a type that is easily absorbed by the body, e.g. sugar or glucose. It is not sufficient merely to allocate additional standard rations as emergency rations. The emergency rations pack should be clearly labelled, and should not be used to supplement normal meals but kept for emergencies only. The packing of such rations must be durable, in order to withstand repeated handling without actually being opened for use (e.g. mess tin). Suggested emergency food types are: glucose tablets, hot drink (e.g. tea, coffee, Complan), muesli-type food, dried fruit, chocolate bars, sweet biscuits, nuts, fuel and matches. It may be advisable to include the food pack with the emergency equipment such as flares, survival bag and emergency stove.

TABLE 5.6 **ONE-PERSON 24-HOUR EMERGENCY RATION**

	Basic item	Weight (oz)	(g)	Alternative	Weight (oz)	(g)
1. Breakfast	Porridge oats	3	90			
	Drinking chocolate	2.5	75			
2. Main meal	Cuppasoup	1				
	Dehydrated prawn curry	2.5	75	or dehydrated stew	2	60
				or vegetable curry	2.5	75
	Dehydrated peas	1.5	45			
	Smash potato	2	60	or pre-cooked rice	2.5	75
	Dehydrated vegetable	2	60			
	Apple flakes	1	30	or apple and bilberry		
				flakes	1	30
3. Snack	Chicken and bacon spread	2	60	or chicken/beef/cheese		
				spread	2	60
	Jam	1.5	45			
	Margarine	1	30			
	Plain biscuits	2.5	75			
	Sweet biscuits	2.5	75			
	Chocolate bars	3.5	105			
	Chocolate toffees	2	60			
	Nuts and raisins	1.5	45			
	Date and dried fruit bars	2	60	or muesli bars	2	60
	Dextrose tablets	1	30			
4. Drinks and	Instant coffee	0.5	15			
sundries	Instant tea	0.25	7.5			
	Oxo	1 cube				
	Dried milk	0.5	15			
	Sugar	1	30			
	Salt	Pinch				
	Spoon	1				
	Small tin opener	1				
	Waterproof matches	1 box				
	Paper tissues	1 pack		or toilet paper	5 sheets	
	Face wipes	1				

FURTHER READING

Axcell, C., Cooke, D. and Kinmont, V. (1986) *Simple Foods for the Pack.* San Francisco: Sierra Club Books Publication.

Bennett-Jones, H. (2002) Base camp health and hygiene. In: *Expedition Medicine*, 2nd edn. London: Profile Books, pp. 93–103. Available from www.rgs.org/eacpubs

Davies, J. and Dickerson, J. (1991) *Nutrient Content of Food Portions.* Cambridge: Royal Society of Chemistry.

Drew, E.P. (1977) *The Complete Light-pack Camping and Trail Foods Cookbook.* New York: McGraw-Hill.

Fleming, J. (1988) *The Well Fed Backpacker.* London: Random House.

Gibb, J. (2002) *Reluctant Cook.* London: Adlard Coles Nautical.

Gifford, N. (1983) *Expeditions and Exploration.* London: Macmillan.

Goodyer, P. and Goodyer, L. (2002) Water purification. In: *Expedition Medicine*, 2nd edn. London: Profile Books, pp. 105–110. Available from www.rgs.org/eacpubs

Gunn, C. (1988) *The Expedition Cookbook.* Leicester: Cordee Books.

Mears, R. (2002) *Bushcraft: An inspirational guide to surviving in the wilderness.* London: Hodder & Stoughton.

Prater, Y. and Mendenhall, R. (eds) (1982) *Gorp Glop & Glue Stew: Favourite foods from 165 outdoor experts.* Seattle: The Mountaineers.

Stroud, M. (1998) *Survival of the Fittest: Understand health and peak physical performance.* London: Random House.

Thomas, D. (1994) *Roughing it Easy.* Cincinnati: Betterway Books.

Vegetarian Society UK Ltd (1990) *The Vegetarian Handbook.* The Vegetarian Society, Parkdale, Dunham Road, Altrincham, Cheshire WA14 4QG. Website: www.vegsoc.org

Wiseman, J.L. (1996) *The Urban Survival Handbook.* New York: HarperCollins.

Wiseman, J.L. (1999) *SAS Survival Handbook.* London: HarperCollins.

6 MAPS, NAVIGATION AND GPS

Peter Simmonds

All expeditions, no matter what their aim, require maps and some form of navigation. Irrespective of the type of terrain, mode of transport or the season of travel, navigation will play a dominant part in the preparation, planning and execution of any form of journey or fieldwork. Expeditions will have designated roles with perhaps a science coordinator, safety officer or medics and it is strongly recommended that you also include of a chief navigator. The navigator should be responsible for the provision of all expedition mapping, ensuring that all participants have good map-reading skills and are equipped with the tools by which to navigate.

MAP PROCUREMENT

Maps will be needed from the very outset of planning to determine how to get to your expedition area and to determine the extent of the area within which the expedition is to operate.

Try to obtain your maps as soon as possible. To do this you will need to identify what you need the mapping for, what scales of map are best suited to the various objectives of your expedition, from where you can source the maps, what the availability is like for your area and how much money is available to purchase maps and what quantity.

Sources for obtaining maps

There are many sources from which mapping is available, but enquiries and research are often needed to establish the best source. Any expedition in the UK will have access to national mapping from almost any walking shop, which can be purchased easily and without dilemma. Expeditions taking place away from our shores, especially in wilderness areas, face greater challenges. It must also be appreciated that the time needed to acquire this mapping could be weeks, or even months, so plan early to avoid disappointment.

Published maps for topography, geology, vegetation cover and soils will probably exist at national and regional levels – whether they are still in stock in the host country is another thing. Try to purchase regional and local maps from a UK map supplier, such as Stanfords or GeoPubs, before going. Tactical Pilotage Charts show elevation, topography and basic vegetation cover, at 1:500,000 scale, and are available for most parts of the world. Regional surveys of soils and vegetation cover have been carried out for many countries by the Food and Agriculture Organization of the United Nations (FAO, Rome): enquire at the HMSO to see if the region of interest has been mapped. You may be able to view maps in the map collection of the Royal Geographical Society (RGS), or review the coverage of a region using the CARTO-NET database of the British Library. A worldwide directory of national earth-science agencies and related international organisations is published by the US Geological Survey (USGS); it gives details of overseas agencies that might hold maps or sets of air photographs. It may also be worthwhile to enquire at the National Cartographic Information Service (NCIS) of the US Geological Survey, or the Cartographic Section of the US National Archive. More detailed maps, at 1:50,000 or 1:25,000, often cover only a small percentage of less developed countries and may date from colonial times. If you can obtain detailed maps, think yourself lucky, but bear in mind that the map data may be over 20 years' old and accuracy may not be as high as UK Ordnance Survey maps.

If your fieldwork is overseas, you should be aiming to carry it out jointly with a team from the host country: this will facilitate access to data. Furthermore, this facilitates the transfer of skills to places where they are most needed, and helps to reduce bureaucratic and logistical problems.

More and more data summarising the features of the world are being digitised and are available over the Internet, from word-processed documents and spreadsheet tables, to scanned air photos and satellite images of entire continents. At the expedition planning stage, the websites of the RGS–IBG Expedition Mapping Unit (www.rgs.org/mapping), the US National Geographic Society and the USGS are particularly useful. Digital maps are now easily obtainable at broad continental and national levels, usually on CDs. These digital maps of developing countries are unlikely to be more detailed than 1:1,000,000, and are therefore of use only where regional studies are important.

Map scales

In the UK we are fortunate to have the world's best quality mapping at a scale that can suit almost all eventualities. Once you depart from the UK, life becomes harder and the quality of mapping can decrease measurably. Choosing the right scale of map to suit your various expedition requirements can be hard, but try to list the preferred options. Maps are generally described as being either "large scale" or "small scale". A large-scale map will normally imply a scale from 1:2500 to 1:100,000 and a small-scale map from there upwards. Small-scale maps are normally used for planning. They are ideal for journeys into and out of an expedition area and are very good for long distance journeys, especially when in vehicles or on animals because on one map the distance represented could as large as 150 km. Large-scale maps are usually used for specific locations and are of better use when travelling on foot, but the distance represented by one map might not exceed 40 km. So more maps will be needed for large areas.

Map costs
Always be prepared to pay for mapping. You might not always get value for money, but that is a risk you may have to take. To overcome this you may wish to see the mapping before you purchase it. When buying your maps take into consideration the exact quantities that you need. Despite the importance placed upon expedition members to keep their maps safe, you can guarantee that the first item misplaced will be a map, so buy enough to cover all eventualities.

Map reliability/accuracy
The Ordnance Survey maps of the UK are arguably the best in the world, of the highest accuracy and contain all the information that you could wish for. The same cannot be said for other countries and you must be able to appreciate the currency and accuracy of the information.

When viewing the maps have a look for the "Print Note" or "Compilation Note" in the map margin. Most maps will tell you when the map was compiled, the age of the information used to create the map and when it was printed. Do not be misled by a print date of 2002 when in fact the map was compiled in 1955! Compare detail between different scale maps, especially for road information because these may still be depicted as a track in some areas when in fact a major road has been built through the area.

The other factor to take into consideration will be the language; since the dissolution of the Warsaw Pact there are many more excellent quality large-scale maps available but annotated in Cyrillic.

Finally, note the date and rate of change of the magnetic variation or magnetic declination because these will change with time and must be recalculated by you.

Satellite imagery
Satellite imagery can also be used to complement maps that you have acquired for your expedition. In today's modern world it could be said that acquiring satellite imagery will be easier than acquiring some aerial photography or even mapping.

The main use for satellite imagery is in a geographic information system (GIS). That is another subject all by itself and is not covered here.

There are many satellites collecting imagery of the world. The types of images that they produce are endless. From a purely navigational point of view it will be easier and cheaper to acquire panchromatic images. This will give you a picture of the world below and will augment your maps extremely well. Colour images may sound nice, but the cost will rise and some detail can be confusing in colour. The more familiar types are LANDSAT 7 and SPOT.

The list of sources for satellite imagery is huge. It is so widely available that it is hard to know where to start. Contacting one of the commercial vendors, such as Info Terra or Space Imaging Europe, will aid you in identifying exactly what is available and the costs involved. They might well be able to offer slightly older imagery at a reduced price. Sources of satellite imagery are given on the RGS–IBG Expedition Mapping Unit website (www.rgs.org/mapping).

The accuracy of a satellite image is based on its ground resolution, i.e. the size of a pixel at ground level. Most satellites these days are producing images to resolutions of 50 metres and better. This is ideal for navigation, because most maps may not have an accuracy of greater than 50 metres.

As there are so many satellites acquiring imagery, the reliability of those images is likely to be better than maps. An image could be taken on a Monday and by the Wednesday you could have a copy of it for use.

Satellite imagery is available in either a hard or a soft copy. It is always wise to procure a soft copy because you can then make multiple copies at your leisure, as long as you have the ability to view it and print it. Free viewers are available and can be sourced through a commercial vendor.

Different types of images cover different sized areas, e.g. LANDSAT 7 comes in a scene 170 km × 185 km. The size will depend on the company supplying the data. A LANDSAT 7, panchromatic 170 km × 185 km scene might cost you around £500. However, there is also free imagery. A joint US/Japanese project called Aster has free imagery for users. It is of a good enough quality for a navigator. The National Imagery and Mapping Agency (NIMA) also has free imagery available on its home-page. Look for its Geo Engine and search from there.

Aerial photography

Another product that can augment your mapping is aerial photography. Aerial photography comes in either monochrome or colour. Again, monochrome is both cheaper and more suited to navigation.

As with satellite imagery there are many sources for aerial photography. Getting it for any western European country will not prove to be too difficult. Further away from that things will get harder; this is mainly the result of security restrictions.

The big problem with aerial photography is that all images are captured using an aeroplane, and are always subject to tilt as a result of side, head or tail winds. Therefore all aerial photographs are distorted, and the image that you see will not represent the ground perfectly. Not all aerial photographs come with a grid annotated on them and objects can often be hidden in shade. New buildings, roads, etc. might have been built since the sortie was flown.

It is easier to get aerial photography as hard copy. Normally this will have a glossy finish. If possible try to obtain the index plot for the sortie which, if available, will show the relationship between individual photos and the ground. This may be vital in wilderness areas because it can be very hard to pinpoint features on the air photo and the corresponding features on the map.

Each aerial photo should come with a strip of information located on one or more edges. The most important information to identify is the photo or print number, the date when the photo was taken, the focal length of the lens used in the camera and the flying height. Examples of those are given below.

Print number	056
Date Time Group	061245ZMAY91
Camera focal length	152.4 mm
Flying height	6000 feet

Determining air photo scale

The easiest method of scaling an air photo is to use the following:

$$F \text{ (camera focal length)}/H \text{ (flying height [AMSL])}$$

Example, based on the above information:

$$F = 6 \text{ inches, } H = 6000 \text{ feet}$$
$$F/H = 6/6000 \times 12 \text{ (to make the same units)} = 1/12{,}000$$
$$F/H = 1{:}12{,}000$$

Note that 152.4 mm = 6 inches (approximately).

Putting the map grid on to an air photo

With care the map grid can be transposed on to a photo by inspection. Care must be taken, because the grid will not plot as a regular shape.

To plot the map grid on to the aerial photograph, start by plotting the four corners of the photograph on to the map. Be as precise as possible. The map grid may then be transposed on to the photograph. Remember to add grid numbers and other information as required.

AIDS TO NAVIGATION

Once the maps have been procured and the detail scrutinised to build a mental

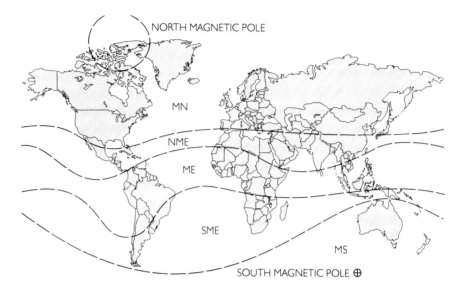

Figure 6.1 *Silva compass magnetic bands*

picture of the terrain, the time has come to consider the navigational tools to recommend to each expedition member.

The compass

Every expedition member must have a compass. The best type is the plastic or lightweight compass, which is cheap and very reliable. It should, however, have the following basic features:

- Oil-filled housing to allow for proper floatation and movement of the compass needle.
- Luminescent stripes for night use.
- Lanyard for keeping it on your body.
- The needle in the plastic compass should be weighted to suit the magnetic attraction in a particular part of the world. A rough guide can be seen in Figure 6.1.
- The two- to three-letter combinations can be found on the rear of plastic compasses. Some companies can provide a "Global" compass that ensures a free-moving needle no matter your location.

Compass bearings

Navigators will use a combination of true, grid and magnetic bearings:

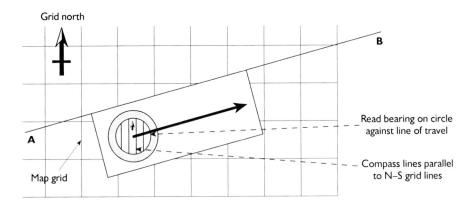

Figure 6.2 *Grid bearings*

- True bearings are measured clockwise from meridians of longitude (graticule) to the desired location on the map.
- Grid bearings (Figure 6.2) are measured from grid north clockwise to the feature on the map.
- Magnetic angles (Figure 6.3) are measured from magnetic north clockwise to the feature as identified on the ground.

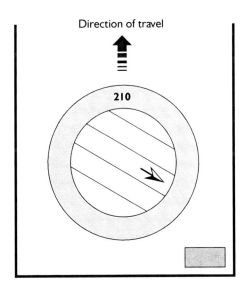

Figure 6.3 *Magnetic angles*

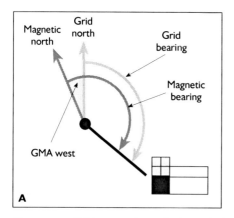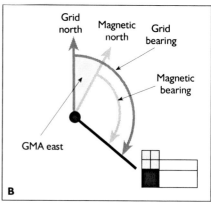

Figure 6.4 *Grid magnetic angle (GMA): (a) magnetic bearing = grid bearing +*
GMA; (b) magnetic bearing = grid bearing − GMA

Grid magnetic angle

Grid magnetic angle (GMA) is the difference between grid north and magnetic north
(Figure 6.4). It is defined as being east or west of grid north. Incorrectly applied it will
lead to incorrect bearings:

- If the GMA is **WEST** then you will need to **add** it to a grid bearing to obtain a
  magnetic bearing.
- If the GMA is **EAST** then you will need to **subtract** it from a grid bearing to
  obtain a magnetic bearing.

Magnetic variation

Magnetic variation (MVAR) or magnetic declination is the difference between true
north and magnetic north. It can be defined as being east or west of true north
(Figure 6.5).

Remember, if you are using the longitude lines on your map, you will be
measuring the true bearing and not the grid.

- If MVAR is **WEST** you will have to **add** it to a true bearing to obtain a
  magnetic bearing.
- If MVAR is **EAST** you will have to **subtract** it from a true bearing to obtain a
  magnetic bearing.

Watch

This is another basic, but vital, tool for navigation. Knowing your time accurately is
important, so ensure that you have a good reliable timepiece in each party. They also

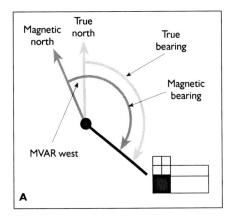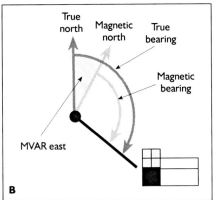

Figure 6.5 *Magnetic variation (MVAR): (a) magnetic bearing = true bearing +*
MVAR; (b) magnetic bearing = grid bearing − MVAR

become vitally important if faced with a survival scenario. The preferred type is an
analogue watch with both hour and minute hands. See "Survival navigation" (below)
for use in a survival scenario.

Altimeter
A good altimeter is the most effective guide to telling you your height/altitude. This
becomes of prime importance if in extremely mountainous or jungle terrain. All
altimeters require calibration and this must be done at a known height. This could be
a known spot height from a map, a local airfield, a known survey point or if none of
the aforementioned is available then from a global positioning system (GPS) receiver.
If you do use a GPS receiver your height accuracy will degrade. Currently the global
height accuracy for GPS is about 20 m.

Pacing
If you are on foot and are travelling in terrain with minimal features, then knowing
how many paces you take to walk 100 m is ideal.
    A single pace is a double stride and is normally counted on the heel of your right boot
hitting the ground. A local athletics track with a 100 m bay is ideal for this. Walk the 100 m
at least twice and walk normally. Try doing this carrying different loads, because your
strides to 100 m will differ depending upon the weight in your rucksack. Once you are
happy with your pacing, plot all the variables on a small piece of graph paper and cover
it with Fablon. It can then be kept with you at all times. Due consideration is then taken
for hilly terrain or muddy ground, but nevertheless it is a tool that is always with you.
    Another tool to aid in pacing is a pedometer. These devices can judge the distance
that you have travelled on foot.

Plot sheets

When navigating in a featureless terrain such as the desert or the Arctic, the mapping will be plain and potentially out of date because sand dunes/snow drifts move continuously.

To help you in these environments it is very useful to plot every move that you make on a piece of graph paper.

This "plot sheet" needs to be scaled and all bearings, distances, spotted features and new tracks can be plotted to update your own mapping. All notes taken should be annotated on a dead reckoning log (explained below).

Global positioning system

This has been intentionally left to last. Do not become totally dependent on it. GPS is a very good aid to navigation, but that is all it is. Batteries will run out, the receiver could get damaged in a small fall and in some terrains you might not even be able to track any of the satellites. Becoming over-reliant on GPS is bad practice, and you should never forget how to use the more classic navigation tools.

The basics

GPS is an American satellite-based positioning system first developed in the mid-1970s. At its core is a constellation of 24 satellites that orbit the earth and transmit information that can be processed by a GPS receiver. This system is capable of providing very accurate positions in a very short time frame. As this goes to print the global accuracy for a standard hand-held GPS receiver is 13 m or less (position only). You may get better or worse than this. The location of the satellites in the sky and the number of satellites being tracked will help to determine this.

However, this accuracy and its relationship to your maps will all depend upon your ability to set your receiver to specific parameters as defined on your maps. Remember that maps came first and that mapping theory is used on GPS receivers.

GPS set-up

The two most important parameters are the map datum and the coordinate format that you wish to use.

If you select the wrong datum on your receiver you can expect to get a position error of up to 1000 m. All modern mapping will specify its datum in the map margin.

The choice of a coordinate format will depend upon your mapping. Be aware that certain countries have their own position format – this could be a bit of an unknown. Whatever format is chosen, remember that GPS can give you only a unique position on the world, so you must conform to the format exactly. All errors are user errors!

Other set-ups that you may wish to consider are:

- Bearing style and bearing units: it is worth setting your receiver to magnetic bearings. Any bearing displayed can then be set on your compass as well!

- Time: all receivers have the ability to be set for local time.
- Navigation units: how you wish distances and speeds to be displayed.

Initialising a GPS receiver

Once you have arrived in your expedition area you will need to initialise your receiver. Nearly all GPS receivers have the option to select the continent, country and sometimes the nearest city. Choose this and then wait for the receiver to display a new position, but be patient as it may take 12–15 minutes to initialise.

Waypoints

A big advantage with GPS is the ability to load a waypoint (map feature) in coordinate format into the receiver and then let the receiver guide you to it. Try to choose features of the map and not just any old grid reference.

Routes

If you are in vehicles or on animals then you can create a route to follow. Individual waypoints are needed for this and, once loaded, these can be imported into a route. Once activated, the receiver will guide you from point to point and usually will inform you on arrival at a point.

This feature should not be used when travelling on foot.

Navigating with GPS

GOTO is a function that when chosen will display a list of all stored waypoints on the GPS. All receivers have a GOTO function. This will prompt you to select a loaded waypoint and then guide you to it. Remember that the information displayed to aid you will be "as the crow flies". GPS cannot predict obstacles such as cliffs, rivers, etc., so be careful. Do not get into the habit of looking at your receiver permanently and thereby becoming blinkered to your surroundings.

Nearly all receivers will have either a compass screen or a highway screen. The easiest to use for foot navigation is the compass screen. If in vehicles or on animals the highway screen may be easier.

Background mapping on a GPS receiver

Many companies now offer the ability to load, from a CD, mapping into specific receivers. This digital mapping should be treated carefully. Outside the UK no digital mapping is available that shows true topographic information, i.e. footpaths, contours, etc. The best use of a package like this is the opportunity to create waypoints via a PC keyboard rather than the keypad of your receiver. These can then be "burned" into the memory of the receiver for use.

Other navigational tools

Other navigational tools that can be easily forgotten but are still vitally important are:

- graph paper: for plot sheets
- pencils: for plotting bearings, completing fixes, completing logs
- protractor: a more precise way for plotting bearings than a compass
- conversion tables: kilometres, miles, mph, kph, etc.
- notebook: recording information
- dictaphone: especially good when using vehicles or animals; saves trying to rewrite notes at the end of a day
- parallel rulers (small): exceedingly good value
- ruler: metric and metal.

Jungle navigation

Jungle navigation can be extremely difficult and frustrating because known promi-nent features are often very hard to find, let alone identify, on the map.

Primary tropical forest is normally easy to walk through, fallen trees being the only obstacle. Route selection should be based on the contours and drainage pattern. In steep terrain keep to ridges (even though they lack water), otherwise the going will be extremely slow. Planning of the route is essential, because visibility is very restricted.

Secondary tropical forest occurs around the edges of primary jungle or in patches where the primary jungle has fallen down or been cleared. The growth can be extremely dense and should be avoided. Your visibility will be severely restricted and routes should be related to the shape of the ground.

Swamp areas should also be avoided. If unavoidable a straight-line route should be planned from a well-defined starting point and ending at some identifiable point. Aim off from this point so that you know that, on reaching the far side, you have to turn left or right to locate it.

Sketches should be made of features that could be used for navigating – river crossings, clearings, etc., including the direction you were facing when doing the sketch.

Clearings should always be sought after, especially if on high ground.

Remember, in the jungle the shadow will point away from the sun.

Survival navigation

Using the sun

The sun can be used to determine the four major points of the compass. Remember the sun rises in the east and sets in the west.

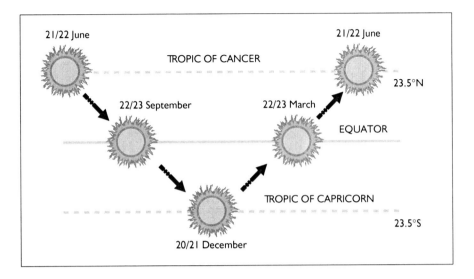

Figure 6.6 *North or south and the sun at midday*

If you are north of the Tropic of Cancer, at midday the sun will give you south. If you are south of the Tropic of Capricorn, at midday the sun will give you north.

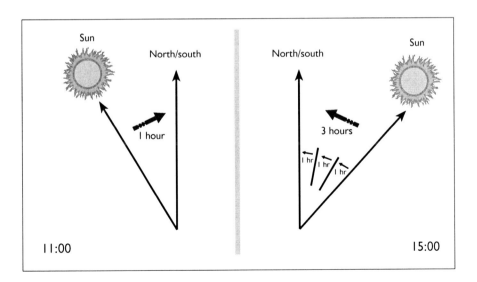

Figure 6.7 *Finding north and south*

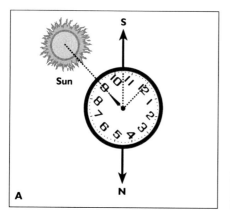
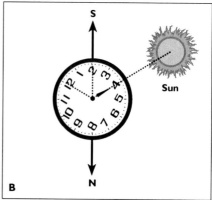

Figure 6.8 *Using an analogue watch in the Northern Hemisphere: (a) morning, (b) afternoon*

North or south (Figure 6.6)

Depending on your latitude and if you are in between the Tropics, the sun at midday could give you either north or south.

If you are going to a country within the two Tropics contact the RGS–IBG Map Room with dates and latitudes (email: info@rgs.org). A definite answer can then be calculated for you.

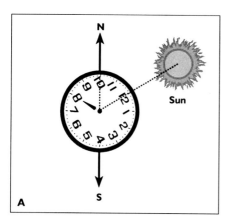
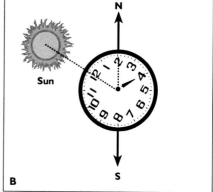

Figure 6.9 *Using an analogue watch in the Southern Hemisphere: (a) morning, (b) afternoon*

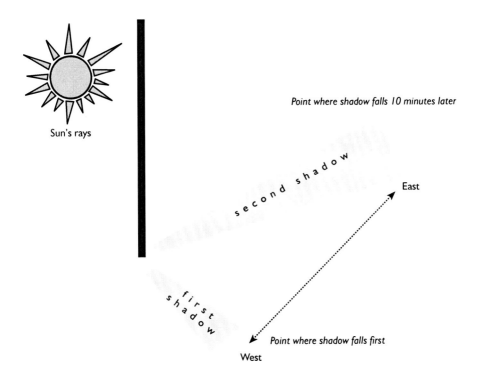

Figure 6.10 *Finding east and west*

Finding north or south (Figure 6.7)
By stating that one outstretched hand at arm's length is equal to one hour, you can determine north or south from the sun using your hands.

To save you from staring at the sun, try this technique with your back to it. Remember, however, that this will also give you the opposite heading.

Using an analogue watch
In the Northern Hemisphere, north of the Tropic of Cancer, lay your watch flat with the hour hand pointing to the sun; south will be midway between the hour hand and 12 o'clock on your watch (Figure 6.8).

In the Southern Hemisphere lay your watch flat with the 12 o'clock pointing to the sun; north will be midway between the hour hand and 12 o'clock (Figure 6.9).

Finding east and west (Figure 6.10)
Place a stick in the ground, which should be flat and clear of debris. Mark the tip of the shadow, e.g. with a stone. A minimum of ten minutes later, mark the tip of the

Figure 6.11 *The stars: Northern Hemisphere. Note that Polaris is shown brighter than it is in reality*

shadow. The line joining the two points is the east/west line, no matter what the latitude or time of day is. This, however, is not a very accurate method.

The stars

The star that best denotes the direction of north at night in the Northern Hemisphere is Polaris (in the Southern Hemisphere Polaris is not visible). There are three main star constellations that can be used to locate it: Cassiopeia, the Great Bear and Orion (Figure 6.11).

The vertical angle of Polaris also gives you your latitude. So the further south your location, the lower in the sky will Polaris be found. The more remote your location, the more stars you will be able to see. Finding specific stars in this environment can be harder.

In the Southern Hemisphere life is less easy because there is no bright star to use as a south indicator. The best way of estimating the location of south is to use the Southern Cross and the brightest star in the Hydrus constellation as your reference.

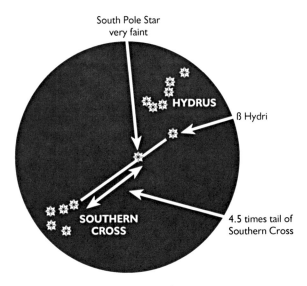

Figure 6.12 *Southern Cross in the Southern Hemisphere*

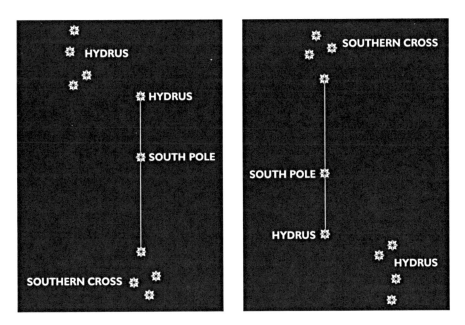

Figure 6.13 *North–south line – how to find in the Southern Hemisphere*

Draw a line from ß Hydrus to the Southern Cross. Imagine that the Southern Cross is a kite; extend the longer axis 4.5 times and that point will be south (Figure 6.12).

When the line joining the tail of the Southern Cross to Hydrus appears to be vertical it is a good indication of the north–south line (Figure 6.13).

Other tools
Another tool you can use in a survival situation is the wind. If you know the direction of the prevailing wind you will be able to see this in features on the ground.

ROUTE PLANNING
Planning a route is critical. The selection of the route must be methodical and follow the following basic guidelines.

Creating a route planner
When creating a route to follow you should always complete a route planner. Never try to memorise the route in your head. A route planner should consist of the following minimum headings:

Leg no. Start position End position Distance Bearing Estimated time Remarks

The following are points to note:

- State clearly the distance units, i.e. kilometres, miles or nautical miles.
- State clearly the bearing type, i.e. magnetic, true or grid.
- The remarks column should specify features to look for or features that you may well cross, i.e. rivers, wadis, saddles, cols, etc.

The start and finish positions of the route
These positions must be in a coordinate format and not just relate to a map feature.

Understanding the contents of the maps you will be using
A simple mnemonic to remember is VAGS:

V – **vertical interval**, or contour spacing: being able to recognise the relief patterns on the map and where necessary the steepness of the terrain.

A – **age of the map**: understanding that the detail on the map might well not represent the features on the ground. The map could be so old that a lot has changed in the intervening years, i.e. track direction.

G – **grid magnetic angle** or **magnetic variation**: understanding the corrections that you will have to apply to compass magnetic bearings to be able to draw/plot those bearings on the map.

S – **scale**: as already mentioned, different scale maps will represent the land in different ways.

Once you have fully digested all of the map information, you can start your route planning. Again you can go about this by using another mnemonic – TRECH:

T – **time and distance**: take into consideration the terrain you are crossing, the effect it will have on the distance that you are realistically likely to cover by your mode of travel in a set time scale. Do not attempt too much in one single leg. Many legs will make navigation easier.

R – **relief and going**: take into consideration the relief and how badly it will affect movement. The well-established Naismith's Rule, based on hillwalking experience, suggests that you should allow, for movement on foot with pack:
  5 km per hour
  + 0.5 hour per 300m of ascent
  – 0.5 hour per 300m of descent
  + 0.5 hour per 300m of steep/difficult descent.
  This provides a reasonable guide when tempered with experience.

E – **ease of navigation**: try to plan the legs of your route that coincide with prominent terrain features. It is worth considering the following:
  Direct and non-direct routes: the fastest route may not be a straight line.
  Handrailing: you may wish to handrail features to aid in your navigation by the use of obvious features such as power lines, pipelines or cliffs.
  Height: do not lose height, especially if on foot and in the jungle.
  Aiming off: if point is on a linear feature, aim to hit it to one side and work towards it.
  Attack points: pick an easier feature and use it as a base to find the end of a leg.
  Catching features: bounce off larger, obvious features to find the end of a leg or route.

C – **cover**: the use of terrain or land features that can provide cover from the elements, i.e. the leeward side of hills if the prevailing wind direction is known.

H – **hazards and safety**: recognition of areas where problems might arise if no alternative is available. We are, after all, explorers! Are there possible escape routes?

ROUTE FOLLOWING

With a route finalised it is now time physically to go and follow it. Do not dive in feet first without first of all setting your map. This can be done by various techniques. You can use your compass to set the map to north. You can set your map by equating features visible on the ground with those same features on the map. And lastly mark on your map your known position.

To complement this it is always worth checking to your rear for back sights. Note start time of each leg. Continue to check your position by carrying out fixes (techniques on this subject are covered later) and keep your wits about you.

If you are in a vehicle, ignore wheel slippage, because it will be minimal. You should concentrate on the extra distance entailed in avoiding obstacles or the drift of the vehicle caused by the driver and side of the steering wheel. A right-hand drive vehicle has been proven to drift off to the right while driving on desert-type terrain.

If you are in any doubt, then stop. Do not guess! Your colleagues might be thankful for a stop for a hot drink or some food while you confirm your position.

Above all, concentrate!

Finding your end of leg

The best method of locating your destination is the square search. This is a proven technique. However, you must always leave someone where you finished before starting the search, because this will be your last known location!

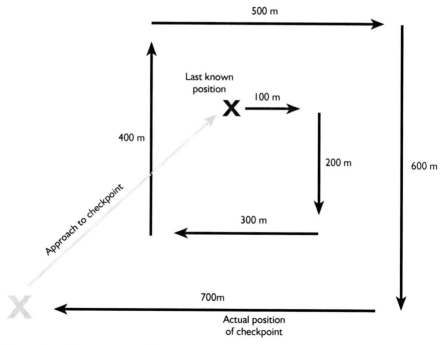

Figure 6.14 *The square search*

On nearing the end of the leg, you should be searching for it. Get a feel for how far you can see to your sides and to your front, e.g. this could be only 100 m.

At the end of the route, stop and leave at least one person there. You then proceed east, no matter the final direction, for the distance that you can search to your front and to your sides. In this example it would be 100 m.

At the end of the 100 m turn 90 degrees and search for 200 m, but keeping your search distance to 100 m.

Continue in this fashion until you locate your checkpoint, target, etc. (Figure 6.14).

Dead reckoning

The principle of navigation by dead reckoning (DR) is simple. From a defined position you travel in a defined direction, for a defined distance after which a new position can be determined.

This sounds good and easy in theory but in practice many errors can occur which will seriously affect your final DR position. The two main errors are in the measurement of distance (e.g. odometer error, poor pacing) and maintenance of your direction (e.g. obstacles, hazards). In order to counteract these errors the navigator must update his or her position regularly by using features that are on both the map and the ground.

There are two main methods available and they are either a multiple fix or a single fix. A modern term for the multiple fix is a resection and if done properly it will define your position very accurately. A single fix does as it describes. It uses only one bearing to obtain a position fix. This makes it less accurate than a multiple fix, but in some terrains you may have only one object that is both visible on the ground and appears on the map.

Multiple fix (Figure 6.15)

1. The size of triangle will depend upon the following:
   (a) compass and pointing errors (poor compass operation, not converting magnetic bearing to suit the map)
   (b) incorrect plotting of detail on map or chart.
2. Your true position will be within the plotted triangle.

Single fix (Figure 6.16)

If you only have one identifiable feature to use, then a single bearing fix should be used:

1. DR position plotted on map from course, speed and distance.
2. True position must be along bearing line from positively identified landmark or feature.
3. Assuming DR log to be correct, most probable estimated position (EP) is shortest distance (perpendicular) from bearing line to DR position.

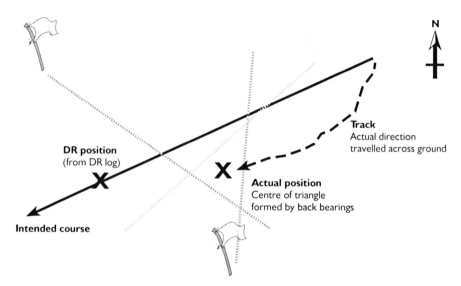

Figure 6.15 *Multiple fix. DR, dead reckoning*

Alternate running fix (Figure 6.17)

Where only one feature is identifiable you can use it more than once to determine your position:

1. Start as you would for a single bearing fix. Mark DR position on map. Take bearing to feature and plot it on map.
2. Travel a fixed distance D on a fixed direction.

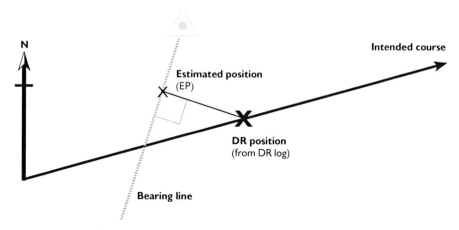

Figure 6.16 *Single fix. DR, dead reckoning*

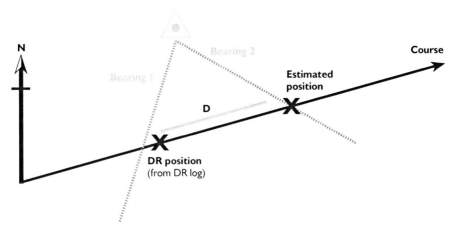

Figure 6.17 *Alternate running fix. DR, dead reckoning*

3. Plot second bearing to feature.
4. First and second positions must lie on bearing lines.
5. Distance D and course travelled can only fit in one unique position and a new EP can be plotted.

DR log

The DR log is filled out by the navigator during the navigation of a route and is a continuous update of the route planner.

The remarks box is where bearings for resections and single bearing fixes are logged (Figure 6.18). DR logs can be designed with your mode of transport in mind.

| Leg | Time | | Time | Bearing | Speedo | | Dist | Remarks |
	Start	Finish			Start	Finish		

Figure 6.18 *Dead reckoning log*

How far can I see in a flat featureless environment?

In flat featureless terrain, it is useful to know the distance to the visible horizon and the following formula can be used:

$$\sqrt{(\text{Height of the person in feet} \times 3/2)} = \text{Distance in nautical miles (nm)}$$

For example, for a 6-foot observer:

$$\sqrt{(6 \times 3/2)}$$
$$\sqrt{18/2} = 3 \text{ nm}$$
$$1 \text{ nm} = 1.852 \text{ km}$$

Which implies that the average person can see out to approximately 5.5 km. Standing on a vehicle can increase visibility to 8–10 km.

EQUIPMENT MANAGEMENT

Maps

The best thing to do with any map is to place it in a proper map case. Try not to cut up your maps, there is so much information in the margins. Get used to handling a few maps in a single case. You may decide that you want to cover your maps with laminate or Fablon. This is all well and good for waterproofing, but not for carrying out any position fixes (the author can testify to this!). The only accurate way of drawing bearings on to a map is by pencil!

Compasses

Remember to keep a lanyard on your compass and hang it safely somewhere. Around the neck is ideal and it is easy to access from there. If you use a prismatic compass, then get a padded case to keep it in until it is required. Keep away from metal objects when taking bearings and do not store your compass by a metallic object. After a period of time this will affect the correct direction of magnetic north. If you do happen to get a bubble in the compass housing, the best way to disperse it is gently to heat it (not over a fire!) and hope that that does the trick. If it persists and affects the movement of the compass needle, then put a small hole into the housing to drain the fluid.

Electrical items

Remember that the life span of a battery will change depending upon the environment that you are in. If you are using lithium batteries the life span can be extended, but the fall-off in power is sharper and quicker. Remember, however, that all batteries should be disposed of properly.

The environment will also affect the way electrical bits work. Placing an equipment item on the dashboard of a vehicle while in a hot climate is just asking for trouble.

Always have proper cases for all electrical items. They can keep heat at bay, keep heat in and also provide protection against falls, etc.

CONCLUSION

So now you should have the basics to enable you to put your mapping, navigation and GPS together for your expedition. Acquiring your expedition maps will probably be one of your hardest procurement jobs. Plan early and you should not be disappointed. You will, no doubt, run training sessions before departure; use these to practise navigation techniques and to introduce the tools that you will be using.

FURTHER INFORMATION

Further reading

Burns, B. (1999) *Wilderness Navigation: Finding your way using map, compass, altimeter and GPS*. Seattle: The Mountaineers Books.

Clark, M. (1993) *Expedition Use of Maps, Air Photographs and Satellite Imagery from Expeditions Planners Handbook and Directory, 1993–1994*. London: RGS–IBG Expedition Advisory Centre.

Ferguson, M. (1997) *GPS Land Navigation: A complete guidebook for backcountry users of the NAVSTAR satellite system*. Boise, Idaho: Glassford Publishing.

HMSO (1989) *Manual of Map Reading and Land Navigation*. London: HMSO.

Kals, W.S. (2002) *The Complete Land Navigation Handbook*. San Francisco: Sierra Club.

Keay, W. (1994) *Land Navigation*. Windsor: Duke of Edinburgh Award Scheme.

McWilliam, N. (ed.) (2003) *Expedition Field Techniques: Geographical information sciences*. London: RGS–IBG Expedition Advisory Centre, in press.

Royal Institute of Navigation (2003) *Guide to GPS for Navigators in Foot*. London: Royal Institute of Navigation.

Sources of information

RGS–IBG Map Collection, 1 Kensington Gore, London SW7 2AR. Tel: +44 207 591 3000, email: info@rgs.org, website: www.rgs.org

The Map Room of the Royal Geographical Society (with the Institute of British Geographers) contains one of the largest private collections of maps and related material in the world. The one million sheets of maps and charts, 2600 atlases, 40 globes (as gores or mounted on stands) and 700 gazetteers comprise the core of the Map Room's material. It includes printed items (on paper, vellum, and cotton or silk) dating as far back as AD 1482, manuscript items from 1716 onwards, aerial photography from 1919, satellite imagery maps and CD-ROMs.

RGS–IBG Expedition Mapping Unit. Website: www.rgs.org/mapping

Fieldworkers use many conventional survey and recording techniques. They can also benefit from new technologies: the global positioning system (GPS) to record positions and to navigate; remote sensing (RS) to provide environmental information; and geographic information systems (GIS) to record, process and display all kinds of spatial data. The Expedition Mapping Unit exists to share expertise in these techniques. The emphasis is on low-cost, reliable, relatively simple approaches – ones appropriate to most non-profit research and conservation endeavours. This website compiled by a volunteer network of fieldworkers has a useful list of web links on a wide range of geographic information resources including printed maps, satellite imagery, GPS and data sets for GIS.

British Cartographic Society. Website: www.cartography.org.uk
The Map Curators' Group of the British Cartographic Society has compiled the 4th edition (2000) of *A Directory of UK Map Collections* on-line only and can be found at: www.cartography.org.uk/Pages/Publicat/Ukdir/UKDirect.html

British Library. Website: www.bl.uk
Bodleian Map Library, Oxford. Website: http://www.bodley.ox.ac.uk

British Orienteering Federation. Website: www.britishorienteering.org.uk

British Schools Orienteering Federation. Website: www.bsoa.org

International Map Trade Association (IMTA). Website: www.maptrade.org

Land Info International. Website: www.landinfo.com
Maintains the world's largest commercial database of digital GIS data products. Covering more than 75 per cent of the Earth's surface, Land Info has an expanding archive of over 250,000 maps.

The Map Shop, Upton-on-Severn, Worcs. UK. Website: www.themapshop.co.uk

National Library of Scotland. Website: www.nls.uk

National Library of Wales. Website: www.llgc.org.uk

National Navigation Award. Website: www.nnas.org.uk

National Remote Sensing Centre Ltd. Website: www.nrsc.co.uk
Satellite images.

Public Record Office, Kew. Website: www.pro.gov.uk

Oddens's bookmarks. Website: http://oddens.geog.uu.nl/index.html
Comprehensive and up-to-date list of map and map-related sites.

Ordnance Survey. Website: www.ordsvy.gov.uk/

Nigel Press Associates Ltd. Website: www.npagroup.co.uk

Royal Institute of Navigation. Website: www.rin.org.uk

Stanfords. Website: www.stanfords.co.uk
International map and guidebook retailers.

United Kingdom Hydrographic Office. Website: www.ukho.gov.uk

SECTION 2
FINANCE AND FUND-RAISING

7 BUDGETING AND FUND-RAISING

Nigel Winser

Fund-raising is without doubt the most arduous part of preparing for an expedition. It will inevitably take up a very large proportion of your pre-expedition time and cause great anxiety. Nevertheless, hundreds of teams are successful in raising sufficient funds each year, and there is no reason why you shouldn't be among them. All that is needed to realise success is thoughtful preparation and a firm resolve. Good luck and humour help.

FINANCIAL MANAGEMENT

The budget
The first draft of your budget is a cornerstone of your planning, so get it down on paper at the earliest opportunity. Do not be afraid of committing yourself, because it represents your thinking at an early stage and provides a basis from which further changes can be made. You will have to adapt your budget as more information becomes available during your preparations, but nevertheless the initial budget will help to ensure that you have a comprehensive list of all the financial components. In addition it will assist in identifying the inescapable "fixed costs" from those that are "adjustable". By altering your fixed costs (e.g. international flights) you will almost certainly affect the overall plan, which may in turn change the size and length of the project. By altering the adjustable costs (e.g. food costs) you are more likely to limit the scope, comfort, efficiency or degree of safety of the operation. There may come a time when the treasurer has to tell the leader that there will not be enough money for the expedition to take place in its planned form. Such cut-off points should be identified in advance and some thought given to the alternatives such as: choosing a closer/cheaper destination, reducing the research programmes, cutting membership or duration and so on. Be sure to allow for inflation – abroad as well as in the UK.

Most expeditions should be working in conjunction with the host country. In many cases there is confusion about the allowances for the host country participants. It is strongly recommended that these expenses be agreed before the expedition leaves.

The budget for an imaginary expedition to the remote Lotogipi Forest is given in Appendix 5, with possible fixed and adjustable costs. The art of the leader is to distinguish between the two and, through consultation and reference to current information, decide which costs belong in which category. In so doing, do keep all the team fully in the picture, so that they understand when their field accommodation gets downgraded!

When drawing up the budget, it is helpful to "guesstimate" a maximum and minimum expenditure for each item and a maximum and minimum income from possible sources. Your budgeting success will be determined by your ability accurately to manage incoming and outgoing funds within continually changing limits. Failure to do this will usually result in team members or their families temporarily having to make up the shortfall, or having to rely on loans or overdraft facilities.

Reducing costs

There are as many ways of reducing costs as of raising the funds to pay for them, limited only by your initiative and the absolute necessity to maintain personal safety in the field. Certain areas relating to safety, particularly flights and insurance, should not be jeopardised by focusing solely on the cheapest available option. This does not mean that you should not investigate the possibilities of reducing prices – just investigate with caution.

A reconnaissance visit to the host country can be extremely beneficial in this respect. You may be able to avoid excess luggage or freight charges by buying food supplies or equipment there, or ordering what you need through the local branch of an international organisation. The reconnaissance will provide detailed field costs of items such as food and accommodation. A host university or institution may be prepared to offer transport, accommodation, loan of equipment or other services in exchange for the opportunity for their students or members to participate in your expedition. Groups within the host country are more likely to be able to provide services than financial assistance, and you are far more likely to get help with logistics through early meetings with local counterparts during a reconnaissance. They are busy people too, so don't expect them to drop everything to run to your aid the moment you arrive. A reconnaissance allows time for them to prepare as well.

International flights, and travel in general, are probably going to be your major costs. Especially in the case of return visits you could investigate the possibility of writing an article for the airline's in-flight magazine (high-quality photographs are essential) in return for concessionary tickets. If you are a large party, you can probably negotiate a group discount.

If you are supported by an organisation, and you ask politely, it may be possible to use its communication services, including postage, telephones, fax, email or photocopying facilities. Schools and university departments will always help if they can, especially if the project has secured official support.

Many outdoor equipment companies are annually besieged by worthy expeditions asking for climbing equipment, rucksacks, compasses, and so on. You may be able to acquire discounts if you are a recognised expedition, but don't expect much to be donated, unless you have something particular to offer, such as publishable quality photographs, or highly experienced members who may be able to field-test new products.

Each expedition is unique, and will have its own set of contacts and ideas. As a rule it should be feasible to cut substantially the actual amount of funds you need to raise by first looking at the options for reducing costs. Having the "best–worst" case budget always to hand (on your laptop) will help you and your team with your planning.

SETTING UP THE FINANCIAL INFRASTRUCTURE

Administration

Even before your hard-earned grants and sponsorship begin to accrue, the methods for administering the money should be established and a treasurer appointed in charge of financial administration. Relatively large amounts of other people's money will be in the care of your expedition treasurer. The expedition must be accountable to both the team members and the sponsors. It is imperative that the accounts are kept with precision and all transactions assiduously recorded throughout the preparatory, fieldwork and post-expedition stages. Given the overall coordinating role of the leader, it is usually unwise for the leader also to be the treasurer.

Although the treasurer controls the mechanisms of incoming and outgoing cash, expenditure should be continuously monitored and agreed by the whole team, especially if expensive items of equipment are involved. While in the field, keeping accounts is extremely difficult, requiring constant vigilance to keep up with daily expenses and changes in exchange rates. Even obtaining receipts can be notoriously arduous in some countries. It is essential to keep the accounts up to date, and to ensure that all team members are aware of the expedition's financial situation. It should be remembered that, in reality, each member of the team is equally responsible for all aspects of their expedition, even if individuals have been allocated coordinating tasks. This is pertinent to the job of budgeting and coordinating funds. The final accounts should appear in the final report and copies of this should go to all those who have contributed to the expedition fund.

Bank help

Make an early appointment to see your bank manager. The manager will be able to advise the best way of distributing funds between accounts as well as the appropriate international services that will be of assistance to you, once you have explained the what, where, when and why of your proposed project.

- Explain your reasons for opening an account and ask advice on allocating the funds between a current and a deposit account. Investing hard-won funds in a deposit account or high-yielding investment account can sometimes be surprisingly productive. Long-term deposits will require an accurate estimate of cash flow to be effective.
- Ask what arrangements can be made in the country that you are visiting. Remember that some countries do not permit funds to be exported and so you might not be able to bring surplus cash back. In these circumstances you will not want to send abroad more than you are likely to spend. Local financial sponsorship is more forthcoming and easier to handle if you have a local account. Consider the various forms of taking money overseas: cash, travellers' cheques, credit card, banker's drafts to local bank. Travellers' cheques are often difficult to cash in remote places, particularly if they are in sterling or from a locally unknown bank. On the other hand, credit cards work in most foreign countries through local banks and shops, but be aware that credit cards are charged at official rates of exchange. You may lose out if you use them in a country with high inflation or where there are better exchange rates from local dealers.
- If you are unsure of which bank to approach, find out which bank is best represented in the area that you plan to visit (e.g. Lloyds International for South America, Barclays International for Africa, etc.). For convenience "up-country" it may be best to use the largest clearing bank of the host country, which has a branch in the town nearest to the expedition area.
- Ask about any local sources of funding. Managers are often trustees of charities such as the Round Table or Rotary Club.

Details to be considered at the outset will obviously vary from project to project, but may include standing orders for personal contributions or initiating contracts with overseas banks or branches. You are unlikely to be offered either substantial loans or other credit facilities if you don't have collateral or a guaranteed income to offset against the loan.

Securing the support of your local bank manager will always pay dividends should you need to negotiate any short-term loans. Don't forget that, if you are a young team, the bank manager sees you as long-term business associates and will be keen to secure your custom. An acknowledgement to your bank in the final report is always a good move.

Cash control

The treasurer's juggling act with the team's finances has to be one of the most unenviable jobs on any project. They should always know the current status of the expedition bank account, expected outgoings and promised additions, and all members must keep within any agreed budget plan. There will be some costs that are fixed to dates (especially flights), whereas others are flexible and so can be set aside until more cash arrives.

Passing information from individual members to the treasurer about their own fund-raising responsibilities and expected costs related to their part of the project is essential throughout. Without this continuous feedback the team runs the risk of having insufficient money at critical times.

Ensuring that sufficient but not excessive amounts of cash are available for daily expenses in the preparatory stage and during fieldwork is an awkward problem. Limiting petty cash to a minimum will help to keep accounts orderly through the use of cheques. Assiduous collection of receipts in exchange for repayment of cash to individuals is essential but difficult, especially in the field.

In the host country you may well appear very wealthy to the local people, so be sensitive about having large amounts of conspicuous cash around your base. There is always a risk of theft, so don't encourage it by inadvertently showing your entire field budget to the whole village. This is an avoidable problem!

PREPARING TO FUND-RAISE

Image and marketing

Your corporate image

It is essential to project the right impression – your image – if your fund-raising efforts are to be fruitful. You will often be competing directly with other expeditions for finite funds, so you will want to create an impression of competent professionalism. This can be done in a variety of ways, which might include printed stationery with recognisable identity, demonstrating that you have the support of knowledgeable people, and a general competence in your appearance and activities.

Producing attractive headed notepaper with the expedition's name, logo, contact address and major supporters can be easily and inexpensively achieved using high-street printers and/or desk-top publishing systems. Despite the minimal effort involved, headed paper immediately gives a professional "corporate" image to all expedition correspondence. An eye-catching design will stand out against a mountain of other applications for assistance on a managing director's desk, and could clinch the vital advantage in swaying the decision on sponsorship in your favour.

The brochure

The expedition brochure is probably the key document for successful fund-raising from grant-giving organisations, trusts, and the world of commerce and industry. A concise, informative and attractive brochure provides evidence of competent preparation on the part of the team, and will rapidly answer the initial questions that will be asked by all those you approach for assistance: Where? Why? When? Who? and the all-important, How much?

The design of the brochure is very important, and should involve input from all team members. This is the single most important document representing the expedition's intentions until the production of the post-expedition report. The cover design should answer the "where? why? and who?" immediately, with the contents addressing each question in more detail and incorporating the budget. A simple, appropriate, clearly identifiable logo can be a visual cue repeated in the prospectus, headed paper, T-shirts and merchandise, and can contribute to a "professional corporate image" for your team. As with the headed paper, desk-top publishing offers inexpensive high-quality brochure-production opportunities, and is usually freely available within universities and schools.

The contents of the prospectus should be factual and accurate as well as innovative and inspiring. Be creative and use stunning photographs. It should be well designed but not lavish and should be something that all members are proud to distribute whenever they have an opportunity. Some of the following might be appropriate to include:

- Describe the project's aim and objectives, clearly.
- Explain how the project came about and why it is worth supporting.
- Give some background on why this work is relevant and timely now.
- A note about collaboration and who else is involved and/or supporting the project.
- Include a really good map, which shows the location and detailed study area.
- Membership details, with pertinent biographical details of all members, especially your local counterparts and home agent.
- Your income and expenditure budget, with details of how you will raise the funds.
- A section on sponsorship and what you can offer individual and corporate sponsors.
- Contact details in UK and in the field – address, telephone number(s), email, website.

Appendix 6 lists some of the suggested headings for an expedition brochure in more detail.

It can be tempting to over-emphasise a particularly evocative aspect of the proposed work in order to cash in. This tendency, otherwise known as the "sexy species concept", can be advantageous but will be relevant only to a few expeditions. Be careful, because interviews and application forms will rapidly expose inappropriate emphasis.

It is also worth considering the production of different brochures for different types of funds, e.g. when applying to scientific establishments, a more detailed prospectus is required, whereas companies may only want a single-page leaflet.

Desk-top publishing is very flexible and can be used to add names of sponsoring companies or organisations to the headed paper or prospectus as they come on board. This attempts to generate positive feedback by showing the company being approached which other companies already support your efforts. By regularly updating the budget to show the remaining total required serves to emphasise existing successes and encourages others to come on board.

If your expedition is to have a well-known patron or patrons, try to find someone who will add prestige and credibility to your efforts in both the UK and the host country. Consider having two separate patrons, one at home and one abroad. You are more likely to gain sponsorship from branches or subsidiaries of international companies in your host country if your patron is a well-known national of that country. This may be particularly important when raising money to pay for your counterparts' expenses in the case of a joint team including members from the host country. Ideally your patron(s) should be sufficiently interested in your project to be prepared to assist you by speaking or writing on your behalf. If they are able to offer fund-raising support with information or participation, all the better, although high-profile people are inevitably busy. Having said this, a patron is not a prerequisite. Many projects successfully raise sufficient funds without a patron. It is up to individual teams to decide.

FUND-RAISING

Why should people give you money?
An organisation or individual may decide to give you funds towards your expedition for any one of a number of reasons:

- Research: where the interest is in the acquisition of new knowledge and the training of field scientists.
- Personal development: encouraging you and your team to learn from the experience of planning a project, travelling and working with people from another nation, and broadening the horizons of all concerned.
- Publicity: businesses may see you as a good source of publicity, perhaps because they are keen to encourage host country participation where they have business interests, or to raise the profile of a particular product.

Before you appeal to any organisation for support you need to be clear why they might give you money, and word your approach accordingly. Don't be disheartened by initial failure to get a positive response. The overall success rate is inevitably low as a result of over-subscription of the well-known sources of funds. As your expedition image and approach become more professional and finely tuned, so your success rate will increase.

In general, teams that are successful exhibit a number of common factors: belief in the value of the proposed project, clear understanding and belief in the team's ability to realise its objectives, a strong commitment to see the project to fruition, a professional approach in all aspects of the expedition, and ingenuity and persistence in large amounts.

Raising funds

Aside from high-level commercial sponsorship and media interest, which are dealt with in Chapter 8, support for your venture is likely to come in either cash or kind from any of the following sources:

* grant-giving organisations
* charitable trusts
* local authorities
* expedition members
* the public.

To attract help from the above, any number of methods can be used. The suggestions are endless, but outlined below are some of the more conventional methods that have been used by expeditions.

It is often the case that modest requests for assistance at the local level, as opposed to single large amounts of money sought from national or international companies, are the mainstay of fund-raising successes. You are more likely to have personal contacts from school and university (governors, teachers, past students), family or friends, and it is easier for small expeditions to get publicity in the local media than at the national level. It tends to be the more experienced, professional expeditions that are able to tackle national and international companies, although this shouldn't discourage you from approaching relevant businesses irrespective of their size.

For anyone who may give you grants, try to find out what you can give them in return. For most funding organisations this will be a report, and you must make your report suitable for the audience concerned.

The categories below show a range of different sources of income for expeditions. When planning your own fund-raising strategy, look at how similar expeditions have successfully managed to raise their funds by looking through their reports.

Grant-giving organisations

There are a number of grant-giving organisations to which you can apply. Several directories exist which give details of these and the RGS–IBG publishes a *Directory of Grant-Giving Organisations for Expeditions* on its website, which is updated annually, and a selection of these is given in Appendix 8.

Each organisation will have its own criteria for giving money. Do your homework and identify only those pertinent to your project. In this way you will save time, resources and the unnecessary disappointment caused by a flood of rejections from applications to inappropriate sources. Remember that some will give only to individuals, and others only to groups or teams of researchers.

Most organisations want applications well in advance and in a certain format. Find out what details are required before you send in an application. Many will ask you to provide references, so you must identify someone (well qualified) who will be prepared to speak on your behalf.

Most grant-giving bodies like to see healthy "personal contributions" from expedition members. This should be in the region of £500 to £800 for UK members and £100 to £300 for local members.

You may be called for interview. If so, you will be thoroughly questioned on your project and it would be useful if the interviewees include people expert in both the scientific aims and logistics. Trying to disguise insufficient background research is unwise. It is better to admit neglect than to try bluffing! Talk to others who have been interviewed before to learn how to perform well at interview.

Support from organisations such as the Royal Geographical Society (with the Institute of British Geographers), Mount Everest Foundation, British Canoe Union, British Cave Research Association, British Ecological Society, Birdlife International, Flora and Fauna International, Scientific Exploration Society or Young Explorers' Trust will give others confidence to follow suit.

Charitable trusts

There are hundreds of charitable trust funds in the UK and some of them support such activities as environmental conservation and medical research or development, which may be relevant to your expedition. A few are concerned with particular regions. Names of the trusts are found in the *Directory of Grant-Making Trusts*, which classifies them by areas of interest. A letter mentioning why your proposal is relevant to the charity is usually sufficient, together with the expedition prospectus. It obviously helps greatly if you know one of the charity's trustees. Enquiries can also be made through your library for details of locally based charities.

Local authorities

The Lothian Exploration Group has found a selection of the following approaches to local authorities successful for school expeditions both in the UK and abroad.

Grants

- From regional council, often given to individuals or for specific schemes and also available to groups.
- From local education department, usually small amounts for specific expeditions

to support in-service training for staff or to assist with staff expenses.
- Sports awards, from departments of leisure and recreation, available to individuals or groups.
- School councils or community associations.

Indirect support
- By loan of equipment from outdoor education centres or schools. This may include buying specialist equipment for the expedition that will later be used by the centre.
- Items that can come out of a special vote, which are at the discretion of the local authorities or their advisers, e.g. stationery, first aid or transport. Inspection of the relevant administrative memoranda will reveal this.

Local government names and addresses can be found in *The Municipal Year Book*, or contact your local education authority (LEA) outdoor education adviser.

From expedition members (personal contributions)
There is a tendency to keep members' contributions too low. Fix them as high as possible to cover at least 50 per cent of the expected expenditure. Do not be afraid of doing this – if people really want to go they will raise the money and, despite the apparently high costs, an expedition is usually excellent value compared with a holiday. Do not under-rate this aspect of the expedition finances; many expeditions are financed entirely in this way. If you manage to accumulate large funds, the contributions can be reduced by refunding money later.

Get this personal income into the expedition account as soon as you can. It will probably be needed to pay some of the early bills, or it can be put on deposit to earn interest. The size of the personal contribution is an important factor when approaching grant-giving organisations – the bigger the better.

Personal contributions should ideally cover: (1) a substantial portion of the air fare; (2) a joining fee towards general expedition overheads; and (3) a per diem to cover food and subsistence for each day in the field. You will need to agree a per diem well in advance, because it is needed for applications to some grant-giving organisations. Remember that it does not cover just the cost of food consumed by an individual, but represents a share of all administration, base camp and local transport costs.

From the public through fund-raising activities
This is the most diverse way of fund-raising and is especially popular with those who have a large volunteer workforce with time on their hands. Some methods have legal restrictions, and some can cost more to run than you expect, but the opportunities before, during and after the expedition are many. Examples include:

- *Sponsored events* such as walks, swims, canoe rolls, parachute jumps, etc. are often profitable. However, there is sometimes a tendency to complete "gimmicky" sponsored events for the press novelty value. This seems to be of limited value. It would seem more appropriate if sponsored events could be useful, so that your local community benefit from your efforts. Litter collecting, tree planting and community service activities offer a wide range of possibilities. This is more likely to predispose people to sponsor you than subjecting yourself to some ridiculous ordeal. Make it clear from the start how much of the proceeds will go to the expedition central fund as opposed to reducing the fees of the person doing the event; 50:50 is probably right in most cases.
- *Raffles* can be especially profitable if local firms donate good prizes, and you have plenty of friends to help sell tickets. However, you must be aware of legal implications of formally organised raffles, which involve paying for a licence from your district council. Work out a rough cost–benefit analysis before you start, to make sure that it is worth the effort and time.
- *Trading activities* can be highly profitable, sufficient to fund an entire expedition, or they can involve serious financial misadventure. Items bought wholesale in bulk, and inscribed with the project name or logo, can be relatively inexpensive to buy, but you have to analyse your market first to see whether there is an existing demand or an opening to create one of your own. Many projects have their own T-shirts printed as part of the team image, as well as to present to counterparts or sponsors. With a particularly eye-catching design, there are always possibilities of selling to interested family and friends, members of pertinent specialist societies and organisations, and possibly the wider general public. Certainly, with the last, the level of organisation required is extensive, with advertisements, marketing and predictive ordering of merchandise. This must be viewed with caution and, because capital outlay is required, it is possible to lose a great deal of money. This type of venture is ideally suited to ongoing projects of several years.
- *Fêtes, jumble sales and bazaars* can be fun and profitable. Try to recruit someone who has experience; otherwise start with caution. There is plenty of scope here for ideas from enthusiastic supporters.
- *Philatelists* sometimes pay for first-day covers which entails franking the envelopes when you are in the field and carrying them the length of the expedition without getting them dirty. This can be profitable if a relevant stamp is issued near the time of the expedition departure. But, be careful, because this sort of project involves a lot of risk capital. Stamps have no post office value once postmarked and it is thus essential to get prepayment from philatelists before buying and postmarking the stamps.

When lecturing to clubs, schools and universities, make sure that you are aware of the travel costs and the fees that you will be paid, otherwise you may end up out of pocket. Do not rely on this method of fund-raising after the expedition is over – it should really be considered only as a last resort. Normally, only famous names command high fees for lecturing. If you want to have your name added to the RGS–IBG's List of Lecturers on Geography and Exploration, to help try to raise some money this way, contact the Expedition Advisory Centre on your return.

CONCLUSION

There are a number of funding sources available for expeditions but competition is fierce. Your fund-raising success will be dependent on your professionalism, individual ingenuity, persistence, confidence and belief in the value of your work. You will be rejected by some of those whom you approach for sponsorship, but use those rejections to encourage your team to find alternative sources of funds. Don't be disheartened; there are almost unlimited ways of securing your funds, and when you have succeeded you will look down from your mountain, look out across the forest, or look up from your microscope and be certain that it was well worth the effort. Good luck!

FURTHER INFORMATION

Charities Aid Foundation (CAF), 25 Kingston Avenue, Kings Hill, West Malling, Kent ME19 4TA. Tel: +44 1732 520 000, fax: +44 1732 520 001, email: enquiries@caf.charitynet.org, website: www.CAFonline.org
CAF is an international non-governmental organisation that provides specialist financial services to other charities and their supporters.

Directory of Social Change, 24 Stephenson Way, London NW1 2DP. Tel: +44 20 7391 4800, fax: +44 20 7391 4808, email: info@dsc.org.uk, website: www.dsc.org.uk
A registered charity providing information and support to voluntary and community organisations worldwide. They run training courses and publish a number of excellent books on fund-raising and a wide range of directories of grant-giving organisations including the CAF's *Directory of Grant-Making Trusts.*

Institute of Charity Fund-Raising Managers, Market Towers, 1 Nine Elms Lane, London SW8 5NQ. Tel: +44 20 7627 3436, fax: +44 20 7627 3508, email: enquiries@institute-of-fundraising.org.uk, website: www.icfm.org.uk
Publish lists of fund-raising consultants and the Code of Conduct for them, and some useful model contracts.

National Council for Voluntary Organisations, Regents Wharf, 8 All Saints Street, London N1 9RL. Tel: +44 20 7713 6161, fax: +44 20 7713 6300, email: ncvo@ncvo-vol.org.uk, website: www.ncvo-vol.org.uk

8 COMMERCIAL SPONSORSHIP AND THE MEDIA

Paul Deegan

Sponsorship can be found everywhere in modern society. From television dramas to professional sport, many commercial companies are prepared to invest money in order to align their product or services with those individuals or organisations that can deliver exposure to the public. It is no different with expeditions. If you can offer primetime television coverage, front cover news stories and an audience of millions, you will have little trouble in persuading one or more sponsors to underwrite the cost of your expedition.

The problem is that, in the 21st century, expeditions do not usually command this level of attention. So it is vital to think realistically about the amount of exposure that your expedition can generate, and then track down a sponsor who wants the type and level of publicity that you are able to offer.

WHICH COMES FIRST: SPONSORSHIP OR PUBLICITY?

The advantage of securing publicity before you chase sponsorship is that you will be able to demonstrate your commercial value to the companies that you decide to approach. The advantage of gaining sponsorship before you announce your plans to the public is that your sponsor will be able to coordinate your publicity campaign and ensure that their brand is aligned with your expedition from the outset. It is worth bearing in mind that, for every £1 a sponsor gives to an expedition, it usually spends the same amount announcing its support for the project to the world.

Given that publicity and sponsorship are difficult things to obtain, my approach is always to go after both simultaneously.

Sponsorship

Pros and cons of sponsorship versus self-funding
Although it is true that a sponsor who underwrites the cost of the expedition

removes at a stroke the threat of financial debt, there can be a heavy price to pay in other ways. To begin with, you will have to make commitments to your sponsor. These might include public appearances, lectures, interviews, regular updates from the field and product endorsement, as well as returning with high-quality photography and video footage for the media. This can all sound very exciting. But promises made in the warmth of a city centre boardroom can be harder to keep during the expedition when you are attempting to concentrate on ambitious scientific or adventurous objectives, not to mention at the end of the venture when you return to full-time study or employment.

Some friends of mine who have experienced life on a sponsored expedition now choose to raise the money for their expeditions through doing overtime at work. Their argument is that this kind of paid employment is guaranteed to generate the necessary money for their expeditions, whereas weeks or even months of fundraising activities often fail to generate a single penny.

Of course, if your expedition budget is so great that months or even years of employment would fail to generate sufficient funding for it, you might have no choice but to chase potential sponsors, or scale down your ambitions so that you are able to pay for it yourself.

Other expeditions strike a balance between these two extremes by paying for some of the expedition out of their own pockets, while attempting to raise the remainder of the money through sponsorship. Indeed, many would-be sponsors want to see some financial commitment from individual members before assigning their own finances to the project.

Types of sponsorship

Having made the decision to go after sponsorship, it is important to decide how much you need and what it is going to be used for. If some of your finances are going to be spent on clothing and kit, it might be easier to approach outdoor equipment manufacturers for product support. Not only are goods always easier to acquire than hard cash, but also the overall amount of money required for the expedition will be reduced. The same approach can be used with many other essential items on your budget, including flights, fuel and food. By employing this tactic successfully, the outstanding balance could end up being small enough to be covered with grants and small contributions from individual team members.

Your expedition's unique selling point

It will come as no surprise that many expeditions look for sponsorship, but few are rewarded. So decide early on what makes your expedition stand out from any other venture. Is it the place you are visiting, the people involved, the science you hope to achieve or the mountain you intend to climb? Whatever your unique selling point (USP), learn how to communicate this to potential sponsors in under 30 seconds over the telephone, or in two or three sentences on paper. Your USP is the hook that you will use to catch your sponsor.

Approaching potential sponsors

Many expeditions spend a great deal of time, effort and money in designing and printing glossy publicity material which describes the expedition's objectives, extols the merits of the team members, and waxes lyrical about the benefits to sponsors of aligning their name and product with the project. Most of this publicity material is then sent to dozens of managing directors with a letter (on expensive headed paper) that begins, "Dear Sir/Madam". The vast majority of this publicity material is duly thrown away by company secretaries, because it is seen by them as being a "round robin" letter, addressed to nobody in particular. Unsurprisingly, the expedition website that is heavily touted in the brochure is hardly ever visited.

Rather than adopting this scatter-gun approach, why not think carefully about which companies might see the greatest advantage in your endeavours? One approach is to draw up a list of a couple of dozen appropriate companies, and then call up to find the name of the person (normally the managing or marketing director) to whom your letter should be sent. Another – even more profitable – technique is to gather all the team members together for a meeting during which a list is made of personal contacts at various local and national companies. You might think that you don't have any contacts, but your immediate and extended family might.

One word of caution: do not underestimate the amount of time that sponsorship work takes up. It's not surprising that many expedition leaders arrive at the airport at the start of the expedition looking tired and worn out.

Different types of return for sponsors

The only expeditions that the public usually gets to hear about are high-profile adventurous projects. However, other types of expedition do attract sponsorship. For example, some companies look for internal rather than external publicity: they might want to use your expedition as a vehicle of focus for employees, or as a test bed for a new piece of equipment. Other organisations might want to raise their profile within your university campus or local area. So just because your expedition is not planning to ascend a hitherto unclimbed mountain, don't lose heart.

Retaining control of your expedition

It can be very easy for a sponsor deliberately or unwittingly to hijack an expedition, and for the leader to allow this to happen for fear of losing his or her benefactor. The hijacking might be visual, with the sponsor asking for each item of clothing to be branded with five or six badges rather than just one. Or the hijacking can be more physical, with members of staff flying into base camp and asking for a tour of the area just as team members are preparing themselves for a hazardous phase of the project.

Try to decide early on how much ownership of the expedition you are prepared to hand over to a sponsor in return for the support that they are offering.

The importance of "thank you"

In addition to delivering on all the agreements made with the sponsor, the expedition leader who wishes to receive funding for future ventures will go out of his or her way to ensure that the sponsor's expectations are exceeded wherever possible. This tactic of under-promising and over-delivering is infinitely preferable to making promises (such as guaranteeing summit photographs with the company's banner), which for obvious reasons cannot always be kept.

Publicity

Pros and cons of attracting publicity

As with sponsorship, publicity can be a double-edged sword. For the expedition leader looking in from the outside, the world of television, radio and newspapers can look very glamorous. The reality is quite different. News-gathering organisations are enormous machines, capable of reducing the most important aspects of your project to a sound bite, and focusing instead on the more light-hearted aspects of your trip, such as ablutions and physical relationships between individual team members. It is also worth bearing in mind from the outset that the greatest amount of media attention you are likely to generate is if something goes wrong during the expedition. Then you might find media interest to be intrusive.

That said, reputable media companies are more likely to treat your expedition with the respect that it deserves, and seek to inform and educate (rather than titillate) their audiences. If you enjoy communicating your experiences to the public, the media are perhaps the most powerful tool for spreading the word of your intentions before departure, and the results of your hard work both during and at the end of the project. Media coverage will also make your expedition a very attractive proposition to potential sponsors.

Of course, it is better to whisper before departure and shout loudly on your return: more than one expedition has set off surrounded by plenty of hype, only to fail to achieve its objective and return with egg not only on its face, but also on that of its unhappy and embarrassed sponsor.

Start small, think big

It is unlikely that an inexperienced expedition team is going to announce its plans to the world at large and be immediately picked up by national or international news organisations. It is worth remembering that the majority of experienced expedition leaders began their media careers by contacting local newspapers and local radio stations.

As well as enjoying a relatively high level of coverage with regional media, the forward-thinking expedition leader will realise that this platform provides an opportunity to learn how to be interviewed. The first time that I did a piece to camera on national television news, I gave silent thanks that I had been interviewed before on a lower-profile programme. I felt the same way when being interviewed on national radio for the first time. For further advice on sound recording and radio broadcasts, see Chapter 37.

Local newspapers and radio are rarely interested in people who do not either live or work in their catchment area. However, if your team members come from different parts of the country, you will greatly increase the number of media outlets available to you.

It is worth bearing in mind that many of the stories that are reported on by the national media are picked up by researchers who trawl through local news stories. That's not to say that you should not bother contacting the national media directly; just don't forget the local news organisations along the way.

Choose the most appropriate stories for different types of media
Think carefully about the stories that you have to offer, and which types of media are most likely to be interested. For instance, a local newspaper will probably want the story of the local climber who has named an unclimbed peak after his grandmother. By contrast, a national television news programme might be more excited about the fact that the same expedition has returned with video footage of a previously unseen mountain range. By generating several newsworthy stories for the media during the life of the expedition, you will be giving producers and editors every opportunity to continue covering your project.

The importance of a well-written press release
A press release is the time-honoured method of communicating a story to the media. By producing a well-written press release, you will be letting the journalist who reads it know that you have an understanding of what they are looking for in a story. The ideal press release will fit on to one side of A4 paper, be double spaced (to allow the journalist to write between the lines), and be sent to a named individual at the news organisation that you are contacting. An example of a successful press release is included in Appendix 7.

What editors and producers look for in photographs and video
When you return from your expedition, one of the first things that the media will ask for is photographs and video footage. Without these, the chance of obtaining post-expedition publicity will be very much reduced. Photographs should be clear, sharp and well exposed. Transparencies (slides) are infinitely preferable to negatives (prints). Always send professionally made duplicates rather than your precious originals.

When it comes to video, try to keep it simple. Straightforward shots of landscapes and events, with plenty of close-ups of people, are much more usable than footage riddled with pans, zooms and jazzy effects. It is also worth bearing in mind that footage shot on a three-chip digital camcorder will produce a quality of image acceptable to most broadcasters. Footage shot on a single chip digital or analogue camcorder will be used only if the subject matter is truly outstanding or unique. For further advice on video and film-making, see Chapter 38.

Pick the best time to release your stories
Timing is everything. If you announce your expedition on the day that England is playing Germany in the World Cup do not be surprised if you receive zero coverage. Although a gripping news story can break at any time and push your expedition off the schedule, you can at least avoid all the planned events, from celebrity marriages to local elections.

Persuading the media to run with your story
This is something of a black art: many expeditions claim that they cannot even get hold of a journalist to listen to their story. To get your expedition in the public eye, the first step to take is to put yourself in the shoes of the person to whom you want to speak. Journalists and editors are constantly bombarded with stories. An expedition that is going to study weather patterns in the Himalayas will come a poor second to a famous footballer being injured weeks before the World Cup, or a government minister being accused of misleading the public. However, an expedition that is going to the world's highest mountain range in order to try to find out if changes in the weather are responsible for annual floods that threaten the lives of millions of people is more likely to entice the journalist into running the story. For further advice on writing for magazines, see Chapter 35.

The next thing to consider is the appropriate time to contact the person to whom you wish to speak. In the case of a monthly magazine, telephone to ask the receptionist which day the publication goes to press, and then call the editor or journalist the week after this date when the office is less frenetic. In the case of a breakfast television programme, fax in your press release on the afternoon before the day you wish to appear, or contact the advance planning department a week or two before the official release of the story. The department will then file your press release for consideration under the relevant day's schedule.

Useful notes for interviewees
OK, so you have managed to get your press release under the nose of the relevant person in one of the news-gathering organisations, and they now wish to interview you. In the case of the print media, this will probably be done over the telephone. Remember that the journalist probably knows next to nothing about where you are

going or what you are trying to achieve, so try to get over just a few key facts that can be understood easily. The journalist does not want to hear your life story. And don't get too upset if, when you see the printed story, it is very short, incorrect or both.

On television, if the interviewer is in the studio, look at him or her rather than at the camera. If the interviewer is not present, you will need to focus on the camera lens. This can feel very strange: the secret is never to stop looking into the lens. (It's worth noting that individuals who glance away while being interviewed to camera are often deemed by the viewing public as being untrustworthy – watch people who do this being interviewed on the television news and you'll see what I mean.)

In all interviews, keep your answers shorter rather than longer. At the same time, avoid answering with just "Yes" and "No" because this is extremely boring. Finally, if you are the expedition leader but do not want to be interviewed, ask the team if anyone else would rather be interviewed and make that individual the expedition spokesperson.

A final tip: a single mention of one sponsor – included as part of an answer to a question – is usually ignored by the media. If you try to mention more than one sponsor in each interview, don't be surprised if your story is dropped.

CONCLUSION

Sponsorship and publicity are viewed by some people as being essential components of a successful expedition. Nothing could be further from the truth. A great deal of time and effort – that could otherwise be spent preparing for the venture or earning the money to pay for it – will need to be invested if an expedition is to attract any form of sponsorship or publicity.

That said, sponsorship does exist for the expedition team that has the tenacity, drive and determination to find it. Furthermore, by persuading the media to cover your expedition, you will greatly increase the chance of securing sponsorship for your project. Good luck!

9 CHARITY FUND-RAISING EXPEDITIONS

Richard Crane and Justine Williams

Fund-raising for charity by expeditions is now well established. It evolved from sponsored sporting events, such as sponsored bicycle rides and the running craze of the early 1980s, which involved a whole new generation of charity fund-raisers, to add to the traditional circle of coffee mornings and village fêtes. Now we have Charity Challenges, which are in effect energetic fund-raising events in an exotic overseas venue.

Most people have discovered that raising money for charity is not an easy option. The concept of fund-raising for a charity is often an add-on to an existing expedition plan, but in reality may require even greater planning, organisation and time than the trip itself. Whether you raise £100 or £100,000 it can give enormous satisfaction.

It's an exhausting business not to be entered into lightly. However, once the team is committed, there can be many benefits, not only to the charity receiving funds from your efforts, but also by providing a tightly defined target that spurs on the expedition as a group, demanding a new set of management skills from those involved.

CHOOSING A CHARITY

There is an almost endless selection of charities from which to choose. They start at the big end with the likes of Save the Children Fund, Christian Aid and Barnado's, going through the medium-sized groups such as ITDG (Intermediate Technology Development Group), and reaching little dynamos such as APT in Gloucestershire, with merely six workers, working on a plethora of projects in the less developed world.

Most will have their own personal preferences to guide them to their choice of charity. For the undecided there are one or two useful points to consider:

1. Are there any convenient parallels between the expedition and the charity? The

public tend to enjoy the thought that the expedition understands a little about the beneficiaries, e.g. a team of canoeists in Norway creates a strong mental image of bobbing around on choppy seas under dark cliffs, and this could quickly translate to raising funds for lifeboats and the RNLI. A tight concept and close link are essential in many fund-raising opportunities particularly in radio, TV or newspapers, because there is so little time/space to convey information to the audience.

2. It is worth asking yourself whether there is some experience that the charity can lend to the expedition, e.g. knowledge of particular countries, even contacts out there. But please remember that most charities are heavily overworked and many of their staff will not have the time or resources to devote to helping your project.

3. You might choose a large charity that is instantly recognisable to your potential sponsors, or conversely choose a smaller group and help raise public awareness about their activities among a larger audience. The fund-raising team of a larger charity may be able to provide greater help with support materials and a formalised system for collecting funds. Alternatively, there is a danger that you might become forgotten among a mass of other projects. So you could find that small is beautiful.

4. Some teams try to link several different charities together and all monies raised are split between them. This could work to the benefit of a smaller charity linked to a more well-known charity. But in general this tends to mean that the approach to donors is, like Hobson's choice, split so that no one gets maximum return, and none of the charities is particularly keen to service you fully.

Despite all the logic, there is no substitute for putting your efforts to the benefit of the charity that you honestly believe to be doing the most appropriate work in its field. When you are sincerely trying your utmost then everything you do becomes so much more fruitful and enjoyable.

FROM THE CHARITY'S POINT OF VIEW

Funds collected by an enthusiastic self-organising team are always welcome. It is a bonus if they also promote the aims and intentions of the charity. Added benefit comes from listing the names and addresses of all donors so that a long-term network of sympathisers can be built up.

However, expeditioners are prone to being rather active imaginative people who can, if undisciplined, become a drain on the time and resources of the charity. The most cost-effective scenario might be a venture that requires no more than a meeting over coffee to discuss proposals, a request after a couple of months for more hand-outs about the charity, and then a small celebration at the end to enable the charity to receive a cheque for a few thousand pounds.

A continual worry to the charity is that a single mismanaged expedition will tarnish their good reputation and public image. If you are raising money using their name, legally you should have their permission to do so. Charities often have strict rules governing the use of their logo and charity registration number and guidelines on how they like to be described on paper and by the media.

FINANCES

Each expedition must fund all its own costs. Charities do not contribute to expedition expenses.

In particular, teams must be explicit about whether they are approaching a sponsor for expedition funding or materials, or a charitable donation, or both. For example, to solicit Karrimor for some good rucksacks for the expedition is a quite different plea from asking a Women's Institute for a gift to the charity (see "Targeted letters" below).

It is advisable to get all donations sent directly to the charity. Their treasurer or fund-raising officer must, of course, be asked first for approval because sorting and filing all the material is time-consuming. In my experience the fund-raising officers will keep a separate log of all donations sent in as a result of your expedition.

METHODS OF FUND-RAISING

All charity events need to have a fund-raising mechanism by which potential donors are approached and monies collected and channelled back to the charity. It is all very well to say you aim to raise a million – it's a different thing to do it. Aim high – but notify the charity of your "confidently achievable" target. This avoids future disappointments all round when you have raised only £500 of a £20,000 target.

Individual sponsor form

This traditional method requires a lot of legwork on the part of the expedition member to go around all his or her friends and relatives. However, it is a cheap safe method that is usually guaranteed to generate a few hundred pounds from a quick local exploit and as much as a few thousand from a long exotic adventure.

Most charity fund-raising departments will be able to provide pro forma sponsorship forms for your use. Do check with the charity first before designing your own form. The sponsor form must be clear and concise, preferably with a splash of colour to grab the initial interest of the potential donors. Double-sided A4 is usually good because it folds easily into an envelope and also two sheets can be put up side by side on a noticeboard to give a full explanation of the trip.

After the headings, an effective format is to use the first paragraph to outline the expedition project, the second for a brief description of the work of the charity and a third to describe how your donations will be used (if you have arranged this with your charity). Following this a few details of expedition members or the field area or some expected problems lend specific interest to the sponsor form. The last paragraph will usually be a plea for donations rounded off by a "thank you for helping" phrase.

The second side needs to be divided up into columns and boxes for date, sponsor's name, address and amount donated. At the top of the page give a one-line description of the charity plus the expedition title and at the bottom of the page don't forget to put the charity's registration number, address and/or your own address. I usually persuade a friend to make a decent donation on the top line to get the ball rolling.

It is virtually always best to collect donations at the same time as donors sign the form because a stupendous amount of effort is required to seek out all signatories at the end of the event which may be up to 24 months ahead. Sponsor forms with your expedition title can also have their second side designed so that individual donors can send it with a cheque or credit card donation direct to the charity. The charity will be happy to receive donations as soon as possible in order to be able to put them to good use immediately. It is also possible to get your sponsors to Gift Aid their donations. This means the charity will receive a further 28 per cent of the donation amount from the Inland Revenue, at no further cost to your sponsor. You will need to get the correct legal wording from the charity to include this on your sponsorship form. But it is well worth doing as you can help raise nearly a third more in donations for your charity.

Informal sponsor form pyramid

Each expedition member persuades his or her friends, relatives and workmates to go around with sponsor forms collecting signatures, thereby spreading the net wider, saving the member legwork but putting some administrative burden on his or her shoulders. The benefit to the charity is an exponential increase in funds.

Formal sponsor form pyramid

This is the traditional method used initially for fun runs and marathons (the Great North Run, the London Marathon) and now copied and extended to overseas charity bike rides as well as many other "mass participation" events. The idea is that the event is open to as many people as possible and each automatically becomes a collector. It is usually quite expensive and time-consuming to organise and is a headache to supervise, but the returns can be enormous. This method could be applicable to an overseas school visit, for which all the pupils in each participating school join in fund-raising.

Targeted letters

These are usually on expedition headed notepaper to people, groups trusts or companies known to be sympathetic to expedition members or to the charity. Before any contacts are approached you must first check with the charity that the names on your list are acceptable to the charity, which may have an existing financial input with some groups, or have an ethical policy avoiding, say, tobacco or logging industries. This targeted method may require a lot of time-consuming research and the production of letters. Some people who claim that it can generate considerable donations have never raised a single penny this way, although I do know some who were successful with this method.

Internet

Promoting your expedition and its charitable aims on the internet need not be time-consuming to set up, and is a good opportunity to link up your expedition website with your chosen charity's website. For examples of how other expeditions have done this look at the links on RGS–IBG Expeditions database (www.rgs.org/expeditionreports). You can also go to www.justgiving.com to simply create your own fundraising web pages. Your sponsors can also donate directly on-line to your event.

Merchandising

This is the profits from selling T-shirts, postcards or knick-knacks (also included in this category is running a raffle for, say, the expedition's teddy bear).

OUT OF SIGHT ...

If your expedition is a lengthy one you need to ensure you are not "out of mind" of your sponsors. Invite all or selected sponsors to a send-off party; send back a "halfway" or regular newsletter for sponsors and the media. Keep your webpage live while you are travelling by keeping an on-line journal. Appoint a relative or friend as the UK base contact who can occasionally check on your fund-raising total with the charity and tell them where and how you are.

POST-EXPEDITION

Funds can be raised from lectures, articles, sale of photos and books, e.g. John Pilkington has donated several thousand pounds to ITDG through royalties from his two books *Into Thin Air* and *The Road to Tartary: Adventures on the Silk Route*. Many people fear that you need a flair for presentation to participate in any of these fund-raising opportunities, but, having sweated in front of a sea of expectant eyes on several occasions, I can assure you that everyone can relate the story of the adventures in their own little segment of an expedition so long as they pluck up a bit of courage and realise that they are the "world experts" with regard to their own endeavours. Remember to include your charity in feedback.

THE MEDIA PULL

This has the potential to draw spur-of-the-moment donations from people whom the expedition never sees. This can be extremely effective but requires plenty of luck to stumble on the exact formula that goads the media to give coverage to the adventure and also to urge the general public to send in donations. Any exposure in a local newspaper is beneficial to fund-raising because in the eye of the general public it gives an unofficial stamp of approval to a project. Try also local radio, and regional and satellite TV. For further advice on working with the media see Chapter 8.

THE WORLD BEYOND FUND-RAISING

Money is only the first step. Communication is equally, if not more, important in the long run.

From our platform as expeditioners we are able to inform others of the aims of our chosen charity and give them details of successful projects. This information will help them to understand why the work of this particular charity is necessary. You could be responsible for stimulating your sponsors to become long-term supporters of the charity, providing valuable donations way into the future.

Finally, it is worth remembering that, from the moment of the first link with a charity to the last celebration party of the expedition, you are all ambassadors of goodwill for your chosen charity.

SUGGESTED READING

Crane, R. and Crane, A. (1984) *Running the Himalayas*. London: New English Library.
Crane, R. and Crane, N. (1985) *Bicycles up Kilimanjaro*. Oxford: Oxford University Press.
Crane, R. and Crane, N. (1987) *Journey to the Centre of the Earth*. London: Bantam Press.
Hobsons (annual) *Sponsorship Yearbook*. Cambridge: Hobsons Publishing.
Institute of Fundraising. *Code of Practice: Charity Challenge Events*. website: www.icfm.org.uk
Pote, L. (1999) *Diary of a Sponsored Bike Ride*. Oxford: Oxford University Press.
Spittler, M. (1989) *Four Corners World Bike Ride*. Oxford: Oxford University Press.
Tomkins, R. (1999) Charity challenge or self-help trip? *Financial Times*, no. 34 035.
Tomkins, R. (1999) Charities spend donations on fundraisers' holidays. Regulator to probe whether sponsored "challenges" break guidelines. *Financial Times*, no. 34 035.
Tourism Concern (2000) Charity Challenges, special edn. In *Focus* 35, Spring. Website: www.tourismconcern.org.uk

SECTION 3
SAFE AND RESPONSIBLE
EXPEDITIONS

10 RISK ASSESSMENT AND CRISIS MANAGEMENT

Clive Barrow

Climb if you will, but remember that courage and strength are nought without prudence, and that a momentary negligence may destroy the happiness of a lifetime. Do nothing in haste; look well to each step; and from the beginning think what may be the end.

FROM SCRAMBLES AMONGST THE ALPS BY EDWARD WHYMPER (1860)

Risk assessment has become a prerequisite for organisers of expeditions and outdoor activities in the UK and overseas, and is now a legal requirement for commercially organised outdoor activities for under-18s in the UK. There is currently no law in this country governing the organisation of expeditions overseas. Many see this as a good thing. Fortunately, the number of serious incidents among participants in overseas expeditions is very small, at 0.3 per 1000 person-days (Anderson and Johnson, 2000).

However, the climate of opinion in the UK is changing in several ways:

- The public is more circumspect about safety and risk as a result of increased media coverage of expedition or outdoor activity accidents.
- As a nation, the UK is adopting a more litigious culture in line with the USA.
- Expectations of safety among the parents and guardians of young people are becoming higher as a result of the introduction of stringent safety procedures and Health and Safety regulations in educational establishments.

Given this risk-adverse climate, planners and leaders of all overseas expeditions should be conducting a systematic, careful and responsible safety management assessment. Risk assessment is the first and perhaps most important part of this. This chapter is intended to provide a brief practical guide to risk assessment, coupled with the key considerations involved in crisis management planning.

RISK ASSESSMENT (TABLE 10.1)

TABLE 10.1 RISK ASSESSMENT: SOME DEFINITIONS

Hazard	A situation or set of circumstances that have the potential to cause harm
Risk	The likelihood of harm potentially caused by a hazard
Risk assessment	The conscious process applied to the identification of hazards and the risks associated with them and the subsequent identification and implementation of a series of control measures to minimise the risk highlighted

Hazard and risk on overseas expeditions

Hazard and risk are inherent in everything we do and the degree of hazard and risk is dependent on the activity and environment in which that activity takes place. In the UK the degree of risk is considerably less than overseas, particularly in less developed countries where the ability to control our environment, coupled with a lack of knowledge of that environment, is proportionately greater. Risk assessment of overseas projects must therefore consider a wider array of hazards, and must always allow for the unexpected (Table 10.2).

The expedition organiser must always be prepared to adopt alternatives and/or completely abandon an activity if the risk assessment suggests that control measures cannot reduce the risk to an acceptable level.

In attempting to qualify and quantify risk, it is important not to worry unnecessarily about trivia. A risk assessment that is too cluttered with minor concerns will be discarded in the field as a bureaucrat's folly, and will be of less value than not doing one at all. Any severe and persistent risk must appear in the risk assessment document, together with appropriate control measures.

Acceptable risk

On an overseas expedition, risk can never be completely eliminated. Indeed, it is through the management of both perceived and real risks that expeditions of all types can have such beneficial effects on the participants. Most expedition organisers speak of reducing risk to an *acceptable* level. This is extremely difficult to define because opinions about acceptability may differ greatly among individuals.

The experience, age, ability and technical competence of the participants on an expedition or overseas project must be considered, because this will affect the level of risk considered acceptable. When considering the concept of acceptable risk, think first of to whom the risk should be acceptable? To whom are you accountable?

TABLE 10.2 HAZARD AND RISK ON OVERSEAS EXPEDITIONS

Hazard	Risk
1. The team	
Health and fitness (including previous medical conditions)	Increased risk of health problems on existing expedition leading to serious illness/death
Attitude and behaviour	Increased risk of ignoring control measures resulting in illness/injury
Experience and training	Lack increases risk in all activities
Personal equipment	Serious injury/illness resulting from inadequate equipment/equipment failure
2. The environment	
Mountains/sea/desert/jungle	Altitude sickness/drowning/heat problems
Climate and weather conditions	Heat- and cold-related injury/death
Wildlife (including insects)	Attack/poisoning through bites/stings/disease
3. Health	
Endemic disease (dengue fever/ Japanese encephalitis)	Serious illness or death
Malaria	Serious illness or death
AIDS/HIV	Serious illness or death
Polluted water	Serious illness
Contaminated food	Serious illness or death
4. Local population	
Political climate	Political instability/coup/kidnapping/ imprisonment (e.g. UK plane spotters in Greece!)
Attitudes to foreigners/cultural differences	Attack/rape/theft/access to drugs
Hygiene/living conditions	Disease
5. Expedition activity	
Trekking/climbing/mountaineering	Altitude sickness/falls from height
River crossing	Serious injury/drowning
Water-based activities (diving/kayaking/sailing)	Drowning/leptospirosis
Underground activities (caving/cave-diving)	Drowning/suffocation/starvation
Equipment failure/inappropriate use	Serious injury/death
Games/sports activities	Injury/incapacitation
6. Travel and camp life	
Transport (public/private)	High risk of serious injury/death
Road/water conditions	Increased risk of accidents
Other road users	Increased risk of accidents
Camp hazards (stoves/fires/flooding/ avalanche/wildlife)	Burns/drowning/suffocation/injury/death
Accommodation/hotels	Fire/electrocution/serious injury/disease/ mugging/attack

Examples might include your peers, participants, parents, school governors, local education authorities, teachers, sponsors, research bodies, etc.

To quantify acceptable risk in the context of your own project or expedition, it is important to ask key individuals and groups what they feel is acceptable to them. Don't ever assume! The greater the challenge and promise of achievement (e.g. first conquest of a new mountain peak), the greater the acceptable risk.

Control measures

Control measures are the backbone of the risk assessment process. They are what the expedition leader initiates to reduce or eliminate a particular risk. Some examples would be as follows:

- providing first-aid training before the expedition starts
- getting immunised before exposure to disease
- preventing bites by disease-transmitting insects.

In most cases, many control measures can be implemented before the expedition as part of the planning process. However, once the expedition or project actually starts there may be many more control measures to consider.

Risk assessment format

There are many variants on the format for a risk assessment. There is no right or wrong way to draw one up provided that the principles are observed. The important thing to ensure is that any staff or team member should be able immediately to see from the risk assessment document the risks identified, the control measures that have been put in place and any further actions required.

The UK Health and Safety Executive refers to the five steps of risk assessment. These are as follows:

1. Identify the hazards and associated risks.
2. Identify who is potentially at risk and how.
3. Identify the precautions or control measures to minimise the risk, including any further action required to reduce the risk to an acceptable level.
4. Record your findings.
5. Review the risk assessment periodically.

This process is clear and straightforward and can be applied to any expedition overseas.

A convenient format for risk assessment is shown in Table 10.3:

TABLE 10.3 SAMPLE RISK ASSESSMENT

Hazard	Risk level	Control measure	Additional action	Review mechanism
Data collection activities Trekking/ river crossing	High	Careful route selection Use of guides Competent, experienced group leaders Use of ropes/ training in river crossing techniques No activity after dark Safety and medical kit carried at all times Group risk assessment before each day's activity	Leader/staff approve activity or, if necessary, halt progress if new risk rendering it unsafe to proceed	Post-expedition report with information about incidents arising, and changes to risk assessment

Involving others in the risk assessment process

Never assume that members of an expedition team will observe or abide by the contents of a risk assessment in which they have had no involvement. The key to effective risk assessment stems from clarity and commitment on the part of all of those who may potentially be at risk. It is strongly recommended, therefore, that team members play a part in compiling the assessment at some stage of the planning process. This risk assessment is an essential part of turning a piece of paper into a living process for managing day-to-day risk on an expedition.

Reviewing a risk assessment

As a result of changing circumstances and environmental conditions, the risk assessment must be reviewed regularly to remain effective. Changes to the assessment on paper are useless if they are not properly communicated to staff and participants, or if staff and participants cannot see a reason for the changes.

Golden rules of risk assessment

Some simple rules for compiling risk assessments that work are captured in the following acronym:

C   Clarify the hazards and risks
R   Reassess and revise it where necessary
I   Involve all participants in the process
S   State it simply in writing
I   If it's too risky – don't do it!
S   Share knowledge and experience

CRISIS MANAGEMENT

The key to crisis management is to put in place planning systems and measures that help to recognise a crisis in the making, to prevent one from happening in the first place, and to handle a crisis effectively if one does occur.

Planning

Crisis management planning should always concentrate on the "worst case scenario". Day-to-day administration procedures may be adequate when all is going according to plan, but these are largely irrelevant when catering for the possibility of a serious expedition incident. Be a pessimistic planner! The expedition planner must always be flexible in a changing world. On the plus side, better communications are constantly evolving and new medical facilities are becoming available worldwide. On the minus side, beware of the political turbulence that seems to be characteristic of recent history, which makes the world a less predictable place.

Legal considerations

As with planning always consider the worst eventuality, e.g. a fatality for which you may or may not be found negligent. If your procedures and systems concentrate on the premise that omission and incompetence are equally dangerous, you are less likely to fall foul of the legal system. Always ask yourself the question, "How would I justify this decision/action in court?"

Expedition organisers, particularly regular providers, should always have some form of written agreement with the participant. This may be an application form associated with a set of booking conditions, or a simple letter of understanding. Either way, if a crisis results in legal proceedings, documents of this sort will play a significant part in establishing responsibility.

Consider also the "duty of care" placed on the organiser of an expedition. With under-18s this also extends to the requirement to act in a supervisory role as a prudent parent would under the same circumstances (*in loco parentis*).

Insurance

The contract or agreement with a participant must make it clear if insurance is included or excluded from the expedition costs. To establish a lack of insurance during a crisis is a disaster in itself, because it is insurance that is generally expected to meet most of the costs of handling a serious incident. During the planning process, the organiser must carefully establish the risks, before attempting to insure against those that the expedition may be faced with. In the less developed world, there are other eventualities that can be insured against which would not appear in the text of a conventional travel policy. These include contingency/war risk, kidnapping, and search and rescue.

Selection of staff and participants

Given that we are trying to avoid a crisis, we must pay attention to the selection of our staff and participants, some of whom may be "an accident going somewhere to happen". Each organiser will have their own systems for recruiting and, in the context of crisis management, the selection of personnel will be hugely significant. Inexperienced staff will not be able to pre-empt a crisis as effectively as those who have sound skills and particularly judgement. Irresponsible team members are more likely to cause a crisis through thoughtless action. There are many ways in which to assess the suitability of candidates for an overseas expedition, including carefully prepared application forms, references, interviews, assessment and selection courses. For the commercial provider, there is always a conflict between the necessity to generate turnover and the potential liability of accepting unsuitable participants. This is a balance that must be addressed if a crisis is to be avoided.

Training and preparation

Having selected a suitable staff and participant team, there is a requirement for induction, briefing and training. The better physically, emotionally and administratively prepared an expedition team is, the less likely it is that an incident of a serious nature will occur. One should never ignore the fact that it is practically impossible to prepare people for every fast ball that the less developed world may hurl, but there is much that can and should be done in preparation. Methods of preparation may include verbal and written briefings, residential courses, and practical outdoor training events or workshop sessions. It is suggested that a combination of these methods of preparation is most effective to strike the optimum balance between skills and theory.

Emergency procedures

Emergency procedures form a focal point in the effective handling of an expedition crisis. The detail and extent of these procedures will obviously vary according to the type of expedition, and the number of projects and areas that they are designed to apply to. In essence, they should allow the staff and/or team members of an expedition to initiate a process that will permit the following:

- immediate care of a casualty/casualties and other involved parties
- evacuation to relevant medical care
- revision of expedition logistics/objectives
- communication with interested parties at home and overseas
- monitoring of casualty/casualties in care
- liaison with families/close relatives
- liaison with insurers/assistance agencies
- facility to supply information to authorities/media/public
- follow-up and review.

Emergency procedures must be written down and communicated in the same way as a risk assessment.

Contacts

An address book of contacts is invaluable in crisis management planning because it establishes an infrastructure of support to the expedition overseas. Planners must consider this requirement from all angles. In the host country, government agencies and British representatives (embassies/consulates) must be alerted to the organiser's plans and advice sought accordingly. British missions abroad are generally contactable 24 hours a day. Other contacts should be sourced to provide advice and assistance with crisis handling. A national agent of contact is recommended who can act as a focal point for communication to and from the UK, particularly when direct communications are limited from the area of a host country in which a team may be operating, and likewise for local contacts (hotels/guides/rescue organisations), communications, medical back-up and important local knowledge.

The contacts portfolio should always be evolving as planning progresses, and the expedition staff/leaders equipped with a full and up-to-date list of helpful contacts with the type of support available before departure from the UK.

The medical umbrella

Although not always the case, an expedition crisis generally involves an accident, illness or injury to an expedition member or members. For this reason, careful attention must be paid to the establishment of a "medical umbrella". This applies not only to immediate medical support to expedition members, but to the entire planning process.

Attention must be paid to the skills of the expedition members and accompanying staff. There must be sufficient first-aid skills among the team to deal with the immediate care of a casualty. In the absence of an expedition doctor, several courses are now available that concentrate on more advanced medicine for remote foreign travel for competent first aiders. Careful selection of the expedition medical kit is also important. This should be put together with the expert advice of a qualified doctor

113

(preferably with expedition experience) and all expedition members instructed in the appropriate use of its contents.

The investigation and enlistment of locally and nationally available medical support form another essential part of the medical umbrella. British missions in the host country often have lists of recommended doctors and dentists in the capital city, but rarely have information about the further-flung outposts likely to be frequented by expedition teams. For this reason, detailed research is necessary to produce a support network of medical contacts in the areas in which an expedition will be operating. Support may come from local aid projects with medical back-up, clinics and dispensaries, local hospitals, or, on a national basis, the GPs and hospitals commonly used by the expatriate population of the country. The list of medical contacts should preferably also include specialists if possible, plus a recommended dentist (often overlooked).

Communications

Without communications, an expedition team is reliant solely on its own ability to handle a crisis. Thus, the poorer the communications, the more competent, experienced and medically trained the team needs to be. The communications infrastructure in the crisis management plan should aim to incorporate as many options as possible. The reliability of communications in the less developed world can be appalling, and thus the more options that are researched and made

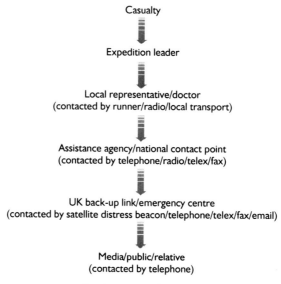

Figure 10.1 *Emergency communication network*

available, the greater the chance of establishing and maintaining links with the outside world.

Essentially, the expedition team relies either on its own communications brought in from overseas (radios, distress beacons, satellite telephones) or on local systems (telephone, runner, telex, local radio communications). In practice, the communications network will comprise some of both, although this will also depend on the nature of the expedition and the size/budget of the organiser. Whatever network is in place, it should allow the team to communicate with the outside world in an emergency. And, after initial evacuation has taken place, the network should allow for two-way communications between the expedition guide/leader and the UK. Figure 10.1 outlines an emergency communication network.

UK back-up

Whatever the size of the expedition, it should have a 24-hour contact in the UK capable of responding and assisting in a crisis. For the smaller or one-off expedition overseas, this may be a family member or colleague who is fully conversant with the expedition medical and contact details for all next-of-kin/closest relatives of all expedition members (including staff). For larger organisations, this back-up may take the form of a duty officer and/or and assistance agency or emergency centre. The function of the UK back-up is to liaise with all the relevant parties in the UK. This may include relatives, sponsoring organisations, insurers, assistance agencies and the press. In addition, this vital link may be in a position to make contact with local support from the UK which, for whatever reason, the expedition in the host country may be unable to contact. The potential scope and extent of this role in a crisis require that the UK back-up be highly capable and responsible, and fully briefed by the expedition organiser.

Sharing experiences

Now that the world has so few frontiers, the likelihood is that, for every expedition to a remote part of the developing world, another has gone before. As a body, organisers of expeditions have a duty to share their experiences with others who follow them in an attempt to avoid recurrent tragedy on expeditions. This can be done in a number of ways from informal conversation between past and present expedition organisers, to formal accident/near-miss reporting such as that set up by the Royal Geographical Society–Institute of British Geographers (RGS–IBG) Expedition Advisory Centre (EAC). For this to happen, the responsibility must lie both with the organiser who has experienced a crisis on an expedition (who should lodge a report with a relevant body such as the EAC/Young Explorers' Trust/British Mountaineering Council/ Alpine Club or similar), and with the planner whose research and initial risk assessment should lead them to such bodies to learn from the experience of others.

FURTHER READING

Anderson, S. and Johnson, C. (2000) Expedition health and safety: a risk assessment. *Journal of the Royal Society of Medicine* **93**: 557–561.
This paper summarises the findings of the RGS–IBG Expedition H&S survey for the years 1995–1997. This paper and more recent data for the years 1995–2000 and beyond can be downloaded from www.rgs.org/medicalcell

Bailie, M. (1997) *Risk Assessments, Safety Statements and all that Guff*. Adventure Activities Licensing Authority (www.aala.org).

Health and Safety Executive (1998) *Five Steps to Risk Assessment*. HSE Books. Tel: + 44 1787 881165, website: www.hsebooks.co.uk
Can be downloaded from www.hse.gov.uk/pubns/indg163.pdf

HSE Adventure Activities Industry Advisory Committee (1999) *Adventure Activities Centres: Five steps to risk assessment*. HSE Books. Tel: + 44 1787 881165, website: www.hsebooks.co.uk

Putnam, R. (2002) *Safe and Responsible Expeditions*. London: Young Explorers' Trust.

Stables, R. (2002) Casualty evacuation. In: Warrell, D. and Anderson, S. (eds), *Expedition Medicine*. London: Profile Books with RGS–IBG, pp. 181–6.

11 EXPEDITIONS: THE LEGAL FRAMEWORK

Rupert Grey

This chapter covers expeditions organised and undertaken by private individuals over the age of 18 with or without the support of third party sponsors. Its purpose is to give a brief outline of the legal framework within which such expeditions operate, and to highlight areas where expedition members should be on guard and, if appropriate, take professional advice.

It does not cover expeditions organised by schools (see Chapter 12), or those organised or funded by other institutions such as universities or commercial organisations; nor, save tangentially, does it deal with different types of insurance (see Chapter 13).

It is offered as a general guide to the legal aspects of organising an expedition; it is not intended to be comprehensive, and it is no substitute for independent legal advice.

STATUS OF EXPEDITIONS

An expedition is not recognised as such by the English legal system. It has no independent legal status; it is simply a collection of individuals who each have the same rights and obligations as any other individual.

You should bear in mind the following:

- an expedition cannot enter into a binding contract in its own name: contracts should be in the name of one, some or all of the members (see paragraph on contracts below);
- the sharing of consequential liability and/or losses is likely to be, and indeed should be, a matter for discussion between the expedition members;
- an expedition cannot sue or be sued in its own name;
- an expedition cannot own property;
- there are no rules governing the relationship between expedition members, unless of course they are agreed between all of them (preferably in writing and signed by each member).

These difficulties can be resolved by forming a company. This will take time and cost money and thus is more likely to suit major or long-term undertakings. The requirement to comply with the Companies Act Regulations (e.g. the filing of annual returns and accounts) is a nuisance but manageable. You will probably require a solicitorÕs advice, and there will be formation fees and associated expenses.

Expeditions are not charitable as such; the scope of charitable status is currently under review, and it may be worth seeking the advice of a solicitor specialising in charitable status as to whether a particular expedition would qualify.

CONTRACTS

Assuming that the expedition is not incorporated, all contracts with third parties such as insurers, photographers, shipping companies and sponsors should be in the name of one or more members of the expedition. Contractual obligations can be created verbally or in writing, and those who enter into them are obliged to comply with those obligations, and are personally exposed to a claim for damages in the event of a breach.

If the membership of the expedition and its purpose are sufficiently well defined, the member who signs an agreement may have an implied right to be indemnified by the other members. But it would be safer to ask all the members of the expedition to sign such an indemnity at the outset.

If significant sums of money are likely to change hands in the course of the expedition it would be foolish not to have such an agreement in place. The agreement can conveniently include other areas of potential dispute such as ownership of property, including photographs, and publishing deals (see below). It would be sensible in these circumstances to invite an independent solicitor to prepare a draft and advise on what it means.

FUNDRAISING

If sponsors provide financial support they will want something back – usually publicity. They, or you, may retain a public relations agent. Licensing the use of photographic images will be an issue, as will the ultimate ownership of equipment, if any, supplied by the sponsors. You should read any agreements covering these issues extremely carefully. If you have any doubt as to the appropriateness of their terms, or as to their meaning, you should take professional advice.

MAINTAINING RECORDS

For the avoidance of argument before, during or after the expedition the following suggestions may be helpful:

- If expedition members delegate tasks to a small group – in effect, a committee – the committee should keep notes of meetings and circulate them to members.
- If the expedition opens a bank account a limit should be placed on the amount that can be withdrawn from it under one signature.
- The task of maintaining records of contributions or donations and expenditure should be delegated to a member.
- Watch out for the taxman. It is conceivable that he will decide to treat donations (in cash or in kind) from sponsors as income in the hands of the recipient, and to tax you accordingly. If significant sums of money are involved you may have to register for VAT.
- The maintenance of accurate records could get you out of trouble. Keep them for at least seven years.

LIABILITY FOR NEGLIGENCE

Accidents are an occupational hazard for expeditions. For some expeditions the taking of risk is an explicit objective, and for others it is an implicit by-product. This approach sits uneasily in a society in which the attribution of blame for accidents is (increasingly) embedded in our culture and legal system. What follows is arguably the most important section in this chapter; the suggestions outlined below are designed to reduce the risk of an accident taking place, and, if it does, to reduce the risk of being blamed for it in a law-suit.

The law, in short, is that members of the expedition owe a duty of care to each other and to all those with whom they come into contact during the expedition. The standard of care is a matter of judgement, but ultimately a person will be considered negligent if they are held by a court of law (who will call relevant expert evidence) to be in breach of the duty of care, and a person has suffered injury or loss as a (reasonably) direct consequence of the breach.

Having regard to the following may give you some protection:

- A leader should be appointed and his/her role and responsibilities defined.
- The qualifications of those with expertise in specific fields, and their role and responsibilities on the expedition, should be accurately described in expedition literature/letters/emails etc.
- The extent to which those responsibilities can be delegated, and to whom, should be agreed.
- The relevant expert should complete a written risk assessment along the lines referred to in Chapter 10.
- The completion of such a risk assessment is no excuse for not continuing to review risks as the expedition unfolds.

- Do not overstate (directly or by inference) the level of expertise of any member of the expedition.
- If persons joining the expedition rely on the specialised knowledge or expertise of members held out as experts, make certain that they have a clear understanding of the scope of the knowledge and expertise relied on.
- Each member of the expedition should confirm in writing that he or she subscribes to the expedition objectives, understands the risks inherent in the nature of the undertaking and agrees to actively participate in steps taken to reduce or control the risk.

It is important that each member of the expedition takes out a policy of insurance to cover the risk that he or she will be found negligent. The fact that the person who is injured is insured for personal injury makes no difference: an insurer who has paid compensation to an injured party is entitled to stand in the shoes of its insured and pursue a claim against the person responsible.

DEFAMATION

The publication, whether written or spoken, of words which damage the reputation of a third party or a member of the expedition will expose the writer or speaker to a claim for damages. This has been known to happen on expeditions. You can insure against the risk of doing so, but it is cheaper to say nothing. Bear in mind that while truth is a good defence against defamation, you may have difficulty proving it in court.

COPYRIGHT

Ownership of copyright in photographs or written material is a topic of considerable importance to expeditions intending to derive income or sponsorship out of articles and photographs. The law in this area is not free of difficulty, but in broad terms the position is as follows:

- Copyright in the written word belongs to the person who wrote the words.
- Copyright in a photograph or video-film belongs to the "person who created it". In practice this will probably be the person who took it.
- An assignment of copyright must be in writing.
- Permission for other expedition members to use photographs does not have to be in writing, but it will avoid arguments later if it is.
- If the photographer wants a credit, he should inform other members in writing.

If it is intended to sell reproduction rights in photographs, film or written material to magazines, newspapers, book publishers or TV production companies, care should be taken over what reproduction rights you are granting them so that the income from subsequent publication or syndication is protected.

All expedition members should be asked to consent in writing to the publication of any material in which they appear or are capable of being identified.

Potential arguments will be avoided if these points are agreed in advance. An exchange of letters will suffice for smaller groups, and for larger expeditions a more formal agreement would be appropriate.

APPLICABLE LAW

As a general rule a contract is interpreted in accordance with "proper law" – i.e. that law with which the agreement and the parties have the most connection. However, you should check the jurisdiction clause, if any, in any contracts which you enter into. Ideally the agreement will state that the contract will be interpreted in accordance with the law of England and subject to the jurisdiction of the English courts. If it invokes another legal system you may not be able to do much about it, but at least you will know in advance.

12 SCHOOL VISITS: THE LEGAL AND PRACTICAL ISSUES

Vanessa Lee

School outings have always been an integral part of a child's education, providing the opportunity to develop key personal and social skills. The experience of outdoor and adventure activities can be invaluable.

Recent tragic deaths of children on school visits have, however, brought the whole topic into focus and debate has been raised as to whether the risks of such visits outweigh the benefits. Following the inquest into the deaths of Rochelle Cauvet and Hannah Black who died in October 2000 after being swept away in a swollen stream in the Yorkshire Dales, Nigel De Gruchy, General Secretary of the National Association of School Masters/Union of Women Teachers (the second-biggest teaching union), said:

> We have reluctantly concluded that until society accepts the notion of a genuine accident, it is advisable for members not to go on school trips.

It was further reported in the *Evening Standard*, 24 April 2003, that Eamonn O'Kane, the present General Secretary, was reiterating the statement and urging teachers to boycott school trips

> because of the growing risk they face from prosecution if pupils are killed or injured in their care.

Is this the end for school outings?

ROLL-CALL OF TRAGEDY

June 1997 – Adi Naseem, 11, drowned on a day trip to an activity centre in Buckinghamshire. Forty-seven children from Feltham Hill School, Hounslow, were allowed to use a swimming pool without lifeguards. Hounslow were prosecuted under the Health and Safety at Work Act and pleaded guilty, receiving a fine of £25,000.

July 1997 – Three children died and 17 were injured when a minibus with no seatbelts plunged 70 feet down an Alpine ravine. The driver was found guilty of manslaughter by a French court and fined £1200. He had never driven a left-hand vehicle.

June 1999 – Gemma Carter, 13, drowned in France on a school trip after becoming separated from the party on a evening swim session. Her supervising teacher was found guilty in France of involuntary homicide and given a 6 months' suspended sentence. The conviction was overturned on appeal in April 2002.

July 2001 – Bunmi Shagaya, 11, drowned in Lake Caniel, France, while on a school trip.

August 2001 – Amy Ransom, 17, was killed when she slipped and fell 500 metres on a mountain during a school trip to Vietnam.

August 2001 – Jason Doulton, 17, drowned on a field trip in South Wales after being swept away by a swollen river.

March 2002 – Amelia Ward, 16, was killed by a falling rock in South Africa while on a school holiday as part of the Duke of Edinburgh Gold Award Scheme. Coroner Paul Knapman, recording a verdict of accidental death, said there had been a "misunderstanding by the man in charge".

May 2002 – Max Palmer, ten, drowned after being swept down river in the Lake District. He was accompanying his mother, a teaching assistant, on a school trip. He was not a pupil at the school.

STATISTICS

The above are just a few of the tragic deaths that have followed school trips. In fact there were seven children killed in 2001 and there have been 47 deaths since 1985.

Understandably, schools, parents and the government are concerned about the future of school visits and how serious accidents and deaths can be prevented. It is important, however, to put the above figures into some perspective. Statistically there is an insignificant chance of a child suffering a fatal accident. On average there have been three deaths per year which gives a fatality rate of 1 in 8 million.

The understandable knee-jerk reaction to stop visits following the death of a child has to be viewed against the overall benefit offered by adventure and outdoor activities on such visits. The immediate focus has to be on the planning of visits and the education and training of all concerned. Accidents are an unwelcome reminder of the responsibilities of schools in organising such visits – responsibilities that translate into legal liabilities.

123

LEGAL FRAMEWORK

The Health and Safety at Work Act 1974 and associated Regulations place responsibility for health and safety with the employer. Employers are responsible for the health, safety and welfare at work of their employees and are under a duty to ensure, so far as is reasonably practicable, the health and safety of anyone who may be affected by their activities – both on and off their premises.

The management of Health and Safety at Work Regulations 1999 requires employers to:

- assess the risks of activities
- introduce measures to control those risks
- tell their employees about these measures.

Disability Discrimination Act 1995

Governing bodies must ensure that their "inclusion" policies address the needs of young disabled pupils wishing to participate in extra-curricular activities, including educational visits.

Health and Safety: Responsibilities and Powers

This was issued by the DfES to all schools in December 2001. It identifies the responsibilities, under Health and Safety legislation, that employers and employees must meet and makes specific reference to educational visits. All visits organised and arranged by schools are covered by these responsibilities, including any visits organised substantially by third party providers.

Who is the employer?

- The local education authority (LEA) is the employer for community schools, community special schools, voluntary controlled schools, maintained nursery schools, pupil referral units and statutory youth groups.
- The governing body is usually the employer for foundation schools, foundation special schools and voluntary aided schools.
- The governing body or proprietor is usually the employer for independent schools.

Employees are also under an obligation to take reasonable care of their own and others' health and safety, to carry out activities in accordance with training and instructions, cooperate with employers over safety matters and inform employers of any serious risks.

Although breaches of these regulations do not confer a direct statutory right of action, there can of course be prosecutions pursued by the Health and Safety Executive (HSE) for breach of the Health and Safety at Work Act and Regulations. In civil actions claimants argue that breaches of the Regulations are evidence of negligence. In common law, teachers have a duty of care to act as any reasonably prudent parent would do – *in loco parentis*.

To assist in complying with the legislation and associated Regulations, the Government published detailed national guidance on school visits in 1998 entitled *Health and Safety of Pupils on Educational Visits (HASPEV): A good practice guide* (see "Recommended reading" at the end of the chapter). This provides advice on planning and organising activities both in and out of school, at home and overseas. Supplementary guidance in support of HASPEV was published in July 2002:

- *Standards for LEAs in Overseeing Educational Visits* sets out good practice for local education authorities in overseeing educational visits carried out by schools. Schools, youth services and others may find the principles set out here useful too (DfES/0564/2002).
- *Standards for Adventure* is aimed at the teacher or youth worker who leads young people on adventure activities (DfES/0565/2002).
- *A Handbook for Group Leaders* is aimed at anyone who leads groups of young people on any kind of educational visit. It sets out good practice in supervision, ongoing risk assessment and emergency procedures (DfES/0566/2002).

Guidance

Although the employer is responsible for health and safety, on a practical level this will invariably be delegated to the school. The employer of course is still ultimately responsible and the supplement to HASPEV is keen to remind of the need for maintenance and compliance, i.e. there should be an audit track making clear who is doing what and confirming compliance with those tasks once delegated.

All employers should therefore provide written guidance for teachers to follow when arranging school visits, which should include laid-down procedures for the approval of certain types of visits. They must ensure that staff are fully trained in their health and safety responsibilities as employees and must be confident that those who are delegated health and safety tasks are competent to carry them out.

It needs to be remembered that for incidents occurring in the UK the HSE enforces Health and Safety law and will normally take action against the employer. Where an employee has, however, flouted an employer's policy or directions in respect to health and safety, the HSE may take action against the employee as well or instead.

Health and Safety of Pupils and Educational Visits (HASPEV)

HASPEV, as the most informative good practice guide in this field, has been adopted and recommended by many LEAs in place of guidance of their own. Others have adapted it for their own purpose. This document and its supplements are invaluable in any criminal or civil proceedings and are relied on either in defence of an incident or to demonstrate breach of good practice. Compliance would be persuasive in the defence of any health and safety prosecution. It is critical therefore that anyone involved in planning or arranging school visits is familiar with HASPEV and/or any other guidance provided by the LEA.

The key to understanding and following HASPEV is to remember that, although it sets out the principles and criteria of good practice to be adopted when planning school visits, it recognises how good practice can vary and much has to be left to the judgement and expertise of those arranging, approving, leading or supervising the trip. Invariably this will be the school staff and/or the governors.

HASPEV provides guidance on the visit from inception to completion including:

- Responsibilities of head teachers, LEA Outdoor Education Advisers (OEAs), Educational Visits Coordinators (EVCs), group leaders, teachers, adult volunteers, pupils and parents.
- Planning of the visit including risk assessments, exploratory visits, financial planning, charging for visits, first aid and other considerations such as equipment, contingency measures, etc.
- Supervision including pupil/staff ratios, parents/volunteers, vetting suitability, supervisor's responsibilities, competence of those leading an adventure activity, head counts and remote supervision, i.e. where pupils may have time on their own while on a visit.
- Preparing pupils for the trip, their participation, equal opportunities, information provided before, during and after the trip, preparing them for remote supervision such as identity cards, knowledge of out-of-bounds areas of activities, emergency contacts if lost, transport, pupils' medical, special and educational needs.
- Communicating with parents: informing them in great detail of exactly what will take place on the trip, parental consent, medical consent, early return, contact between all concerned, i.e. staff, pupils and school.
- Planning transport including legislation, supervision on transport, hiring coaches and buses, appropriate licences and permits required, use of private cars, transport in school minibus, the school minibus driver, maintenance and checks of the school minibus.
- Insurance including confirming insurance in LEAs, the position of the school's insurance and parents – advising them what they have and do not have to do, insurance cover for visits, cancellations and transport insurance.
- Types of visit including adventure activities using licensed providers, adventure activities using non-licensable providers, issues to consider with all adventure activity providers, school-led activities, employment of providers, remote supervision during adventurous activities, coastal visits, swimming in the sea or other natural waters, swimming pools, farm visits, field studies and residential visits.

- Visits abroad including organising your own visit, organising your own transport, using a tour operator and operators based abroad, sources of further advice for school travel abroad, planning and preparation, staffing the visit, preparing pupils for visits abroad, the importance of a briefing meeting for parents.
- Emergency procedures: who will take charge in an emergency, the procedures and framework in place during the visit and at the school base including advice on media contact.
- Model forms: 11 in total dealing with risk assessments, checklists for pupils going on visit, parental consent form for school visit, etc.

IMPROVEMENTS FOR THE FUTURE

It is clear that a better understanding of risk assessments by schools is required. In particular, accidents often occur as a result of a change in circumstances, i.e. weather or an unforeseen event occurring once an activity has actually started. Risk assessments need to be reviewed on an ongoing basis even once the activity has started. A better understanding of the whole topic of risk assessments will increase teachers' understanding that even a relatively safe activity can have the same consequences as more dangerous pursuits. Teachers need to be aware of the type of activities where they might have to call in expert help both before and during the trip.

Where schools are planning visits abroad it is essential that the credentials of any company or centre that is being used are fully investigated. The supplemental guidelines to HASPEV deal with this in some detail, including, for example, receiving confirmation of the qualifications of each individual instructor to be used, together with confirmation of what risk assessments have been undertaken in relation to the activities that will be carried out by third party providers.

When serious incidents occur, it often becomes apparent that either guidance may not be in as much detail as one would hope, or it has simply been passed to the school with little background information or training necessary both to understand and to implement the guidelines. Guidance in the new HASPEV supplements stresses the importance of having both an Outdoor Education Adviser and an Educational Visits Coordinator.

The role of the LEA

The LEA should now designate an Outdoor Education Adviser and other personnel who can support and advise schools in their planning of school trips. In order to undertake its responsibilities for the health and safety of employees and pupils on educational visits, it is good practice for the LEA to:

- define the types of educational visit
- outline a clear system of delegating tasks for approving planning and risk management of these different types of visits
- devise a procedure to ensure that notification of approval takes place at early and appropriate stages in the cycle
- implement a training programme for all members of school staff and governors in the management of educational visits
- maintain and monitor training records and extra qualifications held by school staff
- assess the competence of educational visits coordinators and group leaders
- provide a database of contractors to be used by schools which should be updated via school evaluation forms
- put in place emergency procedures, including 24-hour access, to support schools in cases of extreme difficulty or emergency.

The role of the governing body

Where the governing body is the employer, the governors' responsibilities will be the same as those suggested for the LEA. In addition, it is good practice for all governing bodies to:

- ensure that guidance is available (e.g. from the department and/or LEA as appropriate) to inform the school and influence policy, practices and procedures relating to the health and safety of pupils on educational visits. These should include measures to obtain parental consent on the basis of full information, to investigate complaints and to discuss and review procedures including incident and emergency management systems. Governors may where appropriate seek specialist advice though they should not normally be expected to approve visits
- ensure that the head teacher and EVC are supported in all areas relating to educational visits and ensure that they have the appropriate expertise and time to fulfil their responsibilities
- ascertain what government training is available and relevant
- agree on the types of visit they should be informed about
- be in a position to ask appropriate questions about the visit's objectives and how they will be met. It is not expected that governors should be directly involved in risk assessment in related matters unless they have appropriate competence
- ensure that visits are approved as necessary by the LEA before bookings are confirmed
- help to ensure that early planning and pre-visits can take place, and that the results can be acted upon

- ensure that bookings are not completed until external providers have met all the necessary requirements, i.e. risk assessment training, security etc.
- ensure that the head teacher and the EVC have taken all reasonable and practical measures to include pupils with special educational needs and medical needs on all visits.

The role of the head teacher

It is essential that head teachers ensure that any visits arranged through the schools comply with the Regulations and Guidelines provided by the LEA or governing body and the school's own health and safety policy. Head teachers should ensure that the group leader is competent to monitor any risks that may arise throughout the visit. They should also be aware of their own role if involved in the visit as a group member or supervisor. They should always follow the instructions of the group leader who has the sole charge of the visit.

Paragraph 23 of HASPEV outlines the good practice for head teachers to adopt including ensuring matters such as adequate child protection procedures are in place, all necessary risk assessments and actions have been completed before the visit begins, training needs to be assessed etc. In addition, with the introduction of the supplementary guidance to HASPEV, head teachers are now in a position to delegate tasks to the Educational Visits Coordinator (EVC) and agree who will approve a visit at school level or submit it to the LEA for approval if so required. It is anticipated that the EVC will perform this function with the head teacher countersigning. The task can, however, be wholly delegated where required by the head teacher.

It is important to remember that the role of the head teacher even where tasks are delegated to the EVC remains paramount in the organisation of any trip at school level. Any local authority or head teacher considering their policy in relation to the role of the head teacher should refer both to HASPEV and Paragraph 11 of the Good Practice Supplement.

The role of the Outdoor Education Adviser (OEA)

Traditionally every LEA would have had an OEA but due to budget constraints their function has diminished over the years. The thrust of the new guidance, however, is that an OEA will enable the LEA to monitor what schools are doing. This will help them to ensure, so far as is reasonably practicable, the health and safety of pupils while on school visits.

The guidance suggests that it is good practice to have an OEA or someone who carries out this role as part of their job description. It may, for example, be combined with the position of head of outdoor education centres. Individuals holding this post will have a strong understanding of Health and Safety legislation and in particular be able to advise on the legal responsibilities and powers of LEAs and schools in relation to school visits. Anyone who holds the position of Head Outdoor Education Adviser should be competent to advise on both low- and high-risk activities. It is important, if this role is to work, that the individual holding the position of OEA has sufficient authority in the LEA to effect necessary changes and influence people.

The key aims of the OEA is to provide:

- support to schools
- approval for visits where approval may not have been delegated to school level
- a focal point for schools to contact to give expertise on adventure activities, expeditions, visits generally and specifically overseas visits.

They are also able to provide generic risk assessments to schools and monitor the visits carried out by the LEA schools. Specifically they will be able to ensure the competency and qualifications of school staff, group leaders and Educational Visits Coordinators.

The role of the Educational Visits Coordinator (EVC)

The guidance states that it is now good practice for each school to have an educational Visits Coordinator (EVC). It is important to stress that many schools may argue that they already have somebody who carries out the responsibilities of an EVC, although not with this particular title. In many cases it may be the head teacher, but it can be any other member of staff with an interest in this area with the relevant qualifications and/or experience. The aim of the EVC is to help the school comply with its health and safety obligations while on school visits and generally to act as a single point of contact for the LEA and within the school.

It is also the responsibility of the EVC to support all concerned to ensure that all guidelines for leading activities are followed.

The LEA should work closely with EVCs, providing them with advice, guidance and appropriate training and, particularly, to enable EVCs to access specific training for staff involved in leading or staffing school visits.

The general functions of the EVC are as follows:

- To liaise with the employer to ensure that the visit meets the employer's requirements including those of risk assessment.
- To support the governors and head teacher with any decisions and approval.
- To ensure competent people lead or otherwise supervise a visit.
- To consider and assess the competence of leaders and other adults for a proposed trip.
- To organise thorough induction and training of leaders and adults going on any particular visit.
- To ensure that criminal records and/or disclosures are in place where required.
- To work with the group leader with regards to parental consent and provide

full details of the visit beforehand enabling parents to provide such consent on a fully informed basis.
- To organise and ensure emergency arrangements are in place and contact is available for each visit.
- To keep records of individual visits including near misses, i.e. reports of accidents and "near accidents".
- Continually to review and, where appropriate, monitor practice.

The role of the group leader

The EVC may not be the group leader for each planned visit. A group leader should always be appointed to be responsible for the safe planning and execution of visits.

The group leader will have overall responsibility for the conduct and supervision of the visit, having regard to the health and safety of the group. It is important therefore that the group leader and the EVC are very clear about their respective roles. The group leader should have been appointed or approved by the head teacher or the governing body. In particular the group leader should:

- always obtain the head teacher's prior agreement before any offsite visit takes place
- follow LEA/governing body regulations/guidelines and policies
- appoint a deputy group leader
- clearly define each group supervisor's role and ensure all tasks have been assigned
- be able to control and lead pupils of the relevant age range
- be suitably competent to instruct pupils in all activities and be familiar with the location centre where the activity will take place
- be aware of child protection issues
- ensure that adequate first aid measures are in place
- undertake and complete planning and preparation of the visit including the briefing of group members and parents
- undertake and complete comprehensive risk assessments
- review routine visits and activities and consider whether adjustments need to be made and advise the head teacher accordingly
- ensure all teachers and other members of staff accompanying the trip are fully aware of what the visit involves
- have sufficient information regarding pupils for the visit to assess their suitability to accompany the trip or have their suitability assessed and confirmed
- ensure the appropriate ratio of supervisors to pupils is in place
- consider stopping the visit if there is a risk to health and safety of pupils
- ensure that group supervisors have details of the school contact

- ensure that group supervisors and the school all have a copy of the emergency procedures
- ensure that the group's teachers and other supervisors have details regarding pupils' medical needs, special educational requirements etc.
- observe guidance set out for teachers and other adults accompanying the trip.

The role of other teachers and accompanying adults

All teachers and adults accompanying any visit need to be certain of their roles and responsibilities. In particular, all must be certain that they are to follow the instructions of the group leader regardless of status or standing within the school on a normal day-to-day basis. All are responsible for ensuring the health and safety of all members in the group.

Adults accompanying the visit should be guided in their role by teachers and if possible should not be left in sole charge of pupils except where it has been previously agreed as part of a risk assessment of the activities to be undertaken.

SUMMARY

It may be considered that the following checklist is the absolute minimum that a school should consider to ensure that a trip is safe:

- The school should make full use of both the OEA and EVC at the school in ensuring that the guidance is followed from start to finish, i.e. from preparation through to conclusion of the trip. Full use should be made of the skills and abilities of the individuals holding these positions; the aim of the guidance is to improve awareness of the health and safety risks that can and are faced when organising educational visits.
- The trip should be planned by means of risk assessment, with the aim of reducing any risks to the lowest practicable level.
- The group leader should try to visit the site; if this is impractical, as much information as possible should be obtained to establish that the site is safe – this should include a letter from the venue on its safety procedures, qualifications of staff employed and/or confirmation from other schools that have visited.
- If taking a trip within the UK, it is necessary to ensure that any activity centre is licensed under the Activity Centres (Young Persons) Act 1995. If activities are to be led by members of staff, all must have appropriate qualifications specific to that particular activity.
- There should be a pre-visit meeting with parents advising of the likely risks to be encountered and safety measures that have been taken for the trip. Parents must be asked to sign forms consenting to the trip. Full details of a child's

medical problems should be obtained, together with consent for individual activities and who will be responsible in the event that a child has to be sent home early.

- Pupils must also be fully aware of the aims of a trip and understand the safety precautions that need to be taken with reasons explained as to why such precautions are necessary. All should be fully aware of the emergency procedures in place.
- There should be clear guidance to all staff members and pupils as to who is responsible for each individual activity that is carried out and all must understand their responsibility and obligations.
- Supervision ratios have to be appropriate and consideration given to circumstances where the level of supervision will need to be increased.
- Public liability and, depending on the nature/venue of the trip, travel insurance cover must be in place. This will provide protection, among other things, in respect of any accident to a child and aspects like cancellation and medical costs. Most LEAs carry a significant excess on their Public Liability insurance. This gives them a clear financial incentive properly to manage the risks of such visits.
- Other factors – reputation. The cost of a claim is one thing; however, it is even more difficult to put a price on an LEA's good reputation that can soon be tarnished by the media when something major goes wrong and the "who was to blame" questions quickly materialise. From this last perspective it is important as part of your emergency procedures to consider how you respond to the media. Clear and consistent messaging by the authority in the early stages can save problems later on, not least with your liability insurers.

The above checklist is by no means exhaustive and HASPEV or appropriate LEA guidance should be followed when making arrangements to plan and maintain a safe trip.

The guidance is onerous but necessary. A common theme throughout is to keep all stages of the trip from start to finish fully documented. Planning a school trip is time-consuming and where incidents do occur the rise of the blame culture means many teachers, if things do go wrong, risk the prospect of criminal prosecution and/or risk losing their jobs.

CONCLUSION

In the light of HASPEV and the supplementary guidelines, LEA schools and teachers need to review their existing systems and it is anticipated many will request further training. Funding for this will have to be found. Some LEAs are already investing in this area and offering group training to teachers.

The real emphasis is on the quality of ongoing risk assessments and the ability of group leaders and staff involved in school outings to recognise when such assessments need to be made. This can only be done effectively with comprehensive training.

Understandably there has been much debate, in particular by many of the teaching unions, that involvement in school visits with the amount of work needed is simply not worth it. The reality is that no matter what steps are in place the risks of an accident can never be eliminated. With further training and a better understanding of how to plan school visits, it is hoped there will be a reduction in the number of incidents to an even lower level than at present.

FURTHER INFORMATION

Publications of the Department for Education and Skills (DfES)

Printed copies are available on request from +44 845 602 2260, email: dfes@prologistics.co.uk and they are also available as downloadable web files (www.teachernet.gov.uk/visits).

Advice on organising educational visits can be found in the DfES's 1998 good practice guide *Health and Safety of Pupils on Educational Visits* (DfES/HSPV2/1998). This can be downloaded at www.teachernet.gov.uk/visits together with the following supplements, produced in 2002:

- *Standards for LEAs in Overseeing Educational Visits*
- *Standards for Adventure*
- *A Handbook for Group Leaders*
- *Group Safety at Water Margins*

Health and Safety: Responsibilities and Powers, sent to all schools and LEAs in December 2001 (DfES/0803/2001), can also be downloaded at www.teachernet.gov.uk/visits

Supporting Pupils with Medical Needs Good Practice Folder (DfES/PPY194), can be downloaded at www.dfes.gov.uk/medical/

Guidance on First Aid for Schools (DfES/PP3/34348/698/254), can be downloaded at www.teachernet.gov.uk/Management/guidance/firstaid/

Group safety at water margins published by the Central Council of Physical Recreation (CCPR). Copies can be obtained from DfES Publications on +44 845 602 2260, quoting publication code 0270/2003, or downloaded from www.teachernet.gov.uk/visits or www.ccpr.org.uk

Curriculum Division of the DfES has also been working with HSE, RoSPA and others to produce *Safety Education: Guidance for schools*. This is available on the web at www.teachernet.gov.uk/Management/guidance/safetyeducationguidance// Although the guidance is aimed at schools, there are references to learning about safety in adventure, including an outdoor education case study on page 19.

Department of Health Guide to the Protection of Children Act 1999 is at
www.doh.gov.uk/pdfs/childprotect.pdf

LEA Outdoor Education Advisers' Panel (2002) *Overseas Expeditions: A guide to current good practice* –
for LEAs, city and county councils, and other interested bodies, when producing policies and procedures
to inform and guide their employees.

Adventure Activities Licensing Authority, 17 Lambourne Crescent, Llanishen, Cardiff CF14 5GF. Tel: +44
29 2075 5715, fax: +44 29 2075 5757, website: www.aala.org

In 1996 it became a legal requirement under The Activity Centres (Young Persons' Safety) Act 1995 for
providers of certain adventure activities to undergo inspection of their safety management systems and
become Registered as licensed.

This licensing scheme applies only to those who offer activities to young people under the age of 18
years and who operate these activities in a commercial manner.

Generally, licensing applies to these activities only where they are done in remote or isolated environments, e.g. climbing on natural terrain requires a licence, climbing on a purpose-built climbing wall does
not. A licence is not required for:

- voluntary associations offering activities to their members (e.g. scout groups, local canoe clubs, etc.)
- schools and colleges offering activities only to their own pupils or students
- activities where youngsters are each accompanied by a parent or legally appointed guardian (does
  not include, teacher or youth leader).

The licensable activities include:

- climbing (on natural outdoor features)
- watersports (on most lakes, fast-flowing rivers and the sea)
- trekking (in remote moorland or mountain areas)
- caving and rock climbing.

If you wish to send your child, or a child in your care, to an activity centre/provider you can call the
Adventure Activities Licensing Authority and they will tell you if the centre is registered.

More information about the licensing scheme is available in the Health and Safety publication entitled
Guidance to the Licensing Authority on The Adventure Activities Licensing Regulations 1996 (available from
HSE Books – tel: +44 1787 881165).

Further sources

OCR Certificate in Off-Site Safety Management

Oxford Cambridge and RSA Examinations (OCR) www.ocr.org.uk

This qualification has been designed to recognise candidates' knowledge, understanding and skills in planning and evaluating the safety aspects of off-site activities. Developed in partnership with the British Association of Advisers and Lecturers in Physical Education and the College of St Mark and St John, it
addresses the content of recent legislation in this area and provides an opportunity for candidates to
explore the application of the legislation in practical situations. The qualification is appropriate for those
working in a range of different situations where they are responsible for taking children, young people and
adults out of their everyday environments, e.g. educational visit coordinators, teachers, youth leaders and
university lecturers, who are involved in activities such as geography fieldwork, adventurous outdoor

pursuits and cultural visits. Candidates are required to have attended a course at an approved centre before they can be entered for assessment. Approved centres include:

- RGS–IBG Expedition Advisory Centre. Website: www.rgs.org/eacseminars
- Plas y Brenin. Website: www.pyb.co.uk

The Protection of Young People in the Context of International Visits
Revised 2002. Available from: Wendy Laird, Publications Unit, City of Edinburgh Council, Edinburgh Department, Level 2, Wellington Court, 10 Waterloo Place, Edinburgh EH1 3EG. Tel: 0131 469 3328, fax 469 3311, email: Wendy.Laird@educ.edin.gov.uk

The Coastguard. Website: www.mcagency.org.uk

Child-Safe. Website: www.child-safe.org.uk
A website set up by the Avon and Somerset Constabulary's Child-Safe Project.

Royal Society for the Prevention of Accidents. Website: www.rospa.com
Their website includes *Health and Safety at School Guidance* in the Safety Education section, covering:

- *School Trips – Part 1*: this will enable you to gain further understanding of your legal obligations and responsibilities in connection with school visits and trips.
- *School Trips – Part 2*: covers trips involving activities with a higher risk, and visits to foreign countries.
- *Minibus Safety 1*: this guide will enable you to understand your legal obligations and responsibilities in relation to pupils and staff using the school minibus.
- *Minibus Safety 2*: a useful pre-drive safety checklist and advice for operators, parents and children.
- *Minibus Safety: A code of practice*: www.rospa.com/pdfs/road/minibus.pdf
- *Framework for a School Health and Safety Policy*: some key headings and discussion prompts for developing a whole school health and safety policy.
- *Safety and Disaster Management*.

And factsheets on *Water and Leisure Safety*.

Safe Sport Away: A guide to good planning
Amateur Swimming Association and NSPCC, 2001.
Available from De Brus Marketing services. Website: www.debrus.co.uk
See also www.sportprotects.org.uk

Commonwealth Youth Exchange Council. Website: www.cyec.org.uk
Publications include *Crossing Frontiers: A guide for youth leaders taking groups abroad*. This activity pack aims to introduce key issues through practical, interactive exercises for young groups planning Commonwealth and international exchanges. It is also helpful for leaders of groups who are interested in world development issues and international understanding.

Young Explorers' Trust (YET). Website: www.theyet.org

Safe and Responsible Expeditions
Incorporating *Guidelines for Youth Expeditions*, revised edition 2002. These can be purchased or downloaded from either the YET or the RGS–IBG website www.rgs.org/eacpubs

13 INSURANCE FOR EXPEDITIONS

Mark Whittingham

WHO NEEDS INSURANCE?

Insurance cover is essential for all members of expeditions. There is a need to guard against unforeseen expenses arising from accidents, illness, natural disasters, loss or theft, which could prove to be a financial nightmare for the expedition. Insurance protects leaders and expedition members from claims made against them, and is a special requirement for school and youth groups, which are legally responsible for the members of their party and their actions.

THE BASIC TENETS

Expeditions, almost by definition, seem rarely to have sufficient funds for their true objectives, let alone the "luxury" of insurance. This can easily result in failure to insure adequately. Do not take short cuts by under-insuring. Always bear in mind that, if you cannot afford the premium, you are even less likely to be able to afford the potential loss.

Without doubt, the most important thing to remember when arranging insurance is that the law requires that the person applying for insurance cover disclose all material facts to the insurers whether or not they ask for it. Failure to comply with this fundamental tenet of insurance law can have the effect of completely invalidating the insurance contract. There is no easy guide to what is material, but a simple test is that, if you were the insurer, would you want to have this information to enable you to decide on a fair premium? For instance, if part of your expedition involves white water rafting or mountaineering it is important to declare it.

INSURANCE COVER FOR WAR AND TERRORISM

Following the World Trade Center tragedy, insurance cover has varied greatly from insurer to insurer. Depending on the attitude of the insurance company selected, one of three situations is likely to exist on the insurance policy:

1. total exclusion for terrorism and war
2. full cover for terrorist acts but war excluded
3. full cover for terrorism and war risks; war cover only if one of the five major powers not involved.

Although the third option is the most beneficial there could be additional premium conditions and insurers normally include a 7-day cancellation clause in their wordings which they can invoke.

Definition of terrorism

Whilst there is no insurer consensus the most commonly used definition of terrorism is in the reinsurance (Acts of Terrorism) Act 1993: "Acts of persons acting on behalf or in connection with any organisation which carries out activities directed towards the overthrowing or influencing, by force or violence, of Her Majesty's government in the UK or any other government de jure or de facto."
Insurer policy wordings should be checked very carefully.

LEGAL LIABILITY AND INSURANCE COVER

The litigation resulting from high-profile deaths on school visits abroad makes it vitally important that any leader/individual/teacher leading or running overseas expeditions has adequate public liability insurance which will also provide the indemnity for the costs of defending any legal action that may be taken against them.

The problem with liability cover and overseas expeditions is that outside the UK countries have their own laws, which vary greatly from country to country – an action against an individual or organisation may not necessarily be issued in the UK. It is for this reason that it is important that insurance policies purchased in the UK have worldwide jurisdiction and appropriate geographical cover. On this basis insurers will also defend legal liability claims outside the UK.

English law and liability insurance policies are based on negligence; a person cannot be blamed for genuine accidents where no fault attaches. If no one is at fault no one can be successfully sued for negligence. For a third party to be successful with a claim/legal action, the onus is on them to prove that you have been negligent. To minimise a potential legal liability claim, it is therefore essential to try to minimise any risks that are foreseeable – an unactioned foreseeable risk would allow a third party to pursue a negligence claim.

Full risk assessments are therefore essential. Advice on preparing a risk assessment is given in Chapter 10. In the context of your insurance, the risk assessment document can also be used to:

- "sell" your expedition to any underwriter showing that risk has been

minimised and therefore has the potential for a reduced insurance premium
* provide evidence to defend any third party legal action.

CATEGORIES OF INSURANCE COVER

Insurance policies for expeditions can be divided into various main categories of cover as follows.

Medical and additional expenses

This is a most important insurance cover. It usually covers medical and travel expenses for each member of the expedition following accidental bodily injury or illness. These expenses may vary from a doctor's visit through to major surgery and after-care. The UK has reciprocal national health arrangements with some countries and further details are given below.

This category of insurance should include the following either for an individual visit or on an annual basis:

* emergency assistance, search and rescue and repatriation including air ambulance or air transport costs
* emergency dental treatment
* travel and accommodation expenses for people who have to travel to or remain with or escort an incapacitated insured person
* local funeral expenses or transportation of the body to the UK.

Any medical conditions known to exist before the start of the expedition may not be covered, although this exclusion may not apply provided that the insured person has been without medical treatment or consultation during the previous 12 months. Expedition members who are in doubt about this exclusion should consult their insurance adviser before departure and/or obtain a medical certificate from their doctor stating that they are not travelling against medical advice. This may satisfy the insurance company's requirements.

Professional advice may be available to help with hospitalisation, repatriation or alterations in any travel plans. It is important that, if your travel policy has a 24-hour emergency telephone number for hospitalisation or repatriation, this number be used when an accident or illness occurs. Any action taken by the expedition in the field without consultation with the emergency rescue company/insurers may have to be justified to the company afterwards. A diary of events should therefore be kept.

Do not rely totally on the insurer's emergency assistance rescue company; because expeditions visit isolated and remote areas there are no guarantees that they will have sufficient local resources. Be sure to have a contingency evacuation plan agreed and arranged beforehand. Insurance companies are no substitute for a sound crisis management plan.

If foreign nationals are on the expedition they may need to be repatriated back to their own country instead of the UK. Insurer agreement needs to be obtained for this before the expedition commences. Some countries, particularly those with their own state insurance schemes, may forbid insurance of their nationals outside their own country. Before leaving the UK, the expedition leader should therefore check with their insurance adviser whether or not arranging insurance for foreign nationals from the UK is in contravention of any foreign insurance law.

Cover normally excludes claims relating to HIV-related illness. It is possible to obtain separate "dread disease" insurance for nurses, doctors and health workers where a benefit is payable should a person test HIV positive.

All travel insurance policies have geographical limits. Premiums are lower if cover is just restricted to Europe instead of worldwide; however, careful consideration needs to be given to the insurer's definition of Europe.

No limit of less than £5 million per person should be accepted for journeys to the USA. Travellers are recommended to carry proof of medical expenses cover at all times in the USA to avoid authorities not providing treatment.

Medical treatment abroad
There are over 40 countries outside the EU with which the UK has reciprocal health-care agreements that entitle British visitors to emergency medical treatment. The Department of Health leaflet *Health Advice for Travellers* provides vital information on obtaining emergency medical treatment abroad, and contains details of how to use Form E111, the passport to free or reduced-cost emergency medical treatment in most European countries. This is an important and complex process and the leaflet is essential reading. The leaflet and an application form E111 are available at main post offices. Remember that you must get your E111 stamped and signed by the post office for it to be valid

Always take out adequate health insurance before you travel, even if you are travelling to an EU country covered by the E111. The E111 may not provide adequate cover for all medical expenses

Personal accident
This covers death or disablement after accidental bodily injury. An amount is paid in the event of loss of use of any eye or limb, permanent total disablement or death. Cover should include disappearance, and death or disablement by exposure. The amount paid will be additional to any other personal accident or life assurance that individual members of the expedition have arranged for themselves. Note that personal travel policies do not include weekly benefit amounts for temporary disablement or for disablements not specified, e.g. loss of finger or toe.

Benefits should be payable for disability from *usual* occupation as opposed to *any* occupation. Note that cover should be accidental bodily injury; avoid insurance covers that restrict cover to violent visible or external means.

Make sure that cover is on a 24-hour basis and not just restricted to certain activities, and includes commuting to and from the expedition departure place.

As expedition members can change, make sure that cover is on an unnamed basis for all members as opposed to individuals.

The lower age limit should be carefully checked; the death benefit will be restricted to a nominal amount for minors below the age of 16 years. Some insurers will try to apply the nominal amount to members aged 16 and 17.

If your insurance policy is a group policy for all expedition members, the insurer may try to apply a limit of liability in respect of more than one individual being injured on an aircraft or other conveyance. Larger expeditions should check the policy wording to make sure that any aggregate conveyance limit is adequate.

Public/personal liability insurance
This is one of the most important elements of expedition insurance. All members must have adequate insurance against any legal liability in the event of an incident occurring, which would include liability to other members of the expedition.

The legal necessity for public or third-party liability varies greatly from country to country (care should be taken to comply with local laws). This type of cover should include liability for bodily injury or illness caused to anyone. Cover should also include damage to other people's property, other than property in the care, custody or control of the expedition.

> **Warning:** do not admit liability in the event of an incident, because you may prejudice your insurance cover.

If you are sued for negligence, the cost of professional defence could be considerable, even if you are ultimately found not liable, especially as people are becoming more litigious and recent compensation awards from the courts have been rising. Even a limit of £2 million per person may not be adequate. Local authorities are now recommending that their service providers have no less than a £10 million limit. Ask the insurer for as high a limit as possible.

Leaders should ensure that the policy extends to include actions taken against the leader by a member of the expedition. Leaders of school expeditions and teachers should ensure that the school's liability policy extends to include the teachers'/leaders' liability in full (the cover arranged by the school should include what is known as professional indemnity and officials' indemnity) and in the country to be visited. If the school's insurance cannot be extended to provide this cover, some other form of liability insurance should be arranged. Check with your insurer that cover also extends to cover the expedition organisers.

Cover will exclude mechanically propelled vehicles – this includes waterborne craft and aircraft other than as passengers. Separate liability policies will be necessary for all waterborne craft and motor vehicles.

If hiring a car in the USA/Canada the indemnity limits will be low – separate top-up cover is normally necessary.

Company insurance

If the expedition is being arranged by a company or organisation that has received a fee for its services, additional areas of liability should be considered, e.g.

- breach of copyright if material is to be published
- internet liability – use of email and websites
- directors' and officers'/trustees' indemnity
- personal liability for wrongful acts in the course of duties, including libel and slander liability
- employers' liability and workers' compensation.

In particular, consideration should be given to the Package Travel Regulations 1992. These define a tour operator as prearranging a combination of two or more of the following: (1) transport, (2) accommodation and (3) "ancillary" services. These regulations could therefore potentially apply to schools, universities, coach operators, local authorities, or activity or sports centres. Special insurance cover may be required in the following areas:

- legal liability to third parties for financial losses
- security for passenger payments.

Replacement and rearrangement

You can insure additional travel and accommodation expenses for a replacement expedition member after the death or disablement of an insured person. In addition, this type of insurance would cover the cost of returning the originally insured person to complete the expedition following recovery.

Cancellation and curtailment

This category of insurance provides cover against cancellation for a number of reasons, e.g. compulsory quarantine, jury service, illness or death of the insured person or close relative, or hijack. Cover should also include curtailment, i.e. returning home before completion of the proposed venture or project. It is recommended that insurance cover be arranged as soon as travel expenses, such as air fares, are about to be paid, because this type of insurance provides for recovery of lost deposits in many situations. Cancellation through lack of funds does not constitute a claim. Political risks, such as when a visa is refused, are also likely to be uninsurable.

Very few insurers will be prepared to extend this section to include cover where a formal recommendation is made by the Foreign and Commonwealth Office to leave or not to travel to the country of expedition. This extension is highly recommended for expeditions.

Expedition equipment stores and money

These categories of insurance (which are available only from a very limited number of insurers) provide cover for equipment and stores accompanying the expedition, plus money. It is normal for insurers to ask for details of items above a certain value and these need to be listed. You may find that your sponsors or supporters or university will loan expensive equipment only subject to proof of insurance. This type of insurance is normally subject to certain exclusions, which may include:

- the first amount of each claim (£25 minimum)
- losses not reported to the appropriate authority as quickly as possible
- wear, tear, gradual deterioration, electrical or mechanical breakdown or derangement, atmospheric or climatic conditions
- loss or damage to equipment while in functional use
- breakage of brittle or fragile articles except in certain circumstances (check policy wording)
- loss or damage caused by delay, detention or confiscation by customs or other officials. If it appears to be theft (e.g. by a foreign government official) an explanation of the circumstances to the insurer may result in reimbursement. Expeditions should check on import duties that may be levied on imported goods, or restrictions on bringing certain goods into the country. The expedition may have to prove that it is re-exporting the goods at the end of the expedition.

Equipment and stores sent unaccompanied can be insured under a separate marine cover (see below).

Personal belongings and money of individual members

Expedition members' personal belongings, including spectacles, watches, photographic equipment and valuables, should in most cases already be insured by them on their own personal policies which will need to be extended for the period of the expedition. The cost of insuring personal belongings has risen dramatically of late and large excesses are applied, especially to photographic and video equipment. For these

143

reasons, it is often cheaper to leave this cover to the individuals to arrange. Often over-looked is the single article limit imposed on the majority of policies. The amount varies, but is usually around £250 to £500. If you have a more valuable item, specify it to the insurers, because failure to do so may limit the amount payable in the event of a claim.

Some insurance policies will cover travel and accommodation expenses incurred in replacing a lost passport and many insurers now include the cost of emergency purchase of essential items should your luggage be misplaced for a specific period of time.

Unaccompanied expedition equipment and stores
Cover for this category of insurance is provided under a marine policy regardless of whether the goods are sent by sea or air. A full inventory should be made for both the outward and return trips, and a separate value shown against all items over £250. Difficulties can arise with the type of cover that is required for equipment on loan to an expedition. It is therefore essential that the cover is adequate, because certain "all risks" covers can have "awkward" exclusions. Cover should commence from the date of assembling for shipping to the date of return. Thought should also be given to whether a specific item of equipment is essential for the success of the expedition. If it is, consideration should be given to insuring for the expense of obtaining a replace-ment in some remote part of the world. If a shipping and forwarding agent is used, they may provide insurance as part of their service but you should check exactly what cover is provided, e.g. is there cover after arrival and while in storage awaiting collec-tion? If you are relying on the carrier's insurance ask for a written summary of the cover.

Kidnap and ransom cover
The threat of kidnap and ransom/extortion, not only of personnel, but also of prop-erty, is on a disturbing increase. Although Latin America and the Far East have always been high-risk areas there have recently been an increase in incidents in Europe. It is difficult to estimate exact numbers because only one in ten kidnaps is reported. The following are particularly high hazard risk areas: Colombia, Russia, Mexico, Peru, Brazil, Guatemala, Asia and the Pacific.

It is possible to extend expedition travel covers to insure against kidnap, extortion, wrongful detention and hijack; the insurance should include the following:

- ransom/extortion payment – limit of at least £1 million per person
- loss in transit of the payment
- related expenses, e.g. psychiatric care, personal financial loss
- 24-hour emergency response helpline with specialist crisis consultants
- death benefits.

Insurers will appoint specialist crisis management security consultants who are experienced in defusing the tensions of a kidnap situation and can successfully advise on the negotiating skills required. Usually a consultant can also be deployed to the country to give critical advice. The crisis management team appointed by insurers will work closely with your family and international authorities.

Insurance of vehicles

In Europe
During the past few years many European countries have abolished the inspection of insurance documents at frontier crossings and UK motor insurance policies have been extended to provide cover to any member of the EU, Czech Republic, Hungary, Iceland, Norway, Slovakia and Switzerland.

Motorists visiting Spain will encounter difficulties if they are involved in an accident because the authorities may detain the motorist, or the car, and release can be obtained only against a guarantee or bail deposit for a substantial sum. Insurers issue bail bonds and cover is normally detailed on the reverse of the UK motor certificate. If any part of the guarantee is kept by the Spanish authorities it will be necessary for you to repay such amount to the insurer.

The standard UK motor policy provides cover for sea journeys of up to 65 hours' duration. For longer sea journeys arrangements should be made to insure the vehicle under a marine policy.

Outside Europe
Third-party insurance is compulsory in most countries: exceptions include most Central and South American countries, but insurance is nevertheless recommended.

Why a Green Card is still necessary
The Green Card cover can extend comprehensive cover to certain additional countries. The countries party to this arrangement where Green Cards can be purchased from insurers, or cover can be purchased at the border, are Albania, Andorra, Bosnia, Bulgaria, Cyprus, Estonia, Iran, Israel, Latvia, Macedonia, Malta, Moldavia, Montenegro, Morocco, Poland, Romania, Serbia, Tunisia, Turkey and the Ukraine.

Before leaving the UK you should make enquiries as to what motor insurance is compulsory in each country to be visited and ensure that this minimum cover is arranged before departure or that it can be purchased at the point of entry. Comprehensive motor insurance, although recommended, can be very expensive and is not always available. In many countries the state has a monopoly on insurance and a certain level of motor insurance (usually third party) is compulsory and must be obtained before driving in that country.

Information about the Green Card requirements for a particular country may be obtained from the Motor Insurers' Bureau, Linford Wood House, 6–12 Capital Drive,

Linford Wood, Milton Keynes MK14 6XT (tel: +44 1908 240000, fax: +44 1908 671681, website: www.mib.org.uk).

Both the Automobile Association and the Royal Automobile Club have insurance packages available to members that include vehicle breakdown and repatriation. Expeditions involving extensive road travel should investigate this source of insurance and will generally find that the AA and RAC are able to provide much useful information about driving conditions, etc., in various parts of the world.

Notes

- Driving licence: drivers should carry their UK driving licence with them. The AA or RAC can assist with advice on international driving documents.
- Vehicle registration document: carry the vehicle registration document with you. If you do not own the vehicle, a letter authorising you to use it should be carried.
- Rented vehicles: read the wording of the insurance cover when collecting the vehicle and check for exclusions. There may be some unsatisfactory restrictions that need to be clarified before driving off (e.g. excluding cover if the vehicle is used off the paved highway).

FURTHER POINTS TO CONSIDER WHEN ARRANGING INSURANCE

1. If you hold insurance in your own name (e.g. life, personal accident, all risks) you should notify the destination and details of your expedition activities to your insurers. If you do not, your policy could be invalidated.
2. When relying on an "umbrella" policy (e.g. a school or association policy), check that the cover is adequate. Insurance provided by schools' policies will not usually cover children who left school at the end of the term before the expedition.
3. If you hire local labour, make enquiries about your responsibilities before the expedition starts. In many countries something equivalent to the UK employers' liability insurance, normally known as workman's compensation, may be needed. In most cases, this can be arranged locally, before engaging local labour, and exact requirements can usually be confirmed from the host country's embassy. In addition many expeditions will work with local scientists and helpers who should be included in the expedition's liability insurance, subject to local insurance laws.
4. Read the insurance policy details carefully and explain them to all members of the expedition.
5. Take some claim forms with the expedition to complete while the incident is still fresh in your mind. **It is absolutely essential that any claim be reported to the**

insurer immediately because an insurance policy may time bar a claim if late notified.

6. Be careful to declare separately to the insurer any holiday taken after the expedition has finished. Separate cover may need to be arranged as a separate risk from the rest of the expedition.

7. Check that your policy will not expire if your expedition is delayed beyond the planned return date, as a result of circumstances beyond your control. It may be impossible to contact your insurer from the field.

8. Some insurers will try to exclude any cover arising from "war risks". This should be strongly resisted because expeditions often work in politically sensitive areas. A more acceptable wording is an exclusion of war risks by major powers only. If in any doubt about the stability of the area to which you are travelling, check with the Foreign Office and declare the facts to the insurer in writing for agreement.

CONCLUSIONS

A great deal of time and effort goes into the planning of an expedition and insurance is a vital part of the background organisation. For peace of mind, please ensure that early consideration is given to your insurance needs so that there are no gaps in the insurance cover, and no last-minute panics.

Depending on the type of expedition, insurance can be very expensive. The greater the chance of a claim, the higher the premium. Hence it is strongly recommended that advice be sought on provisional premium levels before finalisation of any budget.

Cheap premiums usually mean inadequate cover with many exclusions – you get what you pay for.

When obtaining a quotation make sure that the price you are quoted is inclusive of insurance premium tax (IPT). The present IPT rates are as follows: personal travel insurance 17.5 per cent, travel insurances as part of employment 5 per cent.

WHERE TO GET INSURANCE

RGS–IBG Expedition Travel Insurance Scheme

Aon Ltd, Richmond House, College Street, Southampton SO14 3PS. Tel: +44 23 8060 7500, fax: +44 23 80 63 1055, email: expeditions@ars.aon.co.uk, website: www.aon.com
Aon have drawn up an Expedition Travel Insurance Scheme for the RGS–IBG that is designed to meet the specialised needs of scientific and educational expeditions. A leaflet describing the scheme can be obtained from the RGS–IBG Expedition Advisory Centre or direct from Aon.

Other firms

Relatively few insurance consultants are qualified to arrange expedition insurance, but among those who have shown an interest in insuring expeditions are:

Campbell Irvine Ltd, 48 Earls Court Road, London W8 6EJ. Tel: +44 20 7937 6981, fax: +44 20 7938 2250, email: ci@netcomuk.co.uk, website: www.campbellirvine.co.uk
In the first instance, please submit brief details of the expedition in writing.

Harrison Beaumont Insurance Services, 2 Des Roches Square, Witney, Oxfordshire OX28 4LG. Tel: +44 870 1217 590, Fax: +44 870 1217 592, email: info@hbinsurance.co.uk, website: www.hbinsurance.co.uk
If obtaining quotations from any other insurance intermediaries, make sure that the insurance broker is a member of the General Insurance Standards Council (www.gisc.co.uk) or (effective in 2004) the Financial Services Authority.

For sporting expeditions
Many clubs and associations have special insurance schemes arranged for their members. These range from mountaineering and hang-gliding to canoeing and caving, and are designed to provide insurance cover for specialist high-risk activities.

Beware: some of these schemes have restricted cover, whereas others may not last because of either experience of bad claims or lack of support.

For mountaineering expeditions
British Mountaineering Council, 177–179 Burton Road, Manchester M20 2BB. Tel: +44 870 010 4878, website: www.thebmc.co.uk
The BMC has an insurance scheme for hillwalkers, climbers and mountaineers, and welcomes new members requiring insurance.

For winter sports, trekking and rafting
Snowcard, Lower Boddington, Daventry, Northants NN11 6BR. Tel: +44 1327 262805, fax: +44 1327 263227, email: enquiries@snowcard.co.uk, website: www.snowcard.co.uk

For outdoor and adventure training instructors
The Institute for Outdoor Learning has developed a public liability insurance scheme specifically for outdoor instructors, offering comprehensive protection against the many risks involved in working as an instructor. Available to both voluntary and professional instructors who are members of the Institute.

For further information contact JLT Corporate Risks Limited, Roebuck House, Brunswick Road, Gloucester GL1 1LU. Tel: +44 1452 511400, fax: +44 1452 511401, email: rachel_richards@jltgroup.com, website: www.outdoor-learning.org

For general insurance
Endsleigh Insurance Services, 3 Kings Street, Watford WD1 8BT. Tel: +44 1923 218438, fax: +44 1923 218458, website: www.endsleigh.co.uk
Have local offices throughout the UK and offer a range of policies designed for the specific needs of the independent traveller.

Note that if an individual is dissatisfied with an insurer's service, the following option is open to him or her:

- Ask the Financial Ombudsman Service to review the case. Their offices are at South Quay Plaza, 183 Marsh Wall, London E14 9SR. Tel: +44 845 080 1800, email: complaint.info@financial-ombudsman.org.uk, website: www.financial-ombudsman.org.uk

If you contact the above body in respect of complaints, this will not affect any legal right of action that you may have.

14 EXPEDITION MEDICINE

David Warrell

Expeditions and fieldwork in remote and challenging places are likely to expose members to greater environmental extremes and to more unusual hazards than do other types of travel. The aims of expedition or wilderness medicine are to improve, through knowledge, planning and skills, the confidence, enjoyment and achievements of the people who participate in these expeditions.

RISKS OF EXPEDITIONS: REAL AND PERCEIVED

The risks of exotic infections, such as a viral haemorrhagic fever, plague, rabies or sleeping sickness, attacks by large or venomous animals and even of meeting cannibals, may loom large in the imagination of expedition members. However, the reality is much more mundane. Travellers' diarrhoea and other gastrointestinal disturbances are now recognised to be the main cause of expedition illness, whereas the leading causes of expedition mortality are falls and other injuries, road traffic accidents, altitude sickness, heat stroke, infections such as malaria, drowning and homicide. Overall, the health risks of participating in a well-planned expedition are similar to those encountered during normal active life (Anderson and Johnson, 2000). However, some expedition activities carry much higher mortality rates: 16 per cent of those attempting to reach the summit of Everest will die, 2.9 per cent of Himalayan mountaineers and 1 per cent of those over-wintering in Antarctica, compared with 0.83 per cent of expedition participants in general, 0.014 per cent of Himalayan trekkers and 0.013 per cent of low-altitude joggers (Anderson and Johnson, 2000).

REDUCTION OF HEALTH RISKS BY PLANNING

Health risk assessment demands consideration of the terrain, altitude, climate, and endemic fauna and diseases of the area to be visited, and the intended aims of the expedition. Much of this information may be available beforehand. During selection of the expedition team, it is important to identify those with special problems (Table 14.1). Depending on the type of expedition, many of these may be accommodated by careful planning. However, the stress of travel in remote areas can destabilise chronic medical conditions and this could, in certain circumstances, cause danger to everyone in the group.

TABLE 14.1 EXPEDITION MEMBERS' SPECIAL PROBLEMS

Pregnancy
Immunosuppression (by drugs or diseases)
Chronic illness (diabetes, epilepsy, asthma, ischaemic heart disease, etc.)
Psychiatric problems
Physical/mental handicap
Alcohol/drug abuse

All expeditions should have a designated medical officer, who, in most cases, will not be medically qualified. All members should attend first-aid training, which, ideally, should be aimed at the particular needs of the expedition. Essential first-aid skills for all expeditions are clearing the airway and resuscitation, controlling blood loss, treating shock, relieving pain and ensuring the safe evacuation of injured people. Prevention of medical problems on an expedition depends on awareness of local diseases (based on up-to-date information from journals, books, websites and telephone advice services), appropriate immunisations and chemoprophylaxis, a pre-expedition dental check-up and, if possible, resolution of known surgical and medical problems well in advance of the expedition's departure. Explicit instructions should be given to expedition members about safe and sensible behaviour: in the use of equipment and techniques; and about food and water hygiene; protection from climatic and environmental hazards; as well as safe sex.

Expedition medical kits need to be much more comprehensive than those carried by ordinary tourists. Local medical back-up must be arranged in advance through the expedition's local agent. Hospitals or other medical facilities nearest to the site of the expedition must be identified, contacted and, if possible, assessed in advance. Emergency evacuation of severely ill or injured expedition members must be antici- pated and planned well in advance and medical insurance cover should be generous to allow for in-country medical care (especially expensive in North America) and, if need be, repatriation of the sick or injured person. Many newer technical aids have improved safety through communication (radio/satellite telephones) and navigation (satellite location systems).

THE EXPEDITION MEDICAL OFFICER

This is an essential and responsible role. The expedition medical officer must take the lead in planning and organising pre-expedition medical education, as well as deciding the location of the base camp, making arrangements for food, and providing for the psychological and pastoral needs of the expedition members. Depending on the particular circumstances, expeditions may feel some responsibility for helping with medical problems of the indigenous peoples of the area. This can be a difficult issue because time, equipment and drugs are always in short supply.

ROAD TRAFFIC ACCIDENTS

It is astonishing that people who have spent much time and money in preventing illness during an expedition should, on arrival at the destination, entrust their lives to untried crazy-looking drivers and unsound vehicles. The risk is much greater in less developed countries, where there has been an epidemic increase in road traffic accident fatalities over the past 20 years and where 85 per cent of these deaths now occur (*British Medical Journal*, 2002). The risk of accidents can be reduced by avoiding driving at night outside cities, ensuring that the driver is not tired or under the influence of alcohol, antihistamines or other sedative (or recreational) drugs, avoiding driving alone, watching the driver for signs of fatigue, taking regular breaks and checking the basic functions of the vehicle (steering, lights, brakes, tyres, etc.) before setting off. Using seatbelts reduces the risk of death by 65 per cent.

IMMUNISATIONS/VACCINATIONS

The current wave of dangerously misinformed criticism of immunisations (MMR, Gulf War syndrome, etc.) must not discourage travellers from this most effective form of disease prevention. Do not assume that everyone has received a standard childhood course of immunisations (in the UK: diphtheria, pertussis, tetanus, mumps, measles, rubella [MMR], *Haemophilus influenzae* b [Hib], meningococcus C). Even if the traveller received a childhood primary course, boosters will be needed for diphtheria, tetanus and polio (eliminated from the Americas and Europe but still present elsewhere) after 10 years. Other basic immunisations recommended for travellers to almost every less developed country are BCG (for tuberculosis/leprosy), and those for hepatitis A, typhoid and rabies. Special immunisations for travellers to certain parts of the world include yellow fever (equatorial Africa and Latin America), Japanese encephalitis (Asia and New Guinea), meningococcus A (meningitis belt of sub-Sahelian Africa and new epidemic areas) and tick-borne encephalitis (central Europe and Scandinavia). Yellow fever is the only immunisation for which a certificate is a statutory requirement for travellers from and to endemic areas (Monath and Cetron, 2002), e.g. you will not be allowed to fly from Ecuador to Brazil without a valid yellow fever

Figure 14.1 *Distribution of meningococcal meningitis in Africa*

immunisation certificate. Recent deaths from yellow fever in tourists to West Africa and Latin America emphasise the continuing importance of this immunisation. Cholera vaccine is no longer recommended by the World Health Organization because its adverse effects outweigh its usefulness, although a new oral vaccine is promising. The risk of hepatitis A, acquired from infected food/water, in less developed countries ranges from 300/100,00 to 2000/100,000 unprotected travellers per month of stay. Active immunisation is safe, effective and durable, and there is no longer any justification for short-term protection with immunoglobulin. Epidemic meningococcal meningitis occurs in the cool, dry season (December–February) most years in countries of the sub-Sahelian "meningitis belt" of Africa (from Senegal and the Gambia in the west to Sudan in the east) (Molesworth et al., 2002) (Figure 14.1). Travellers to this area, and to other new sites of epidemics, should be given meningococcal group A + C (or ACYW) vaccine. The meningococcal group C vaccine now given to children in the UK does not provide adequate cover in these areas.

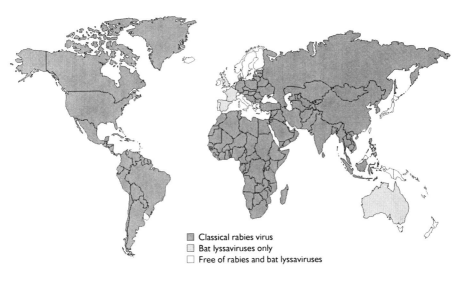

Figure 14.2 *Global distribution of rabies and the rabies-related bat lyssaviruses, 2003*

Pre-exposure immunisation against classical rabies and the European and Australian rabies-related bat lyssaviruses (Figure 14.2) is being used increasingly in travellers. (These bat lyssaviruses are related to classical rabies virus and produce clinical effects identical to classical rabies in infected people.) Although the risk of transmission is low, the lack of effective treatment for rabies encephalitis and the fear engendered by a dog bite justifies immunisation now that safe and potent vaccines are available. Cost can be reduced if an ampoule of vaccine is divided among ten vaccinees, each being given one-tenth of the dose by intradermal injection.

Plague and anthrax vaccines cause serious side effects and, if there is real risk of infection, antibiotic prophylaxis or post-exposure treatment should be considered (doxycycline for plague, ciprofloxacin for anthrax). Japanese (B) encephalitis (Figure 14.3) and European tick-borne encephalitis vaccines should be considered in travellers to the endemic areas, especially during the seasons of transmission. Hepatitis B is a risk for medical staff whose work involves contact with human blood, to those receiving unscreened blood transfusions in some less developed countries and to those who take the high risk of unprotected sexual activity and intravenous drug abuse.

Typhoid is still prevalent in many less developed countries, especially in the Indian subcontinent. Effective injectable and oral vaccines are available that do not have the serious side effects associated with the old "TAB" immunisation.

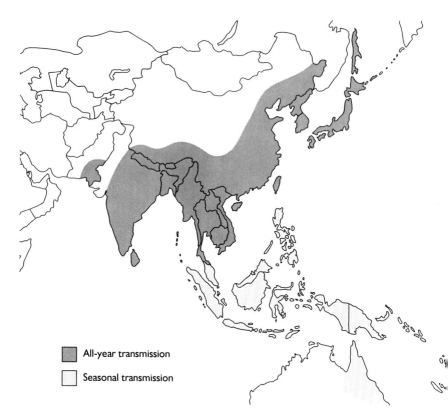

All-year transmission

Seasonal transmission

Figure 14.3 *Geographical distribution of Japanese encephalitis, by endemic countries and regions of South-east Asia, 2003*

INFECTIONS STILL PREVALENT IN SOME TROPICAL/LESS DEVELOPED COUNTRIES

Travellers' diarrhoea

This is by far the most common health problem experienced by expedition members. Many different kinds of food- and water-borne organisms can cause acute and debilitating diarrhoea, usually associated with colicky pain and prostration, and sometimes with vomiting, fever, bloodstained motions (dysentery) and even kidney failure. Enterotoxogenic *Escherichia coli* bacteria are responsible for about 50 per cent of cases. Other important infections are giardiasis and cryptosporidiosis (in which there is explosive watery diarrhoea, abdominal distension, nausea, weakness and passage of exceptionally foul-smelling gas), salmonellosis (especially from undercooked

chicken, eggs and milk products), amoebic and bacillary dysentery and cam-
pylobacter infections (blood in the stool), and viruses. Travellers' diarrhoea is very
rarely fatal but can ruin an expedition.

Prevention of travellers' diarrhoea: food and water hygiene
Drinking water should be filtered, boiled, treated with sterilising tablets or commercially bottled. Beware of ice in drinks because this is frequently made from tap water.
The rule for eating is "cook it, peel it or forget it", but this rule can be difficult to
enforce, without causing offence, when receiving hospitality. Especially hazardous
are salads (even peeled tomatoes), which may have been fertilised with human faeces,
raw egg products such as mayonnaise, undercooked chicken (pink at the bone) or
eggs, milk or cheeses (which also carry the risk of brucellosis, listeriosis and cam-
pylobacter infection), rare or frankly raw meat (relished in France, the Middle East
and Ethiopia) and ice cream. Deep fried food is safer than grilled food and "barbe-
cued" usually means raw in the middle.

TABLE 14.2 PREVENTION OF TRAVELLERS' DIARRHOEA

Food and water hygiene

1. "Cook it, peel it or forget it!"
2. Drink only water that is boiled/filtered/chemically sterilised/bottled.
3. Beware of ice cubes.
4. Avoid unpasteurised milk and milk products – cheese, ice cream, etc. – and
   raw eggs.
5. Avoid shellfish and crustaceans, even if boiled.
6. Consider prophylactic antibiotics, e.g. ciprofloxacin.

As a result of the wide range of possible causes, prevention with even a broad-
spectrum antibiotic ("kills all known germs") such as ciprofloxacin (and other
fluoroquinolone drugs) will be only partially effective. Early treatment with
ciprofloxacin (500 mg) after passing the first loose stool has proved effective. Other
drugs such as doxycycline or co-trimoxazole are less effective.

Travellers' diarrhoea: treatment
Repeated and copious diarrhoea and vomiting rapidly dehydrate the victim. Patients
should be encouraged to rest and to keep drinking clear fluids (frequent small sips to
reduce the risk of vomiting). In severe cases, oral rehydration salts should be added.
These contain glucose to promote absorption of minerals. Very severe cases will
require intravenous fluids. It is best to avoid solid food, which may stimulate

vomiting and further colicky pain and diarrhoea ("gastrocolic reflex"). Ideally, the victim should rest quietly in bed but, if travel or exertion is unavoidable, diarrhoeal symptoms can be damped down with codeine phosphate (Imodium or Lomotil). It may be possible to swallow and retain anti-vomiting drugs such as Stemetil or metoclopramide. Otherwise, these can be given by suppository (through the anus into the rectum).

TABLE 14.3 TREATMENT OF TRAVELLERS' DIARRHOEA

1. Rest; take small sips of clear fluids frequently.
2. Oral rehydration salts.
3. Avoid solid food.
(4. Palliate diarrhoea and vomiting).
5. Take antibiotic (ciprofloxacin) immediately.
6. If symptoms continue for more than 48 hours, are very severe or there is blood in the stools, SEEK MEDICAL ADVICE.

The vast majority of attacks of traveller's diarrhoea will resolve spontaneously after 12–48 hours of conservative treatment. More prolonged and very severe symptoms require medical advice or a trial of ciprofloxacin (500 mg once a day for 3 days) or, if the symptoms suggest giardiasis (see above), a dose of tinidazole (Fasigyn) 2 g (repeated 1 and 2 weeks later) or metronidazole (Flagyl) 250 mg three times a day for 5 days (avoid alcohol!).

Malaria

One of the world's major killing diseases, malaria is also the most common cause of life-threatening illness in travellers. Each year, about 2000 people arrive back in the UK with malaria, three-quarters of them with life-threatening *Plasmodium falciparum* malaria, 7–16 of whom will die of the disease (Figure 14.4). The other three kinds of human malaria, *P. vivax*, *P. malariae* and *P. ovale*, cause an unpleasant feverish illness but very rarely kill. Malaria is transmitted by night-biting anopheline mosquitoes throughout most parts of the tropics except in islands east of Vanuatu in the Pacific and the Caribbean islands (except for Haiti/Dominican Republic).

Symptoms of malaria

All four kinds of malaria cause high fever with shivering and shaking, severe headache, pains in the neck, back and muscles, prostration, nausea and diarrhoea. Sometimes the episodes of fever, lasting a few hours, are interspersed with 24- to 48-hour intervals of feeling almost well. *Plasmodium falciparum* malaria can cause

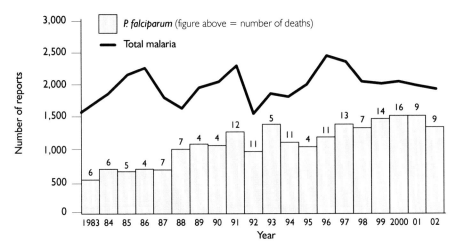

Figure 14.4 *Number of imported malaria cases to the United Kingdom reported to the Health Protection Agency Malaria Reference Laboratory 1983–2002; and number of fatalities each year from* P. falciparum *infections*

unconsciousness and fits (cerebral malaria), jaundice, bleeding, black urine (black-water fever), severe anaemia and other dangerous effects. Symptoms of malaria start no sooner than 7 days after the infective mosquito bite and usually up to a few months, but sometimes longer, after the traveller has returned home.

Diagnosis and treatment of malaria

The diagnosis is made by examining a specially stained blood smear under the microscope or using dipsticks (rapid antigen test). The treatment of malaria has been complicated by the development of resistance to many of the established antimalarial drugs. Chloroquine (Nivaquine) is still effective for the three milder types of malaria, but can be relied upon to cure *P. falciparum* malaria only in Central America and Haiti/Dominican Republic. Elsewhere, treatment is with atovaquone–proguanil (Malarone), artemether–lumefantrine (Riamet) or quinine. If an expedition member develops acute fever in a malarious area, but is too far away from medical support to allow laboratory diagnosis of malaria, a full course of any one of these three drugs can be taken as a therapeutic trial (standby treatment). People who are vomiting and cannot keep their tablets down should be treated by intravenous infusion (quinine), intravenous injection (artesunate), intramuscular injection (quinine/artemether/artesunate) or rectal suppository (artesunate/artemisinin). A patient with *P. falciparum* malaria can deteriorate rapidly and so suspected cases should be evacuated to the nearest hospital.

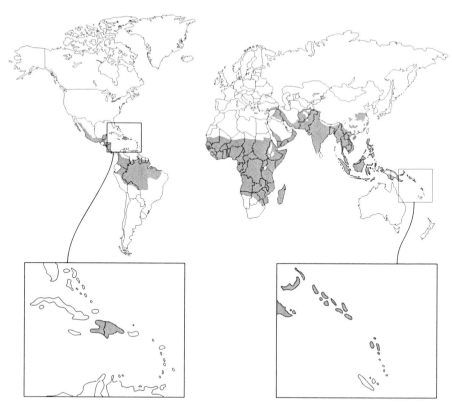

Figure 14.5 *Malaria is endemic in almost all parts of the tropical world as far north as southern Turkey, as far south as north-eastern South Africa, as far west as Mexico and as far east as Vanuatu in the western Pacific*

Prevention of malaria (see Bradley and Bannister, 2001)
It is important to find out if the precise area of the expedition is malarious so that proper precautions can be taken. Night-time exposure (camping, animal collecting) carries a high risk of infection, while pregnant women and people who have lost their spleen or are especially susceptible should, if possible, not enter malarious areas. To reduce the risk of mosquito bites, wear sensible clothing after dark (light-coloured, long sleeves and long trousers), and apply repellents (containing "DEET") to exposed areas. In the sleeping quarters, mosquitoes should be killed by knock-down insecticides and excluded by insecticide (permethrin)-impregnated mosquito nets. Taking drugs to prevent malaria (chemoprophylaxis) is never completely effective. Advice should be taken about which drug is effective in a particular area. In drug-resistant areas, such as Africa, the Amazon Basin and South-east Asia, mefloquine (Lariam),

doxycycline (Vibramycin) or atovaquone–proguanil will be needed. The risk of unpleasant side effects from mefloquine has been exaggerated by the media, but between 0.1 and 1 per cent of people, especially women, may become depressed, dizzy, nauseated and unsteady, and may suffer from nightmares as a reaction to this drug. It is wise to start mefloquine 4 weeks before leaving on the expedition, to allow a switch to another drug in the small minority of people who will develop side effects. Antimalarial drugs should be continued for a full 4 weeks after leaving the malarious area, except in the case of Malarone, which is continued for only 7 days. No prophylactic drug will work unless it is taken regularly, as prescribed, and continued for this period. Feverish illnesses that develop after return from the expedition should be taken seriously! It is important to see a doctor and to mention the risk of malaria.

People who suffer from epilepsy or psychiatric illnesses should not take mefloquine and those with epilepsy or psoriasis should avoid chloroquine. Chloroquine can cause severe itching in dark-skinned individuals.

TABLE 14.4 MALARIA PREVENTION

1. Avoid being bitten: sleep under a treated mosquito net; use insecticides, repellents, sensible clothing and behave sensibly.
2. Take preventive drugs: mefloquine (Lariam) or other drugs, depending on the particular geographical area.
3. Carry a course of standby treatment.
4. SEE A DOCTOR AND MENTION MALARIA – if you develop a feverish illness within a few months of returning!

Dengue
This mosquito-borne virus infection is very widespread in tropical countries and continues to extend its range throughout the tropics, including large conurbations (Figure 14.6). Adults experiencing their first attack will develop fever and severe pains in the head, back, muscles and joints ("dengue" means break bone fever). The only treatment is rest, painkillers and antipyretics. Use paracetamol or codeine phosphate but avoid aspirin and non-steroidal anti-inflammatory drugs such as ibuprofen (Nurofen). After a few days the fever seems to be getting better but there may then be a relapse with the appearance of a red rash and sometimes bleeding. However, in residents of tropical areas, particularly children, a second attack of dengue, with a different type of dengue virus, can cause fatal shock and bleeding. Increasing numbers of travellers are catching dengue in Indonesia, other parts of Asia, the Caribbean and Latin America. There is no vaccine (Almond et al., 2002) and the only way to prevent infection is to avoid mosquito bites. Unfortunately, the stripy-legged,

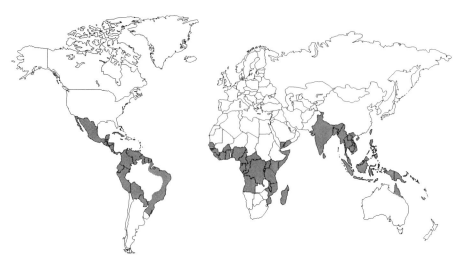

Figure 14.6 *Dengue fever ("break bone" fever)*

dengue-transmitting Aëdes mosquitoes do not confine their biting to night time.

Schistosomiasis (bilharzia)

This infection with a flatworm (fluke) is acquired through contact with freshwater from lakes and sluggish rivers, usually by bathing or washing with water taken from

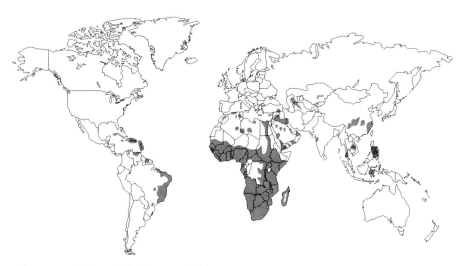

Figure 14.7 *Bilharzia (schistosomiasis)*

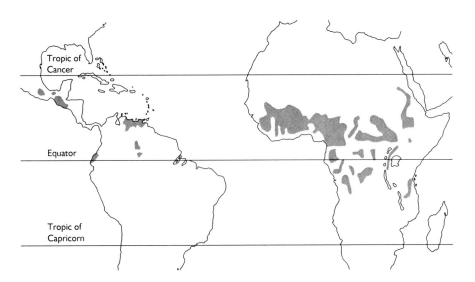

Figure 14.8 *River blindness (onchocerciasis)*

these sources. The water is contaminated by humans or baboons that have defaecated or urinated into it. The intermediate snail hosts that live in the reeds are then infected. Snails release tiny larvae into the water, which burrow through the skin of bathers, causing "swimmer's itch", experienced soon after contact with infected water. However, most cases of "swimmer's itch" are caused by kinds of bird and animal schistosomiasis that do not go on to cause infection in humans. Some people infected with schistosomiasis develop an acute fever with rash a few weeks after infection. Later symptoms include passage of cloudy or frankly bloodstained urine or bloody diarrhoea. Bilharzia is one of the most common travellers' diseases dealt with in travel clinics in western countries. Expedition members usually get worried when they get back from their trip and remember bathing in infected lakes or they hear that another member of their expedition has been diagnosed as having schistosomiasis. Diagnosis is confirmed by finding eggs in the stool, urine or lining of the rectum (diagnosed by "rectal snip") or by a blood test. Treatment is simple, safe and effective with one or two doses of praziquantel (Biltricide). Prevention is by avoiding skin contact with freshwater sources in the endemic countries of Africa, the Caribbean, South America, Middle East and South-east Asia (Figure 14.7). Local advice may be misleading. Lake Malawi, officially declared free of bilharzia for many years, has been the source of many imported cases of bilharzia.

River blindness (onchocerciasis)

This infection of skin, lymph nodes and eyes occurs in parts of Central and South

America, Africa and Yemen (Figure 14.8). It is caused by a filarial roundworm transmitted between humans by viciously biting little black flies which breed in fast-
flowing streams, rivers and waterfalls. Skin changes (itching, roughness, thickening,
decreased or increased pigmentation, and loss of elasticity) and eye damage are
caused by tiny larvae (microfilariae) produced by the adult worms which live in
lumps (nodules) beneath the skin, especially around the waist and hip joints. Diagnosis is by microscopic examination of a skin snip and by blood tests. If there are
nodules on the head or visual or eye symptoms, slit-lamp examination of the eye by
an ophthalmologist is essential. Treatment with ivermectin is effective but may cause
a flare-up of symptoms and so must be supervised in hospital. The only partial
protection against infective black fly bites is the wearing of light-coloured clothing
(long sleeves and long trousers) and the application of DEET-containing repellents
to exposed areas of skin.

Sleeping sickness (African trypanosomiasis)

There is a resurgence of sleeping sickness in Angola, Central African Republic,
western Uganda and other countries of equatorial Africa. There have been some
recent cases in the game parks of northern Tanzania (Serengeti) (Figure 14.9). Vora-
cious, blood-sucking tsetse flies, slightly larger than house flies, transmit the
causative trypanosomes (protozoan parasites) between humans and, in eastern
Africa, between humans and game animal reservoirs (e.g. antelopes and bucks). A
boil (chancre) may develop at the site of an infective bite, followed by fevers,
headache, enlargement of lymph nodes in the base of the neck and, eventually, when
the brain is invaded by the parasites, development of the characteristic sleepiness that

Figure 14.9 *Sleeping sickness (African trypanosomiasis)*

gives the disease its popular name. Trypanosomes may be found in the blood, fluid from lymph nodes or the cerebrospinal fluid bathing the brain. Treatment is difficult and dangerous using toxic drugs. Prevention is by avoiding endemic areas and tsetse fly bites.

Typhus

The various kinds of typhus bacteria are transmitted by mites, ticks, fleas or lice in different parts of the world. African tick fever (typhus) is particularly common in travellers to game parks in Central and Southern Africa (e.g. Kruger). About 7 days after finding an attached tick, severe fevers, headaches, nausea and muscle pain develop and, at the site of the infective bite, a boil comes up which eventually develops a blackish scab (eschar); there is a generalised, reddish rash. Other kinds of typhus may be similar, with abrupt fever, generalised rash, a local eschar and other severe systemic flu-like symptoms. Prompt treatment with a tetracycline such as doxycycline can produce dramatic relief of the symptoms.

Worms

Infections with roundworms (nematodes), flukes or flat worms (trematodes) and tapeworms (cestodes) are enormously prevalent among the inhabitants of many parts of the tropics. Infection with hookworms and *Strongyloides* results from walking bare-footed in areas contaminated with human faeces. The infective larvae can penetrate the skin of the feet. Most of the other worm infections are acquired by ingesting eggs that have been deposited in faeces or by eating the intermediate hosts of the parasites such as fish containing larval forms. These infections can result in a variety of symptoms: anaemia and weight loss in the case of many of the worms that infect the gut; jaundice and enlargement of the liver in the case of liver flukes; coughing up blood-stained sputum in the case of lung flukes; and development of itching rashes and transient cough or asthma in the case of nematodes, the larvae of which migrate in the skin or lungs. The person infected with *Ascaris* may pass a worm (about the size of an earthworm) in their stools or, in the case of tapeworm infection, some wriggling segments of the worm. Depending on the kind of worm infection, diagnosis can be achieved by finding characteristic eggs or larvae in the stools or sputum, or by locating adult worms by x-ray or ultrasound imaging techniques. Effective drugs include praziquantel for flukes, thiabendazole, mebendazole or albendazole for nematodes, and niclosamide for gut tapeworms.

Creeping eruption (cutaneous larva migrans)

People who walk barefooted or lie in the sand in coastal regions of Central/South America, Africa and South Asia may be infected by larvae of animal hookworms. Having penetrated the skin, these parasites are unable to develop further because they are in the wrong host. They therefore crawl around aimlessly under the skin, provoking intensely itchy, sore and reddish serpiginous tracks which may become secondarily infected by scratching. These lesions can creep several millimetres to a few centimetres each day. Treatment is by daily applications of an ointment made by grinding up a 0.5 g tablet of thiabendazole and mixing with 5 g petroleum jelly.

HIV/AIDS and other sexually transmitted infections
Seroprevalence of HIV has exceeded 30 per cent in some African countries, and other sexually transmitted infections, including gonorrhoea, syphilis, chancroid, herpes, venereal warts and hepatitis B, are highly prevalent in many less developed countries, especially in prostitutes, bar girls, "beach boys" and other "professional sex workers". As foreign travel seems often to be associated with a relaxation of usual sexual inhibitions and prohibitions, expedition members should be warned explicitly about the risks of unprotected sex. Although condoms are widely available as part of HIV-prevention programmes, expedition members should carry and use their own supplies of good quality condoms. Immediate medical advice should be sought if there is a purulent discharge from the penis or vagina and if ulcers develop in the genital area or at any other possible site of genital contact.

Potential dangers of blood transfusion
In countries where screening of blood donated for blood transfusion is not possible or is unreliable, there is a risk of a variety of infections of which HIV, hepatitis viruses, HTLV-1 (the cause of tropical spastic paraparesis), malaria and Chagas' disease are the most important. To reduce this risk, some expeditions carry bags of intravenous fluids that can be used as a temporary substitute for blood in the treatment of bleeding and shock. Other equipment included in "AIDS-prevention kits", which might prevent a blood-borne infection, are disposable hypodermic needles (for countries where injection needles are still reused), syringes in case a blood specimen is needed for laboratory tests, and intravenous cannulae and giving sets for the administration of intravenous fluids.

Viral hepatitis
This is a common acute infection in which there is fever with shivering, headache and other pains, weakness, loss of appetite, nausea, vomiting and pain, and tenderness over the liver in the right upper part of the abdomen. As jaundice becomes noticeable in the eyes and skin, the urine becomes very dark and the stools become very pale.

Infection with hepatitis A is through contaminated water or food and is prevalent in less developed countries. Symptoms start between 3 and 5 weeks after infection. It is easily and effectively prevented by immunisation. Hepatitis B and hepatitis C are highly contagious. They are spread by blood contamination of needles, by blood transfusion and by sexual intercourse. Both infections can be chronic, resulting in progressive liver damage, cirrhosis and eventually the development of liver cancer. Effective vaccines are available for hepatitis B but not yet for hepatitis C. Other methods of prevention include avoiding unprotected sex, avoiding any skin penetration by potentially infected needles or other sharp instruments (including body piercing, acupuncture, tattooing and of course the sharing of needles by intravenous drug abusers), avoiding even conventional medical procedures if the practitioner is unable to ensure hepatitis-free conditions, and avoiding contamination by the blood of an infected person – even the sharing of a toothbrush.

COMMON NON-TROPICAL INFECTIONS

Sore throats and respiratory tract infections

Acute sore throat may be part of a generalised viral respiratory infection such as influenza, or caused by bacteria, most commonly streptococci, or from infectious mononucleosis (glandular fever). It may be accompanied by fever and painful, enlarged glands in the neck. Complications include tonsillitis (the tonsils on one or both sides are enlarged, red and covered with flecks of pus), local abscess formation in the throat, such as a quinsy, which may threaten to obstruct the upper airway, sinusitis (profuse, purulent nasal catarrh that may suddenly stop flowing, pain in the face, tenderness over the cheeks or forehead) and middle-ear infection (otitis media; earache, purulent discharge from one ear). There is no reliable way of distinguishing the different causes of a sore throat just by examination. However, if the lymph glands elsewhere are enlarged (e.g. in both armpits and both groins), glandular fever should be suspected and the patient should, on no account, be treated with ampicillin/amoxicillin because this can cause a severe rash. Provided that the patient is not known to be allergic to penicillin, the simplest treatment is to give a 7-day course of penicillin V (phenoxymethylpenicillin) or cloxacillin. Penicillin-allergic people can be given erythromycin. Gargling with water in which an aspirin has been dissolved or sucking anaesthetic lozenges may improve the symptoms.

Chest infections (bronchitis, pneumonia) cause fever, cough, bringing up greenish-yellow sputum (phlegm) and sometimes breathlessness and a sharp, localised chest pain (pleurisy), worse on breathing in or coughing. A stethoscope is useful for detecting signs in the infected lung. Treatment is with antibiotics such as amoxicillin or erythromycin or a cephalosporin or clarithromycin.

Painful red eyes (conjunctivitis)

If one or both eyes become red and painful with a purulent discharge so that the lids stick together at night, an infection of the outer membranes of the eye (conjunctivitis) and/or a local infection of one of the eyelash follicles (stye) is likely. A topical eye ointment such as chloramphenicol or tetracycline should be applied regularly. The eye can be irrigated with sterile (boiled) tepid water.

A piece of grit may become lodged above the cartilaginous tarsal plate of the upper eyelid, causing days of soreness and misery. It may be removed only by everting the tarsal plate, a very useful skill that can earn many grateful patients.

Pimples, boils and other bacterial skin infections

These are very common, especially at the sites of injury, on the feet where sites of friction or abrasion have become infected, or at the sites of insect bites. Lesions should be kept as clean as possible and covered with light, non-adherent dressings. A topical antiseptic should be applied, such as povidone–iodine. If the pustule, boil or carbuncle has developed a yellowish head, or if the local area is tensely swollen and fluctuant, an attempt should be made to drain the pus by lancing with a sterile needle or scalpel blade. This can cause immediate relief. A course of antibiotics may be needed, especially if the local lymph glands are tender and enlarged and there is fever.

Urinary tract infections

The symptoms are frequent, urgent, painful urination with local burning. If the infection is severe there may be generalised symptoms such as fever with shivering, nausea and vomiting, and pain and tenderness in the lower back (loins) on one or both sides. The urine may look cloudy, dark or frankly bloodstained and may have a fishy or other strong unpleasant odour. Treatment is with antibiotics such as trimethoprim, amoxicillin or ciprofloxacin. It is important to drink a lot of fluid.

Vaginal discharge with local itching is commonly caused by thrush (*Candida*, a yeast) especially in women taking antibiotics (such as doxycycline for malaria prophylaxis). Treatment is with clotrimazole (Canesten) cream or pessaries.

Athlete's foot, dhobi's itch and other fungal infections

These are very common problems on expeditions. Athlete's foot is prevented by meticulous attention to foot hygiene, washing with antiseptic soap between the toes regularly and thoroughly drying the feet, which should be kept as well aerated and dry as possible by wearing open sandals without socks. Antifungal powder such as miconazole (Daktarin) can be used. Dhobi's itch is a reddish, irritating rash that may develop blisters and weep. It occurs in moist, occluded areas such as the groin under the scrotum, in the armpits or under the breasts. Washing with antiseptic soap, thorough drying, application of antifungal powder (miconazole) or creams (clotrimazole, econazole or ketoconazole), and maximal aeration are the best treatments. Patches of fungal infection may develop on any part of the body, especially in humid climates. They are often circular or annular with an irregular, scaly, reddish border. A trial of antifungal cream is the best treatment.

ENVIRONMENTAL DISEASES

High-altitude sickness
Rapid ascent from sea level to 11,000 feet (3500 metres) causes acute mountain sickness (AMS), a reaction to hypoxia, in more than 50 per cent of people, while rapid ascent to 16,000 feet (5000 metres) causes AMS in most people. The symptoms, which develop within 36 hours, include headache, lassitude, fatigue, loss of appetite, drowsiness, weakness, dizziness, palpitations, breathlessness, nausea and vomiting. Sleeping is interrupted by irregular (Cheyne–Stokes) breathing. AMS can be prevented by slow acclimatisation. The symptoms are reduced by taking Diamox (acetazolamide) 250 mg 12 hours before ascent and then 250 mg twice a day for 5 days. The most common side effect of acetazolamide is tingling in the fingers and toes. Sedative drugs and the contraceptive pill should be stopped because there may be an increased risk of venous thrombosis. Heavy, physical exercise should be avoided during the 2–5 days after arrival at altitude.

Two severe life-threatening forms of high-altitude sickness are recognised: high-altitude pulmonary oedema (HAPO), which causes breathlessness, coughing up frothy sputum, blueness (cyanosis) and drowsiness, and high-altitude cerebral oedema (HACO) in which there is headache, confusion, drowsiness, double vision and unsteadiness. Treatment of HAPO is with oxygen and rapid descent to lower altitude or the use of a portable hyperbaric chamber. If this is not possible, nifedipine (Adalat) and acetazolamide can be used. An experimental method of preventing HAPO is to inhale salmeterol (Sartori et al., 2002). For treatment of HACO, oxygen, rapid descent or use of a portable hyperbaric chamber is also essential, together with dexamethasone, furosemide or acetazolamide.

Motion sickness
This can be prevented by taking hyoscine hydrobromide by mouth (e.g. Kwells), which is effective in 30 minutes and lasts for 4 hours, or by a skin patch (Scopoderm) which takes up to 8 hours to act but lasts for 72 hours.

Jet lag
Air travel, east or west, across two or more time zones, commonly causes daytime tiredness, disorientation, memory loss, a feeling of unreality, loss of appetite and other gastrointestinal symptoms resulting from disruption of the diurnal rhythm. Excessive alcohol consumption during the flight adds "hangover" to these already unpleasant symptoms. Jet lag can be minimised by sleeping during the flight and, at the appropriate time, after arrival. A short-acting sleeping pill such as zopiclone, zaleplon or temazepam can help. The use of melatonin remains controversial but it may speed recovery from jet lag if taken on arrival, just before going to sleep and, in advance of travel, on waking (westbound travel) or at 2 pm (eastbound travel).

Exposure to light can also help: on waking (eastbound travel) or at the end of the day (westbound).

Heat illnesses and sunburn

When the body's heat-losing mechanisms fail, the body temperature rises with sometimes disastrous results. This is most commonly the result of exposure to environmental heat and high humidity (such as during a heatwave), especially in people undertaking prolonged physical exercise, wearing inappropriately heavy clothing. Heatstroke is a severe form of heat illness in which the increase in body temperature affects the brain, causing confusion, loss of consciousness or fits. This can also be caused by drugs such as Ecstasy. Heat illness should be suspected if a member of the expedition becomes unexpectedly weak, lethargic and tired, complaining of muscle cramps, with mental changes, headache and any impairment of consciousness. The patient should be quickly removed to shade or a cooler place, their clothes should be removed and they should be vigorously cooled by being sponged all over with water and fanned. This is a medical emergency and so medical help should be summoned immediately.

Exposure to the sun, even when it feels cold as at high altitude, can cause acute effects such as sunburn and prickly heat, whereas long-term exposure can cause skin cancers. Sunburn is prevented by wearing a broad-brimmed hat and adequate clothing and by applying sunscreens protecting against UVB and UVA to all exposed areas of skin.

Hypothermia, cold injuries and frostbite

The dangers of cold must be considered and prevented in expeditions to a variety of environments. Hypothermia can develop insidiously and the dangers of inadequate shelter, inadequate clothing, wind, being wet or immersed in water, undernourished and forced to be immobile must all be recognised. Low ambient temperatures may be predictable on geographical grounds, at altitude and at certain seasons, but unseasonal cold snaps and the night-time fall in temperature in many desert regions may catch expeditions unawares. Exposed areas of the face and the extremities, hands and feet, are especially vulnerable to frost-nip and frostbite. Thawing or rewarming of frostbitten parts should not be attempted until the victim has reached a warmer environment where medical care is available.

Allergic and atopic diseases

Expedition members who suffer from chronic allergic/atopic diseases, such as asthma, hay fever and eczema, should take adequate supplies of their usual medications. Those who have suffered anaphylactic attacks from nuts, shellfish and other foods, and from stings by wasps, hornets, bees, ants, etc., should carry self-injectable adrenaline (e.g. EpiPen or Anapen) and make sure that they and other expedition members know how to use this equipment effectively. Contact reactions to plants, animals, insect bites, etc. are common and should be treated with topical crotamiton (Eurax) and corticosteroid ointments (e.g. betamethasone [Betnovate]) and antihistamine tablets (e.g. chlorpheniramine [Piriton] or promethazine [Phenergan]).

Attacks by animals

Wild animals, such as the big cats, bears, wolves, hyenas, elephants, hippopotamuses, rhinoceroses, camels, buffaloes and wild pigs, have all been known to attack and kill humans. Domestic cattle and dogs can also be dangerous. Large wild animals must be respected and avoided unless you are travelling in a vehicle. Attacks by the big cats are especially likely between dusk and dawn. In the water, hippos, sharks and crocodiles can kill. Take local advice about the resident dangers before walking, swimming or camping. Teeth, claws, tusks and horns can produce devastating injuries, blood loss and fractures, with a high risk of contamination from a range of germs including tetanus and rabies. First aid involves control of bleeding, closing gaping wounds with dressings and evacuating the casualty to medical care. Broad-spectrum antibiotics should be given.

Rabies

In most parts of the world (see Figure 14.2), there is a risk of transmission of rabies or rabies-related viruses by bites of wild mammals or domestic dogs and cats. Pre-exposure immunisation is recommended (see above). All bites (including human bites) should be thoroughly cleaned (scrubbed with soap under a running tap), irrigated with clean water and then treated with a strong antiseptic such as alcohol or povidone–iodine. If there is a risk of rabies, a course of post-exposure immunisation should be started immediately and rabies immune globulin infiltrated around the wound. Those who have been immunised against rabies in the past require only two booster injections of vaccine.

Venomous bites and stings

Snake bites are best avoided by wearing proper boots, socks and long trousers, especially in undergrowth and sand, using a light after dark and avoiding high-risk activities such as attempting to handle snakes or snake-shaped animals and putting hands into holes or vegetation. The important first-aid treatment of a snake bite is to keep the bitten limb absolutely still with a splint or sling and to move the patient to medical care on a stretcher as soon as possible. Firm bandaging of the entire bitten limb with a long, crêpe or elasticated bandage may delay absorption of neurotoxic venoms (e.g. mambas, coral snakes, kraits) until the patient reaches a hospital. Most traditional first-aid methods (tight tourniquets, incisions, suction, electric shocks, snake stones, etc.) are dangerous and useless. The decision whether or not to give antivenom, the only antidote against snake venom, should be made by a medically trained person. Fish stings can be treated by immersing the stung part in uncomfortably hot but not scalding water. Jellyfish stings are treated with vinegar (box jellyfish in Australia) or baking soda (Atlantic jellyfish). Scorpions can be revealed with an ultraviolet lamp. Their very painful stings are treated with local anaesthetic. Leeches are very common in the rain forest and in freshwater. DEET applied to the skin, socks, boots and trousers is a partially effective deterrent.

FURTHER INFORMATION

RGS–IBG Expedition Medical Cell
Helping to improve health and safety is a key part of the RGS–IBG Expedition Advisory Centre's work.

Members of the RGS-IBG Expedition Medical Cell (www.rgs.org/medicalcell), chaired by David Warrell, advise the RGS–IBG on all medical matters relating to fieldwork in remote and challenging environments, to ensure the associated risks of participating in such activities are kept to a minimum.

An ongoing Survey of Expedition Health and Safety has been carried out by the Expedition Advisory Centre since 1995 to help improve the effectiveness of the work of the Medical Cell. All expeditions are encouraged to contribute to the survey.

Medical advice for expedition planners is given in the edition of *Expedition Medicine* edited by David Warrell and Sarah Anderson (Profile Books, London, 2002). The Medical Cell also develops and maintains information sheets/guidelines on specific topics. Information is available on *First Aid Training, Children at Altitude, Guidelines for Acclimatisation on Mount Kilimanjaro* and *Heat-related Illness*.

Regular seminars and workshops on matters of expedition health and safety include: a weekend Wilderness Medical Training course, and courses leading to the Certificate in Offsite Safety Management.

The Expedition Advisory Centre also helps expeditions recruit medical personnel for expeditions through its Register of Personnel available for expeditions, and publicises opportunities for medical professionals to participate in expeditions and fieldwork overseas through its *Bulletin of Expedition Vacancies*.

Useful addresses and websites
BCB Limited, Morland Road, Cardiff CF24 2YL. Tel: +44 292 046 4464, fax: +44 292 048 1100, email: bcb@bcb.ltd.uk, website: www.bcb.ltd.uk
First aid kits and emergency medical supplies.
Blood Care Foundation, PO Box 588, Horsham RH12 5WJ. Tel: +44 1403 262652, fax: +44 1403 262657, email: bcfgb@compuserve.com, website: www.bloodcare.org.uk
Emergency blood supplies.
British Association for Immediate Care (BASICS), BASICS Headquarters, Turret House, Turret Lane, Ipswich IP4 1DL. Tel: +44 870 165 4999, fax: +44 870 165 4949, email: admin@basics.org.uk, website: www.basics.org.uk
British Dental Association, 64 Wimpole Street, London W1M 8AL. Tel: +44 20 7935 0875, website: www.bda-dentistry.org.uk
British Medical Association, BMA House, Tavistock Square, London WC1H 9JP. Tel: +44 20 7387 4499, fax: +44 20 7383 6400, email: info.web@bma.org.uk, website: www.bma.org.uk
Centre for Tropical Medicine, University of Oxford, Nuffield Department of Clinical Medicine, John Radcliffe Hospital, Oxford OX3 9DU, Founding Director (Emeritus): Professor David Warrell. Tel: +44 1865 220968, fax: +44 1865 220984, email: david.warrell@ndm.ox.ac.uk
Department of Health (Medicines Division), Market Towers, 1 Nine Elms Lane, London SW1 5NQ. Tel (weekdays 09.00–17.00): +44 20 7273 0000, (other times): +44 20 7210 3000, fax : +44 20 7273 0353
For UK drug export certificates: email: info@mca.gsi.gov.uk, website: www.mca.gov.uk

EXPEDITION MEDICINE

Diving Diseases Research Centre, The Hyperbaric Medical Centre, Tamar Science Park, Research Way, Plymouth PL6 8BU. Emergency tel: +44 1752 209999, fax: +44 1752 209115, email: enquiries@ddrc.org, website: www.ddrc.org

East Africa Flying Doctors Society (AMREF), 11 Old Queen Street, London SW1H 9JA. Tel: +44 20 7233 0066, fax: +44 20 7233 0099

The Fleet Street Travel Clinic, Dr Richard Dawood, 29 Fleet Street, London EC4Y 1AA. Tel: +44 20 7353 5678, fax: +44 20 7353 5500, email: Info@fleetstreetclinic.com, website: www.fleetstreetclinic.com

Health Literature Line, The Library, Department of Health, Shipton House, London SE1 6LH. Tel: +44 800 555777, fax: +44 1623 724524, website: www.equip.nhs.uk

Phone for individual copies of material produced by the Department of Health. If more copies are required, fax or write

Hospital for Tropical Diseases, Mortimer Market, Capper Street, Tottenham Court Road, London WC1E 6AU. Tel: +44 20 7387 9300/4411, healthline: +44 9061 337733, fax : +44 20 7388 7645, website: www.thehtd.org

InterHealth, 157 Waterloo Road, London SE1 8US. Tel: +44 20 7902 9000, email: Info@interhealth.org.uk, website: www.interhealth.org.uk

Long-term advice and treatment for aid workers and expatriates.

International Health Exchange, 134 Lower Marsh, London SE1 7AE. Tel: +44 20 7620 3333, fax: +44 20 7620 2277, email: info@ihe.org.uk, website: www.ihe.org.uk

Maintains a register of health professionals wanting to work in less developed countries, and runs training courses on primary health care and refugee community health

John Bell and Croyden, 50–54 Wigmore Street, London W1V 2AU. Tel: +44 20 7935 5555, fax: +44 20 7935 9605, website: www.johnbellcroyden.co.uk

Pharmacy and medical supplier.

Lifesystems Limited, 4 Mercury House, Calleva Park, Aldermaston RG7 8PN. Tel: +44 118 981 1433, fax: +44 118 981 1406, email: mail@lifesystems.co.uk, website: www.lifesystems.co.uk

First aid and emergency dental kits.

Liverpool School of Tropical Medicine, Pembroke Place, Liverpool L3 5QA. Tel: +44 151 708 9393, fax: +44 151 708 8733, website: www.liv.ac.uk/lstm/lstm.html

London School of Hygiene and Tropical Medicine, Keppel Street, London WC1E 7HT. Tel: +44 20 7636 8636, fax: +44 20 7436 5389, website: www.lshtm.ac.uk

London School of Tropical Medicine Malaria Reference Laboratory. Tel: +44 20 7636 3924; +44 9065 508 908 (24-hour), website: www.lshtm.ac.uk/centres/malaria

Medical Advisory Service for Travellers Abroad (MASTA) Travel Clinics. Tel: +44 1276 685040, email: enquiries@masta.org, website: www.masta.org

MedicAlert Foundation International, 1 Bridge Wharf, 156 Caledonian Road, London N1 9UU. Tel: +44 20 7833 3034, fax: +44 20 7278 0647, email: info@medicalert.org.uk, website: www.medicalert.org.uk

National Poisons Centre. Tel: +44 870 600 6266 (for clinically complex cases), website: www.doh.gov.uk/npis.htm

Nomad Traveller's Store and Medical Centre, 3–4 Wellington Terrace, Turnpike Lane, London N8 0PX. Tel: +44 20 8889 7014, fax: +44 20 8889 9529, email: sales@nomadtravel.co.uk, website: www.nomadtravel.co.uk

Travel pharmacy. Medical kits made to order at a low cost.

Royal College of Nursing, 20 Cavendish Square, London W1G 0RN. Tel: +44 845 772 6100, website: www.rcn.org.uk

Royal Society for the Prevention of Accidents, Edgbaston Park, 353 Bristol Road, Edgbaston, Birmingham B5 7ST. Tel: +44 121 248 2000, fax: +44 121 248 2001, email: help@rospa.co.uk, website: www.rospa.co.uk

SP Services (UK), Unit D4, Hortonpark Estate, Hortonwood 7, Telford TF1 7GX. Tel: +44 1952 288999, fax: +44 1952 606112, website: www.999supplies.com
Emergency medical and rescue supplies.
TALC (Teaching-aids At Low Cost), PO Box 49, St Albans AL1 5TX. Tel: +44 1727 853869, fax: +44 1727 846852, website: www.talcuk.org
Trailfinders Travel Clinics, (London). Tel: +44 20 7938 3999, (Glasgow) +44 141 429 0913
UIAA Mountain Medicine Data Centre. Website: www.thebmc.co.uk/world/mm/mmo.htm
Wilderness Medical Training (WMT), The Coach House, Thorny Bank, Skelsmergh, Kendal LA8 9AW. Tel./fax: +44 1539 823183, email: enquiries@wildernessmedicaltraining.co.uk, website: www.wildernessmedicaltraining.co.uk
World Health Organization, WHO, Avenue Appia 20, 1211 Geneva 27, Switzerland. Tel: +41 22 791 21 11, fax: +41 22 791 31 11, website: www.who.int
Publishers of the WHO Weekly Epidemiological Record, Global Epidemiological Surveillance and Health Situation Assessment, International Travel and Health

Useful web addresses

British Travel Health Association: www.btha.org/site/index.php
Centers for Disease Control (CDC), USA: www.cdc.gov
Department of Health: Advice for Travellers: www.doh.gov.uk/traveladvice
E-Med: www.e-med.co.uk
Fit for travel: www.fitfortravel.scot.nhs.uk
International Society for Infectious Diseases: www.promedmail.org
For disease alerts.
International Travel Health Association: www.istm.org
Public Health Laboratory Service: www.phls.org.uk
Excellent malaria guidelines.
Travel Health Online: www.tripprep.com
Travel Screening Services: www.travelscreening.co.uk
The Travellers' Health website: www.travellershealth.info
Has news and links to over 200 travel health-related sites.

Further reading

Almond, J., Clemens, J., Engers, H. et al. (2002) Accelerating the development and introduction of a dengue vaccine for poor children. *Vaccine* **20**: 3043–6.
Anderson, S.R. and Johnson, C.J.H. (2000). Expedition health and safety: a risk assessment. *Journal of the Royal Society of Medicine* **93**: 557–62.
Auerbach, P.S. (ed.) (1995) *Wilderness Medicine. Management of wilderness and environmental emergencies*, 3rd edn. St Louis, MO: Mosby,
Backer, H.D., Bowman, W.D., Paton, B.C., Steele, P. and Thygerson, A. (1998) *Wilderness First Aid. Emergency care for remote locations*. Boston, MA: Jones & Bartlett Publishers.
Bouchama, A. and Knochel, J.P. (2002) Medical progress: heatstroke. *New England Journal of Medicine* **346**: 1978–88.
Bradley, D.J. and Bannister, B. (2001) Guidelines for malaria prevention in travellers from the United Kingdom for 2001. *Communicable Diseases and Public Health* **4**: 84–101.
British Medical Journal (2002) Road traffic accidents. *British Medical Journal* **324**, 11 May: 1107–10, 1119–54.
Dawood, R. (2002) *Travellers' Health*, 4th edn. Oxford: Oxford University Press.
Forgey, W.W. (2000) *Wilderness Medicine Beyond First Aid*, 5th edn. Guilford, CT: Globe Pequot Press.

Gibbons, R.V. and Vaughn, D.W. (2002) Dengue: an escalating problem. *British Medical Journal* **324**: 1563–6.

Hatt, J. (1993) *The Tropical Traveller*, 3rd edn. London: Penguin Books.

Molesworth, A.M., Thomson, M.C., Connor, S.J. et al. (2002) Where is the meningitis belt? Defining an area at risk of epidemic meningitis in Africa. *Transactions of the Royal Society for Tropical Medicine and Hygiene* **96**: 242–9.

Monath, T.P. and Cetron, M.S. (2002) Prevention of Yellow Fever in persons travelling to the tropics. *Clinical Infectious Diseases* **34**: 1369–78.

Sartori, C., Allemann, C. and Duplain, H. (2002) Salmeterol for the prevention of high-altitude pulmonary edema. *New England Journal of Medicine* **346**: 1631–6.

Shlim, D.R. and Solomon, T. (2002) Japanese encephalitis vaccine for travellers: exploring the limits of risk. *Clinical Infectious Diseases* **35**: 183–8.

Steedman, D.J. (1994) *Environmental Medical Emergencies*. Oxford: Oxford University Press.

Stich, A., Abel, P.M. and Krishna, S. (2002) Human African trypanosomiasis. *British Medical Journal* **325**: 203–6.

Warrell, D. and Anderson, S. (eds) (2002) *Expedition Medicine*, 2nd edn. London: Profile Books. Available from www.rgs.org/eacpubs

Werner, D. (1993) *Where There is No Doctor*. London: Macmillan Education Ltd.

15 MINIMISING ENVIRONMENTAL IMPACT

Nick Lewis and Paul Deegan

Why do wilderness areas and the routes to them continue to suffer so much ecological damage despite the ever-increasing level of environmental awareness? Given the broad range of information available these days, it seems hard to believe that expeditions can be organised without some prior knowledge of the potential impacts that they may cause. The very fact that so many expeditions are planned with little or no regard for the environment suggests that what is needed is a handful of simple tenets that all groups could adopt, regardless of their destination or activity.

This chapter outlines some of the things that can be done during an expedition to avoid causing more environmental impacts than are absolutely necessary. By combining a couple of fresh ideas with a common-sense approach, we have come up with "Five Golden Rules", which if followed will significantly help to reduce the environmental impacts generated by expeditions.

THE NOTION OF MINIMAL ENVIRONMENTAL IMPACT

The only expeditions that have no impact are those that exist only on paper and never leave home. It is impossible to go anywhere and not have an impact – the footprints we leave, the very air we exhale are the least impacts that we will have. It is therefore vital that we accept that we will cause a certain level of environmental impact. The trick is to minimise it. But what is a minimal level of impact and how do we measure it?

To attain a truly minimal impact, it is vital to take an integrated approach by looking beyond the expedition destination and the usual problems of litter and footpath erosion. We also have to consider our impact in the cities, towns and villages of our host country, places where the most far-reaching and destructive impacts of any expedition may be most keenly felt.

Many people reading this book are likely to consider themselves different from the run-of-the-mill package holiday tourist. However, in many of the areas that we travel through on an expedition, we create exactly the same type of environmental impacts as every other visitor. This usually includes waste arising from servicing the expedition, such as from hotels, restaurants or guesthouses. This may seem to be a byproduct of any tourism, but that's the point – if your expedition is to have a minimal impact, it must take these factors into account and plan accordingly. In other words, don't think that, because you are an *expedition*, it absolves you from all the problems associated with normal tourism. Far from it; more care needs to be taken. Therefore, we need to consider all the activities associated with our proposed expedition and view all the potential impacts that may arise.

Rule 1: keep your expedition small
The impact of an expedition is closely related to its size. The more members it has, the more impacts it can potentially have. It's simple mathematics really – more airplane seats to carry you there and more vehicles needed for transport lead to greater fuel emissions; more loads to carry and more boots on paths result in increased soil erosion.

Minimising the number of team members will:

- reduce the cost of the expedition
- decrease the transportation requirements (resulting in a smaller demand for fuel)
- minimise the quantity of supplies that the expedition requires, whether it's the purchasing of scarce provisions in a village or the importing of food from the home country
- lower the number of local porters used (reducing sewage and firewood problems)
- decrease the amount of waste and sewage produced – waste is usually the most significant impact that expeditions leave behind.

Rule 2: appoint an environmental manager
Expedition planning consists of research into the objective: arranging permits, sorting out travel arrangements, raising finances, and organising gear and food. The environmental aspect of the expedition should also be addressed at this stage. Designating a team member to be the environmental manager is a positive first step.

As relatively few destinations request visitors to comply with environmental requirements or legislation, most expeditions need to educate themselves about the environmental issues in their chosen area. It is the job of the environmental manager to find out what these are. The list of subjects that could be researched include:

- Environmental regulations and permit requirements of the host country.

- Specific environmental concerns of the expedition area, such as path erosion, sensitive ecosystems and waste accumulation.
- Transport options to and from the destination. Can existing public transport be used rather than private vehicles to transfer the expedition to the roadhead?
- Accommodation options: try to use hotels or guesthouses that participate in local environmental initiatives.
- Environmental policies of expedition agencies (including guides, trekking companies and boat charters).
- Waste management options: investigate waste-handling structures available in the area. If there is none, consider repatriating your waste. (Waste management options are probably the single most important role that the environmental manager will cover.)

The most effective environmental managers work closely with the other members of the expedition at the planning stage. To take just one example, there are many possible considerations to take into account when choosing food and kit. The diligent environmental manager will want to work alongside the person(s) responsible for provisions and group equipment in order to:

- buy food that has minimal packaging, and then remove any wrappers that are not necessary to preserve the food. All cooking instructions can be written on to a single sheet of paper;
- bag food into separate day packages. The day bag becomes a handy waste receptacle when the contents have been eaten;
- use a liquid fuel (such as gasoline or paraffin) for cooking. A pressurised stove that uses a liquid fuel burns hotter than bottled gas, and has no resulting empty cartridges requiring disposal;
- take comprehensive repair kits and learn how to use them, rather than carrying a number of expensive spare stoves and tents. Remember that poorly maintained products can result in damaged equipment being abandoned in-country. It is worth bearing in mind that less equipment results in less weight, with less to go wrong.

These are just some of the areas that the environmental manager can research; the job can be as focused or encompassing as one wishes it to be. For example, the environmental manager may decide to keep a constant record of possible environmental impacts. A simple account of the expedition's activities can then be incorporated into the final expedition report. A more comprehensive study may include a before-and-after photographic account of places visited, monitoring of waste products produced (types/volumes/weights), and a brief study of any waste-handling infrastructures present in the areas visited.

Rule 3: assess the environmental impacts

When the environmental manager has assembled all the available information, he or she will be in a position to identify and assess the:

- main environmental sensitivities and constraints of the expedition area (e.g. nearby wetlands, protected areas or species)
- environmental aspects of the project that may result in potential impacts (e.g. overland transport, anchoring of vessels, generation of waste).

Environmental impact assessments

Much has been written about the assessment of environmental impacts of expeditions and, unfortunately, much of it is either inaccurate or wrong. Expedition brochures are littered with jargon such as "preliminary study", "audit", "impact assessment", "baseline survey" and "initial evaluation". This problem is compounded by the fact that these phrases have different definitions depending on the context in which they are being used and the legislation of the country being visited.

So for the sake of simplicity, let's define here what an environmental impact assessment (EIA) should be for an expedition:

An assessment – conducted at the planning stage – of the potential environmental impacts that the expedition may cause throughout its course.

The purpose of the EIA is to identify those key areas where the expedition may cause environmental damage. This then allows appropriate contingency plans to be drawn up before the expedition begins in order to prevent any damage being done. Assessments can be simple or elaborate, but, for an expedition, it's important to keep it straightforward.

A simple EIA can consist of a list of bullet points or numbers ranking the most likely impacts in order, with a simple contingency plan outlined for each. An example is presented in Figure 15.1.

This basic form of EIA may be all that is necessary for a small (two-person)

HIGHER SIGNIFICANCE

LOWER SIGNIFICANCE

1. Hazardous waste (e.g. batteries): return to home country
2. Sewage: designate burial pit away from water sources; burn toilet paper and bury with sewage in active layer of soil
3. Domestic waste: minimise packaging before departure; remove all waste from field area and dispose in organised landfill site
4. Path erosion: stick to designated routes; minimise number of porters
5. Vehicle emissions: use public transport

Figure 15.1 *Simple identification, ranking and mitigation of impacts*

expedition. Larger expeditions may need a more elaborate way of identifying and assessing their impacts. A more comprehensive EIA can be conducted by assessing the impacts qualitatively using the following criteria:

- Activity: describes the impacting activity.
- Duration: specifies the duration of the impacting activity.
- Output: names the specific aspect of the impacting activity, e.g. disposal of food waste is one of the outputs of cooking at base camp; fuel spillage is an output of refuelling stoves.
- Nature: identifies the type of impact caused by the activity, e.g. water contamination from poor waste disposal or fuel spillage.
- Scope: pinpoints the geographical area affected by the impact, be it local, regional or continental.
- Persistence: estimates the duration of the impact and whether it is likely to be short term (minutes/hours), medium term (days/weeks), long term (months/years), permanent or unknown.
- Intensity: classifies the overall severity of the impact in relative terms (low, medium or high).
- Probability: evaluates the likelihood of the impact occurring, in relative terms (low, medium or high).
- Significance: rates the overall importance of the impact, assessed in relative terms (low, medium or high). This depends on all the factors previously described plus additional variable factors such as the sensitivity of the environment.
- Type of effect: this assesses whether the impact will have a direct, indirect or cumulative effect. The effects of particular impacts may depend very much on later events.
- Mitigation: indicates possible contingency plans for minimising unavoidable impacts.

All these criteria can be integrated into a simple table to assess the expedition. An example is presented in Table 15.1.

Although this comprehensive approach requires some attention to detail, it is important not to make it unnecessarily complicated. Whatever method you choose to use, the most important point is that the task gets done. Remember, it is very difficult to prevent impacts from occurring unless you know what they are.

It may be difficult or even impossible to foresee what environmental damage will occur in an area in future years. Consequently many impacts tend to have direct, indirect and cumulative effects, e.g. fuel spillage into a watercourse will have a direct toxic effect on the aquatic ecosystem; this water may in turn indirectly pollute food sources for humans and animals if used for irrigation sources. Long-term spillages

TABLE 15.1 EXAMPLE LAYOUT FOR ENVIRONMENTAL IMPACT ASSESSMENT TABLE

Activity	Overland journey from port of entry to roadhead and return	Journey to base camp and return	General expedition tasks
Duration	8 days (total)	10 days (total)	45 days (duration of expedition)
Output	Engine emissions	Walking	Waste head torch batteries
Nature of impact	Air pollution; dust	Erosion of footpaths	Contamination of soil and groundwater resources
Scope	Local–regional	Local	Local–regional
Persistence	Long term	Long term to persistent	Long term to persistent
Intensity	Medium	Low	High
Probability	Medium–high	High	Low
Significance	Medium–high	Low–medium	High
Type of effect	Direct, cumulative	Direct, indirect, cumulative	Direct, indirect, cumulative
Mitigation	Use public transport; use well-maintained hire vehicles; stick to designated routes and speed limits.	Reduce number of journeys to minimum; stick to designated routes	All waste batteries to return home with expedition for proper disposal

may have a cumulative effect whereby the affected ecosystem diminishes.

Rule 4: draw up and use an environmental management plan

On the basis of your assessment, an environmental management plan (EMP) can now be drawn up for the expedition. The EMP should detail how the mitigation measures outlined in the EIA are going to take place. The EMP will include:

- The expedition's environmental statement – what is it you are trying to do from an environmental point of view or, perhaps more importantly, what are you trying to avoid doing?

TABLE 15.2 EXPEDITION MANAGEMENT PLAN FOR DRAVOT–CARNEHAN KAFIRISTAN EXPEDITION, PAGE 1 (MEMBERS: D. DRAVOT, P. CARNEHAN, B. FISH)

Activity or action causing potential environmental impact	Impacts	Mitigation	Responsible party	Monitoring	Timing
Overland journey to and from Jagdallak (gateway town)	Air pollution from engine emissions	Use public transport. If hiring, use only well-maintained vehicles	Dan	Observation, preliminary check	Journey to and from roadhead
Stay in Jagdallak to and from base camp	Waste production in Jagdallak	Minimise time spent in town; use of non-packaged goods	Dan	Check with suppliers	6 days
Walk-in from Jagdallak to base camp	Soil erosion	Stick to designated route; minimise loads and number of porters	Dan	Verbal reminder and observation	Walk-in
	Destruction of wood resources	Use of petrol stoves; provide porters with adequate clothing	Bill	Observation and liaise with head porter	Walk-in
General base camp activities	Destruction of wood resources	Use of petrol stoves	Bill	Observation and liaise with cook	Duration of stay at base camp
	Soil erosion and visual impacts from tent platforms	Stick to designated route; replace stones upon departure from base camp	Dan	Check before departure	Duration of stay at base camp

Contamination of soils, groundwater by fuel spills	Proper fuel storage and refuelling practices, use of spill mats and drip trays	Bill	Regular checks	Duration of stay at base camp
Contamination of soils, groundwater by sewage	Stored in blue chemical barrels and removed to Jagdallak for disposal in municipal sewerage system	Peachey	Regular checks	Duration of stay at base camp
Visual impact, soil and groundwater pollution from garbage	Non-combustible material separated and returned to UK for disposal, combustible material burnt under controlled conditions in Jagdallak	Peachey	Regular checks	Duration of stay at base camp
Contamination of soils, groundwater and wildlife by hazardous wastes (batteries, waste fuel, explosive materials)	No disposal in expedition area – all hazardous wastes separated from other wastes and returned to UK for proper disposal	Peachey	Regular checks	Duration of stay at base camp

- A summary of the environmental problems that may be already present in the expedition area, as well as those that may be generated by your visit.
- The type of environmental permits or applications required.
- The mitigation measures to be conducted.
- The waste management options that you have in place.
- The arrangements for local employment.
- The monitoring programmes to be conducted.
- The reporting requirements on your return.
- Who is responsible for implementing the various measures?

The EMP should be short, succinct and easy to understand; try fitting it on one side of A4 paper and make sure that it is followed! An example is shown in Table 15.2.

Rule 5: consider the environmental impact on gateways
Expeditions are about going to wild places, but we have to go through a lot of urban places to get there. Almost every expedition uses a settlement as a springboard in order to enter the wild place that it intends to visit. This town or village is likely to be the place where multiple expeditions pick up supplies, sort out transport and maybe hire porters. At the end of the expedition, team members may return to this settlement in order to relax and sort things out before starting the long journey home. Unfortunately these settlements are often the places where expeditions choose to deposit their waste products.

As they act as essential access points to and from our destination, let's call them "gateway" towns. Popular gateways include:

- Huaraz and Cuzco in Peru
- Pokhara and Lukla in Nepal
- Skardu and Gilgit in Pakistan
- Leh and Manali in India
- Talkeetna in Alaska
- Tasiilaq in Greenland
- Chalten and Calafate in Patagonia.

Many gateways are in remote locations and suffer from poor transport infrastructures. This leaves remote communities extremely vulnerable to many impacts, e.g. it might not be possible to remove waste products. This forces poor waste disposal in the gateway towns and can lead to soil and groundwater contamination.

What can be done to minimise our impacts on gateways?
Pragmatically speaking, waste needs to be transported to somewhere where it can be dealt with properly. So it is important to determine the location of effective waste-

handling structures along the route that your expedition is taking. If such structures do not exist, repatriation of your waste back to your home country may be the only viable solution. Depending on what wastes you decide to ship home with you, this need not be a burdensome task or one that generates endless amounts of bureaucracy and paperwork. Waste materials such as food packaging, paper, plastic and depleted batteries can be returned easily, especially if you took these items with you at the start of the expedition.

Shipping hazardous wastes (such as fuel or sewage) will require compliance with international regulations. However, many small expeditions have been successful in achieving this. By doing so, they have proved that expeditions can leave only a minimal impact.

CONCLUSION

Remember, it is vital that expeditions consider every aspect of their activities to identify the potential environmental impacts that they may cause. The "Five Golden Rules" highlighted in this chapter are designed to help you to achieve this:

Rule 1:   Keep your expedition small
Rule 2:   Appoint an environmental manager
Rule 3:   Assess the environmental impacts
Rule 4:   Draw up and use an environmental management plan
Rule 5:   Consider the environmental impact on gateways.

Taking responsibility for environmental matters on expeditions is often regarded as a thankless task. Nevertheless, it's the one legacy that not only affects the wilderness and the local community, but also all future expeditions who travel into the region. Who knows, that next expedition may be yours!

FURTHER INFORMATION

Useful web addresses

Conservation of Arctic Fauna and Flora (CAFF): www.caff.is
Leave No Trace: www.lnt.org
Tourism Concern: www.tourismconcern.org.uk
Green Globe: www.greenglobe21.com
International Porter Protection Group: www.ippg.net
Mountain Tourism Guidelines: www.thebmc.co.uk/world.htm
UNEP World Conservation Monitoring Centre: www.unep-wcmc.org

SECTION 4

EXPEDITION LOGISTICS

16 MOUNTAINEERING AND TREKKING EXPEDITIONS

Roger Payne

Whether you are considering the "last great" mountaineering challenge or thinking about your first climbs in the greater ranges, you would be wise to ensure that you have accumulated a solid base of mountaineering and alpine experience closer to home first. The British Mountaineering Council can provide details of local climbing clubs, and details of courses including those run by the national centres for mountain sports at Plas y Brenin and Glenmore Lodge.

Having acquired sufficient experience, the first problem is to identify a suitable objective. The best way to approach this is to consider carefully the different areas in which you wish to travel or climb. Select an area that does not pose too many political and logistical problems. As Alan Rouse said, "It is nice to get to the top first time round." For example, some valleys in Nepal have very good access with relatively straightforward approaches to the peaks, whereas remote Karakoram passes and peaks have proved completely elusive to inexperienced groups! Consider the likely weather and conditions: Alaska and Patagonia can produce conditions as extreme as their positions are relative to each other. Do not make your first trip to the greater ranges your last!

Fortunately, mountaineers have a habit of recording their explorations so there is a vast amount of information available to would-be expeditioners. Apart from the obvious British magazines, some European titles and the Japanese *Iwa to Yuki* (which includes English summaries) are worth consulting. However, the best source of information is the various journals: the Alpine Club's own journal, *The Alpine Journal*, is excellent while the *American Alpine Journal* tries to give a comprehensive worldwide summary and can be considered definitive for South America. The *Canadian Alpine Journal* is excellent for Arctic North America, and *The Himalayan Journal* and *Indian Mountaineer* are very good for those regions. You can try contacting contributors to the various journals for detailed information, via either the journal editor or the club of which they may be members. These journals are all available in the Alpine Club Library, which is one of the best collections of mountaineering literature in the world. Other useful collections can be found in the smaller specialist libraries listed at the end of the chapter. Past expedition reports are often an excellent source of information. The Royal Geographical Society (with the Institute of British Geographers or IBG) is the repository for reports from British expeditions including those supported by the Mount Everest Foundation. An increasing number of traditional-style guidebooks are being produced for trekkers and mountaineers.

In addition to housing its excellent library, the Alpine Club has an online Himalayan index listing most known peaks over 6000 metres in Asia (excluding countries of the former Soviet Union). In recent years, the Alpine Club has organised an annual symposium on climbing in the greater ranges. These tend to concentrate on specific mountain areas and are held in November each year at Plas y Brenin. The proceedings of these are sometimes published. Your local library may also have one or two useful reference books and, of course, you can order other books through them.

The British Mountaineering Council (BMC) has a range of information sheets covering most mountain areas to help in the early stages of planning. In addition, if you have problems after doing your initial research you can always contact the BMC office for advice or an opinion. The BMC keeps copies of reports from BMC-supported expeditions and some other reference works for visitors. The Council also has comprehensive insurance schemes available with worldwide, extended period and expedition cover for mountaineers.

All the Himalayan countries, and to a lesser extent most other less developed countries, have rules and regulations governing access to their mountains. It is essential that you obtain the latest up-to-date regulations and adhere to them no matter how frustrating they may seem at times. If you think dealing with bureaucracy will be a problem for you, choose instead an area with unrestricted access, such as in much of North America. Check with the appropriate Embassy or High Commission for the current situation, rules and regulations, or (where they exist) the national mountaineering authority.

Depending on your circumstances and the nature of the trip you may be eligible for various grants (e.g. the Alison Chadwick Memorial Fund for women climbers, the Nick Estcourt Award for teams tackling difficult peaks and the Eagle Ski Club for ski mountaineering expeditions) but the two main grant-giving bodies for mountain expeditions are the Mount Everest Foundation (MEF) and the BMC. The basic criteria for these grants are that you are attempting first or first British ascents of peaks or routes in a new style, and that you have the necessary experience to stand a reasonable chance of success. The MEF also provides grants to those doing research in high mountain areas. Both grants are administered via the Honorary Secretary of the MEF from whom further details of eligibility and application forms are available.

The UIAA (Union Internationale des Associations d'Alpinisme), which is the world body for mountaineering, produces guidelines to promote good ethical practice in developing sustainable mountain activities. At an early stage in your planning you should acquaint yourself with the Kathmandu Declaration, the Ethical Code for Expeditions and the International Mountain Code, all of which are available from the BMC or from the UIAA website.

Also, as part of your preparation it is essential that you consider likely medical problems, and accident and emergency procedures. The UIAA Mountain Medicine Data Centre produce a range of excellent information sheets covering problems concerned with acclimatisation, cold, injury, etc. In most circumstances in remote or high mountain ranges you will need to be prepared to effect your own evacuation in the event of an accident.

It is always worthwhile consulting others with expedition experience as part of your expedition planning. If you are considering training in Britain as part of your preparation, make sure that those who are providing the training are properly qualified or experienced. In Britain there are various levels of qualification that are recognised in Europe and internationally. For further guidance you can contact the BMC's Training Administrator or the Mountain Leader Training Board (MLTB).

Depending on the country that you propose to visit there are different services available to transport expeditions to and from their base camps. This often involves the expedition hiring local people to transport goods and equipment. Some countries have systems for registering porters and local guides etc., and set terms and conditions for their employment. In 1997, the International Porter Protection Group was formed to look after the safety and welfare of the trekking porter at work in the mountains after a series of tragic deaths. You should make sure that you are aware of their guidelines and follow them when hiring local labour.

Opportunities for inexpensive travel and access to unclimbed peaks and unexplored areas have never been greater. Careful preparation, good teamwork and previous experience help to ensure successful and enjoyable expeditions. In your enthusiasm to travel and climb in the greater ranges do not neglect basic principles such as an environmentally aware approach and respect and fair treatment for local people. It has been said many times before but it is worth finishing on here: "Leave nothing but footprints, take nothing but photographs, change no one but yourself".

USEFUL INFORMATION

Useful addresses

Alpine Club, 55 Charlotte Road, London EC2A 3QT. Tel: +44 20 7613 0755, email: sec@alpine-club.org.uk, website: www.alpine-club.org.uk

British Mountaineering Council (BMC), 177–179 Burton Road, West Didsbury, Manchester M20 2BB. Tel: +44 870 010 4878, email: office@thebmc.co.uk, website: www.thebmc.co.uk

International Porter Protection Group. Website: www.ippg.net

Mountaineering Council of Ireland, House of Sport, Longmile Road, Walkinstown, Dublin 12, Ireland. Tel: +353 1 450 7376, email: mci@eircom.net, website: www.mountaineering.ie
Mountaineering Council of Scotland, The Old Granary, West Mill St, Perth PH1 5QP. Tel: +44 1738 638227, email: info@mountaineering-scotland.org.uk, website: www.mountaineering-scotland.org.uk
Mountain Leader Training England, Siabod Cottage, Capel Curig, Conway LL24 OET. Tel: +44 1690 720314, email: info@mlte.org, www.mltb.org
UIAA Associations d'Alpinisme, Postfach, CH-3000 Bern 23, Switzerland. Tel: +41 31 370 1828, fax: +41 31 370 1838, email: office@uiaa.ch, website: www.uiaa.ch/
UIAA Mountain Medicine Data Centre. Website: www.thebmc.co.uk

Training centres
Glenmore Lodge, Aviemore, Inverness-shire PH22 1QU. Tel: +44 1479 861256, website: www.glenmorelodge.org.uk
International School of Mountaineering, Club Vagabond, 1854 Leysin, Switzerland. Tel: +44 1766 890441, email: ism@alpin-ism.com, website: www.alpin-ism.com
Plas Y Brenin, Capel Curig, Gwynedd LL24 0ET. Tel: +44 1690 720214, website: www.pyb.co.uk

Specialist libraries
Alpine Club Library, 55 Charlotte Road, London EC2A 3QT. Tel: +44 20 7613 0755, email: info@alpine-club.org.uk, website: www.alpine-club.org.uk
Graham Brown Memorial Library, c/o Scottish National Library, Edinburgh
Fell and Rock Climbing Club Library, Lancaster University
Alan Rouse Memorial Library, Central Library, Surrey Street, Sheffield S1 1X2. Tel: +44 114 273 4760
Rucksack Club Library, Manchester Central Library
Yorkshire Ramblers Club Library, Leeds Central Library

Specialist publishers and distributors
Cicerone Press, 2 Police Square, Milnthorpe, Cumbria LA7 7PX. Tel: +44 15395 62069, email: info@cicerone.co.uk, website: www. cicerone.co.uk
Cordee, 3a DeMontfort Street, Leicester LE1 7HD. Tel: +44 116 254 3579, fax: +44 116 247 1176, email: info@cordee.co.uk, website: www.cordee.co.uk
Omnimap, 1004 S. Mebane St, PO Box 2096, Burlington, NC 27216, USA. Tel: +1 336 227 8300, email: custserve@omnimap.com, website: www.omnimap.com
Excellent source for obscure maps.
West Col, Goring, Reading, Berkshire RG8 9AA. Tel: +44 1491 681284

Principal national mountaineering authorities
American Alpine Club, 710 Tenth Street, Suite 15, Golden, CO 80401, USA. Website: www.americanalpineclub.org
Chinese Mountaineering Association (CMA), 9 Tiyuguan Road, Post Code 100763, Beijing. Tel: +86 10 671 23 769, fax: +86 10 671 11 629, email: cmaex@ina.com.cn, website: www.cma.com.cn
Greenland: Danish Polar Centre, Strandgade 100 H, DK 1401, Copenhagen, Denmark. Tel: +45 32 88 01 00 or 32 88 01 20, fax: +45 32 88 01 01, email: dpc@dpc.org, website: www.dpc.dk
Indian Mountaineering Foundation, Benito Juarez Road, Anand Niketan, New Delhi 11002. Email: indmount@vsnl.com, website: www.indmount.org
Pakistan Alpine Club, 509 Kashmir Road, R.A. Bazar, Rawalpindi, Pakistan. Tel: +92 51 927 1321, email: alpineclub@meganet.com.pk, website: www.alpineclub.org.pk

Mountaineering Association of Tibet (CTMA), No 8, East Linkhor Road, Lhasa, Tibet. Fax: +86 891 36366
Mountaineering Association of Xinjiang (CXMA), 1 Renmin Road, Urumqi, Xinjiang, China. Fax: +86 991 2818365
Mountaineering Federation of Russia, Lushnetzkaj naberzhnaj 8, 119871 Moscow, Russia
Nepalese Mountaineering Association, PO Box 1435, Nag Pokhari, Naxal, Kathmandu, Nepal. Tel: +977 143 4525, email: office@nma.com.np, website: www.nma.com.np

Grants

Alpine Ski Club Awards. Email: grants@alpineskiclub.org.uk, website: www.alpineskiclub.org.uk
Eagle Ski Club. Website: www.eagleskiclub.org.uk
Nick Estcourt Award, 24 Grange Road, Bowdon, Altrincham WA14 3EE
Andy Fanshawe Memorial Trust, c/o Ed Douglas, 181 Abbeydale Road South, Sheffield S7 2QW. Tel: +44 114 236 5589, website: afmt_adm@hotmail.com
Fred Harper Memorial Trust, Weston Cottage, West Way, Crayke, York Y061 4TE
    For the advancement for public benefit of education and training in the skills necessary for safe mountain walking and climbing.
Mount Everest Foundation, c/o Hon. Secretary: Bill Ruthven, Gowrie, Cardwell Close, Warton, Preston, Lancashire PR4 1SH. Website: www.mef.org.uk
Polartec Challenge. Website: www.polartec.com

Reference books and further reading

Alpine Club (1984) *Lightweight Expeditions to the Great Ranges.* London: Alpine Club.
Barton, B. and Wright, B. (2000) *A Chance in a Million? Scottish Avalanches.* Scottish Mountaineering Trust. Distributed by Cordee Books, Leicester, www.cordee.co.uk
Bollen, S. (1991) *First Aid on Mountains.* Manchester: British Mountaineering Council.
Brain, Y (1998) *Bolivia – A climbing guide.* Leicester: Cordee Books.
Clarke, C. (2002) High altitude and mountaineering expeditions. In: Warrell D. & Anderson S. (eds), *Expedition Medicine.* 2nd edn. London: Profile Books. Available from www.rgs.org/eacpubs
Cleare, J. (1979) *The World Guide to Mountains and Mountaineering.* New York: Mayflower Books.
Collister, R. (1991) *Lightweight Expeditions.* Crowood Press. Distributed by Farringdon Books, Colchester, Essex.
Daftern, T. (1983) *Avalanche Safety for Skiers and Climbers.* London: Diadem, www.diadembooks.com
Deegan, P. (2002) *The Mountains Traveller's Handbook: Your companion from city to summit.* Manchester: British Mountaineering Council. Highly recommended. Available from www.rgs.org/eacpubs
Echigo, K. (ed.) (1977) *Mountaineering Maps of the World.* Vol. 1: *Himalaya.* Vol. 2: *Karakoram-Hindu Kush.* Tokyo: Gakushu Kenkyu-sha (Gakken).
Fyffe, A. and Peter, I. (1991) *The Handbook of Climbing.* London: Pelham Books.
Hackett, P.H. (1980) *Mountain Sickness.* American Alpine Club. Distributed by Cordee Books, Leicester, www.cordee.co.uk
Higgs, D. (1990) *Mountain Photography.* London: Diadem.
Houston, C. (1987) *Going Higher: The Story of Man and Altitude.* New York: Little, Brown.
Kapadia, H. (2001) *Trekking & Climbing – Indian Himalaya.* Leicester: Cordee Books.
Kelsey, M. R. (1990) *Climbers and Hikers Guide to the World's Mountains.* Kelsey Publishing, 456 E. 100N, Provo, Utah, USA 84606. Tel: +1 801 373 3327.
Langmuir, Eric (1995) *Mountaincraft and Leadership: a handbook for mountaineering and hillwalking leaders in the British Isles.* Aviemore, Scotland: Scottish Mountain Leader Training Board.
Maier (1999) *Trekking in Russia & Central Asia.* Seattle: The Mountaineers.
Mason, K. (1987) *Abode of Snow.* London: Diadem.

Mehta, S. and Kapadia, H. (1991) *Exploring the Hidden Himalaya*. London: Hodder and Stoughton. Especially good reference for Garwhal, Kishtwar and Kulu.

Moran, M. (1988) *Scotland's Winter Mountains*. Newton Abbot, Devon: David and Charles.

Neate, J. (1987) *Mountaineering and its Literature*. 2nd edn. Milnthorpe, Cumbria: Cicerone Press.

Neate, J. (1987) *Mountaineering in the Andes: a source book for climbers*. London: RGS–IBG Expedition Advisory Centre.

Neate, J. (1990) *High Asia: an illustrated history of the 7,000 metre peaks*. London: Unwin Hyman.

O'Connor, B. (1989) *The Trekking Peaks of Nepal*. Marlborough, Wiltshire: Crowood Press. A guide to eight trekking peaks with maps and full logistic details.

Pollard, A.J. and Murdoch, D.R. (1997) *The High Altitude Medicine Handbook*. Oxford: Radcliffe Medical Press.

Razzetti, S. (1997) *Nepal Trekking and Climbing Guide*. Leicester: Cordee Books.

Salkeld, A. (1999) *World Mountaineering: The world's great mountains by the world's greatest mountaineers*. London: Mitchell Beazley.

Seiser, V. and Lockerby, R. (eds) (1990) *Mountaineering and Mountain Club Serials*. Lanham, USA: Rowman and Littleford.

Steele, P. (1999) *Medical Handbook for Walkers and Climbers*. London: Constable and Robinson.

Vallotton, J. and Dubas, F. (1991) *A Colour Atlas of Mountain Medicine*. Aylesford: Wolf Publishing.

Wilkerson, J. (1975) *Medicine for Mountaineers*. Seattle: The Mountaineers. Distributed by Cordee Books, Leicester, www.cordee.co.uk

17 DESERT EXPEDITIONS

Tom Sheppard

ETHOS AND OPERATING ENVIRONMENT

Reliability

Expeditions should be about total reliability, about planning the risks and problems out of the project at the planning and training stage. Nowhere is that more important than in the desert where the consequences of getting it wrong are unlikely to be cushioned by living off the land or getting assistance from local people. Detailed planning

Figure 17.1 *The beauty of the desert is often overwhelming. But the desolation emphasises the imperative need for reliability (© Tom Sheppard)*

Figure 17.2
(Top) Sharp rocks will damage tyres. Use full road tyre pressures and gentle driving. (Centre) Tracks offer a mixture of surfaces; with speed limited to 40 mph, a medium tyre pressure can help. (Bottom) Soft sand (very fine here too) needs a gentle right foot, use of diff locks (not in evidence here!) and lowest tyre pressures combined with severely limited speed
(© Tom Sheppard)

Figure 17.3 *Transverse corrugations are among the most malevolent types of terrain to which it is possible to subject a vehicle. "Track" tyre pressures (see below) are best allied to a "harmonic" speed which transmits less vibration to the vehicle body (© Tom Sheppard)*

of day-to-day operations (not to be confused with inflexibility), total self-sufficiency and thorough contingency planning are essential.

The desert environment

Those already embarked on planning a desert expedition will not need reminding, but for those still uncommitted it is worth remembering the astonishing variety of terrain, topography and climate that deserts encompass. Gravel plains, sand dunes, sand sheets, low rounded hills, harsh rocky landscapes, mountains, stony plains, boulder-strewn outwash fans as well as endless nondescript vistas of tough grass tussocks all offer different obstacles and challenges to travel by vehicle.

Problems for vehicles

Surface roughness, surface unevenness and surface strength will be the main problems for vehicles. Stones and rocks, often with sharp edges or large rounded shapes, will demand careful, sympathetic driving and tyres inflated to the maximum. The unevenness of tracks – undulations or sudden potholes or, worst of all, the widely encountered transverse corrugations – will call for sensitive control of speed and usually an "on–off tracks" tyre pressure tied to a moderate upper speed limit. Typically this might be 1.8 bar and 40 mph for a light 4 × 4 such as a Defender 90.

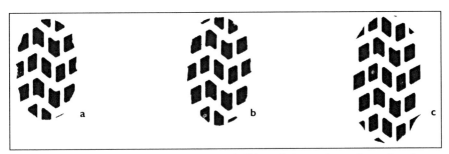

Figure 17.4 *Typical tyre footprints at (a) road, (b) track (40 mph maximum) and (c) emergency soft (15 mph maximum) pressures.*

Corrugations will demand a "harmonic speed" (probably 25–35 mph) which reduces the apparent effect of these features, but they will still be giving the suspension a pounding and at the same time make braking and steering far less effective than on normal tracks.

Reduced surface strength presents as soft sand on dunes, in wadis, on cut-up tracks, or unpredictable patches on sand plains. Here tyre pressures may well have to be reduced to 1.5 bar or even less to enlarge the tyre footprint and benefit flotation,

Figure 17.5
Reinflation of tyres deflated for sand is essential if hard, rocky conditions are encountered or higher speeds used. The TruckAir is a fan-cooled, high-capacity electric pump (© Tom Sheppard)

Figure 17.6 *Bogging in soft sand – and de-bogging – soon becomes a standard procedure. Reverse if you can, deflate the tyres to emergency soft, dig sand away from wheels, place sand mats behind rear wheels, reverse back on to firm ground, recce on foot the best way out (© Tom Sheppard)*

but must be accompanied by speed reduction to 15–20 mph – less if extreme deflation down to 1 bar is used. Reinflation must follow as soon as possible after the bad patches are covered. Failure to do so will cause tyre overheating with delamination, and will destroy covers; remember that this could be all four at once. Check the on-road tracks (40 mph maximum) and emergency soft (15 mph maximum) pressures for the tyres on your vehicle. Contact the tyre manufacturer when your dealer gives you a blank stare.

Problems for people
Despite high temperatures, dry desert climates are subjectively easier on the human body than tropical ones; details follow below. Effective acclimatisation takes about five days. So take things gently at first to give your sweat glands (and therefore cooling mechanisms) a chance to get up to speed. Sunburn can strike hard and quickly from the moment of your arrival – and all the time thereafter – so be meticulous with the use of protective creams and long-sleeved shirts and long trousers. It is even worth taking cotton-backed gardening gloves to protect hands and wrists. See below for more on clothing, water intake, etc.

CLIMATE

Temperatures
Desert temperature extremes of -5 to +50°C are bandied about by dramatists but these tend to be one-offs plucked from the relatively high-altitude Hoggar Mountains in winter to low-lying In Salah in summer. Far more precise, the Michelin planning maps also provide details of monthly daily maxima, overnight minima, and rainfall figures for a host of different African and Middle Eastern locations including desert areas.

Radiation and humidity
Less widely noted is radiation under clear-air conditions. This can be very strong (incoming, solar) during the day and strong (outgoing, from sleeping humans) at night. Typical diurnal shade temperature variation is of the order of 20°C but can feel more when standing in the sun or sleeping with no reflector over you – be it upper cloud, dust or a "space blanket" (see "Camping gear" below). Humidities, however, are very low, permitting most efficient evaporative cooling to be derived from

Figure 17.7 *Strong solar radiation from clear sky plus physical work digging out vehicles can bring on sunburn and dehydration very quickly. Be on your guard. Drink little and often and enough for your urine to be a normal colour. Allow time to acclimatise (© Tom Sheppard)*

sweating, so high temperatures will not feel as enervating as they would in rain forests. Similarly, exiting a sleeping bag on a still, ultra-dry morning will feel almost bracing compared with the penetrating damp cold of even a UK autumn.

Wind
Probably as trying as any temperature extreme are the diurnally variable moderate-to-strong winds in desert areas, often springing up for 2 or 3 hours before sunset to irritate anyone trying to cook, do end-of-day paperwork and map checking – or even trying to lay out a sleeping mat. Less frequently these winds can, over about 22 knots, raise sand and dust to above head height and prove the wisdom of comprehensive polythene bagging of such equipment as cameras. Sometimes these winds will rage overnight, leaving sleeping bags awash with fine sand. They will deposit fine sand into everything. Passing local sand storms can also call for rapid battening down of equipment left outside. All these winds may leave, to a greater or lesser extent, dust hanging in the atmosphere for days at a time, reducing nocturnal radiation cooling and dulling photographic light during the day.

Rain
Rain is rare, though less so in recent years – possibly due to climate change. It damps down the dust, firms up soft sand, cleans dust off rocks and foliage, and results in a riot of small plants erupting within a few days. But rain run-off can also cause flash floods of devastating force in wadis. Many are the tales of a wall of water coming down a wadi to sweep camping gear and even vehicles and people before it. Be aware of the potential when choosing a camping site.

MAP RESEARCH
Advance knowledge of the area that you plan to visit will pay handsome dividends when you reach it. Anywhere in Africa and as far east as Muscat, it would be fair to regard the Michelin 1:4 m maps – Sheets 741, 745 and 746 (with enlarged areas 743/744) – as the definitive planning maps. Road types and surfaces, rest-houses and hotels in remote regions and the all-important fuel availability are regularly updated with 2-yearly new editions. For transit purposes these maps will probably suffice, although the traveller will be missing a lot of interesting detail available on larger-scale maps.

The type, scale and standard of maps available to scales larger than the Michelin 1:4 m vary considerably from country to country and a starting point for subsequent research in the UK will be Stanfords in London's Long Acre, where in-country published road maps will be available plus some topographical maps. It is worth being discerning at this stage. Familiarity may have led you to accept the outstanding cartography of the British Ordnance Survey and the French Institute Geographique National (IGN) as the norm, but nationally published road maps for many overseas

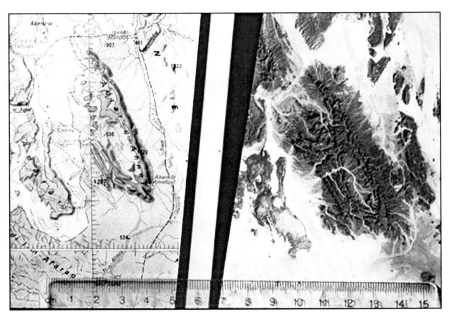

Figure 17.8 *Even the cheapest, no-geo-referenced satellite shots can hugely improve on coverage of even good maps – here a 1:500 k IGN sheet. Fortuitous in deserts is the dark depiction of rock and mountains, helping them stand out as terrain indicators against the light-coloured sand or gravel (© Tom Sheppard)*

countries look like cartoons in comparison, not helped by often depicting planned routes or projects as *faits accompli* superhighways. Nor are many overseas maps updated regularly. Even in the thrusting, go-ahead Gulf, *de facto* city and local roads are way ahead of the maps.

The cost and facilities required for major national surveys derive mostly from military, ex-colonial or geopolitical alliance sources. Thus the British mapping in Oman is superb, the 1960s French IGN 1:200 k, 1:500 k and 1:1 m maps of Algeria are excellent and the Russian maps of Libya are very good – albeit of varying accuracy, and saddled with a unique grid and only Cyrillic notations. Many of the American TPC and ONC maps are hopelessly vague and inaccurate, but are improving as they are updated. Few of these maps are stocked by map shops but may be available to special order with a long wait and high costs. Därr in Munich, like Stanfords, has a good selection and comprehensive catalogue.

Enormous accuracy and detail for enormous money sums up the satellite image market but by careful selection of older images at low resolutions (say, 30 metre pixels), often in black and white, your needs can be tailored to your budget. Draw the

199

distinction, however, between geo-referenced images and straight prints. The former will have had latitude and longitude and/or UTM (Universal Transverse Mercator) grid superimposed on the image – but this is high-priced work done "by hand" and is only as accurate as the best available terrestrial maps, which will have been used as a reference. Costs and availability are improving all the time.

Remember, too, that global positioning systems (GPS) give you a position on a grid – not a position on the ground. Its accuracy as a fixing aid is only as good as the map on which it is used and the accuracy of the grid on that map. Check the grid-on-map accuracy in the field on a prominent and unmistakable landmark.

When measuring distances across country on a map, be sure to apply a "terrain factor". Thus on a gravel plain actual distance covered by the vehicle will probably be up to 1.1 to 1.15 times the distance measured on the map; in bush it may be 1.3 and weaving through varied low hills 1.4–1.5. Getting from A to B over dunes you could travel anything up to 2.0 times the map indication.

EQUIPMENT

Weight and bulk
An overriding consideration to bear in mind when planning equipment is weight. You will read elsewhere in this book never to exceed a vehicle's permitted gross weight. Desert expeditions are extremely demanding logistically. As a result of the distances involved and the lack of replenishment points, fuel and water will usually comprise well over half the vehicle payload. As a result of this and calculations of the logistical "cost" of each person in terms of water and supplies, you may well already have had to limit crew to two people per vehicle (assuming that archetypal "light 4 × 4s" such as a Defender 110 or Toyota are being used). Weight must therefore be the prime consideration for any equipment assigned to a particular vehicle. Do not fall into the trap of putting equipment aboard until you run out of space. Varying load densities and general averages almost invariably mean that a maximum load limit is reached before the load "bulks out".

Packing and lashing
All equipment must be secured to the vehicle and lashed down on to old carpet or rubber mats. Track conditions will be such that severe vibration and shaking will be the norm so think "strap-down" at all times. Packing should be based on the assumption of sand blowing constantly through the vehicle – not an unusual situation. Lidded containers and polythene bags should be the norm. Have the high-density cargo (such as the fuel and water jerry cans) up front against the bulkhead behind the driver and passenger so that the overall load is spread as evenly as possible between the vehicle axles, and not concentrated over the rear axle alone.

Figure 17.9 *Looking aft from driver's seat, cargo lashed for a desert trip. Fuel and water cans are at the front against the bulkhead; lightweight plastic food storage boxes are lidded and lashed. Note fire extinguisher mounted near back door where cooking takes place (© Tom Sheppard)*

Camping gear

To save weight, don't take a tent; a further bonus is the breathtaking beauty of the desert sky at night. Do take a low camp bed, lightweight garden lounger or some such device to keep you off the ground away from the attention of scorpions trying to escape the cold. A sleeping mat on top of this will prevent heat loss downwards. A graded combination of light-to-medium sleeping bag, Gore-Tex bivvy bag, cotton inner sheet bag and long-johns (cotton–acrylic long underwear to use as pyjamas) will cover the varying night temperature conditions at most times of the year. In winter an additional fibre-pile inner may be needed. All this can be used in warm conditions by sleeping in the cotton sheet on top of, then inside the bivvy bag as the night gets colder. Unroll and sleep on top of the sleeping bag within the bivvy bag and it is ready to get into if the temperature really drops as it sometimes does around 4 am. Lock the vehicle, sleep with the key, and keep a water bottle, torch and mini-flare handy. Useful for cold conditions when the sky is very clear is an aluminised "space blanket" to drape over the bivvy bag to reflect body heat. A few desert areas may require the use of a mosquito net and, by sleeping alongside the vehicle, a single-point attachment can be made to the rain gutter.

Clothing

Outer layers
Something to keep the wind out, something to keep the rain off and something to keep the dirt off are the basics of outer clothing for desert trips. The first two are pretty obvious, rare as rain may be, and the waterproofs can be the very lightweight stuff from the economy end of the outdoor clothing catalogues – something that will roll up small for a corner of one of your bags. Depending on what you have brought for inner layer clothing, the rainproof can also be the windproof for use in the field.

Overalls – a large, white, loose-fitting boiler suit is ideal – are excellent for keeping your normal clothing clean when working on the vehicle or doing chores around camp such as re-fuelling and moving boxes and cans. Desert trips are naturally not blessed with generous washing and laundry facilities so prevention is the best approach. With even minimal care, the desert also being an inherently very clean environment, the overall itself can usually be kept clean enough to use at the end of the day as an around-the-camp item in which to cool off. Also part of the working gear will be working gloves to prevent abrasion, cuts and greasy dirt during wheel changing or hoisting cans of diesel.

Day to day
No points are awarded for the ragged explorer image of old and, although modern technical clothing favours the baggy look for sound practical reasons, there is no excuse for it not to be clean or for you not to have a set of "presentables" in your bag too. Appearing at customs posts, campsites or hotels looking like a vagrant does no one any favours. Be careful in Arab countries not to offend local sensibilities by under-dressing – especially if you are a woman. Wear long trousers and cover your arms in any centre of population.

For the clothes themselves, the principle of loose-fitting, well-ventilated garments is obviously applicable with, at least at the start of the trip, long sleeves to keep the sun off. Plenty of pockets, with coin- and key-proof closures, preferably zips on trousers or small Velcro patches on shirt pocket flaps, will ensure that valuable objects stay in your pockets rather than falling into the sand during digging or other activities. It is worth tagging keys with a small length of brightly coloured cord so if they do fall into loose sand you will spot them at once. Regatta and Wynnster do "working trousers" in 70/30 polycotton (with many zipped pockets) that keep an ideal balance between smartness and practicality.

Coolmax (an ingeniously cross-sectioned polyester fibre with a large surface area to encourage wicking) and nylon (polyamide) microfibres are names to look out for in choosing hot climate clothing. The latter should preferably have a hydrophilic chemical treatment to enhance wetting and wicking such as DuPont's Vaporwick (The North Face ventilation shirt) or Aquadry MMS (Craghoppers' Barkhan shirt).

A decade ago, the wearing of nylon against the skin in hot climates would have been unthinkable but the microfibre and wicking enhancement has produced a hard-wearing, comfortable, very quick-drying material ideal for expeditions.

Coolmax-rich trekking socks and lightweight, flexible walking boots will take care of the feet. Select a boot with a low heel cup because ankle flexibility is important when driving. Clarks produce high-street leisure footwear – light, breathable, flexible yet strong enough for rock scrambling – that is ideal for desert expeditions; their FastWalker is a classic. For the other end of the body take a cotton sun hat with (unlike the dreaded baseball cap) a brim wide enough to keep the back of your neck and the tops of your ears in the shade.

COOKING, FOOD, WATER

Cooking
A good, simple, field-maintainable, multi-fuel stove such as the MSR Dragonfly and a 5-litre can of kerosene will keep an expedition going for 6–8 weeks. Trying to run any multi-fuel stove on diesel makes a good promotional exercise for cold food. The ratio of cleaning time to cooking time is about four to one. Leaving your MSR stove fully assembled and mounted on a board to stow in a box is a luxury that a vehicle-based expedition can afford and thus preclude the messy disassembling of it after use with attendant drips of fuel.

Choice of food is common sense except to emphasise that plenty of breakfast cereal fibre, tinned fruit and vegetables should be included. Use Ryvita or similar when local bread is not available.

Water
Water supplies in population centres in Libya, Algeria and Tunisia are fit to drink from the tap. Take a microfilter such as the Katadyn as a back-up, however; it is expensive but will remove parasites, protozoa, bacteria and viruses. You don't want to add gastric infections to any water crisis that you may encounter. Consumption will depend on workload but planning figures (per head, per day) would be:

- night/day temperatures 5/35°C – 5–10 litres (less if you are skinny)
- night/day temperatures 25/45°C – 8–15 litres per head per day (less if you are skinny)

A general ballpark figure for moderate workloads of 7.5 litres per day is safe to generous for most conditions. Usage takes a leap above 42°C ambient.

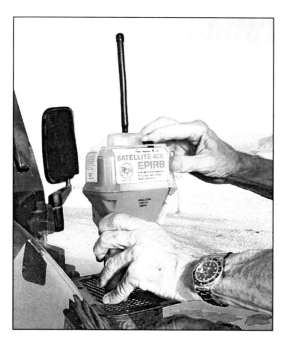

Figure 17.10
*EPIRBs are registered in the
name of the user and will send
a position-tagged emergency
signal to a worldwide ground
station network via a satellite
when triggered. Battery life is
usually 48–60 hours in use, 7
years in storage. For smaller
units and more detail, see*
Vehicle-dependent
Expedition Guide, *Section 5.6.
(© Tom Sheppard)*

SAFETY AND RESCUE AIDS

Epirbs and sat-phones

With the infrastructure already orbiting the earth, satellite-based emergency rescue aids are worth considering. A Sarsat EPIRB (Emergency Position-Indicating Radio Beacon – about £600–800) is registered on purchase and, thereafter, activation will transmit a signal to a satellite which will take a fix and relay your position to a world-wide network of stations. Although, as a prime marine-oriented rescue aid, its back-up is cast iron, it is an all-or-nothing device with no ability to distinguish between "I am trapped under a rolled vehicle and can't move" or "I will need rescue but I have plenty of water so don't rush".

A satellite phone, however, enables you to contact specific agencies or a base party and even to get advice about particular problems. The Inmarsat or iridium-based worldwide phones cost in the region of £1200. However, the Thuraya satellite, commissioned late 2001 (now with back-up satellite), is geo-stationary approximately over the Egyptian–Libyan border with a footprint extending from the western Sahara to India and north to the UK and Germany, and yields ideal coverage for most African and Middle Eastern desert areas. Call charges are considerably lower than the world-coverage networks and the instrument and its set-up cost around £700.

Figure 17.11 *Desert terrain can encompass enormous variety. Expect the unexpected. Always put the welfare of your vehicle first, drive it with care. Recce difficult terrain on foot if in any doubt (© Tom Sheppard)*

Flares

Miniflares – like the EPIRB, these are obtainable from yacht chandlers – are invaluable small-package items that can be issued to individuals, typically to parties going out on foot. If you do equip your party with these, be sure to agree a particular time when they will be used, e.g. 8 pm when it is dark and they will be seen; a set time when all eyes can be trained in the expected direction of discharge.

FURTHER READING

Bagnold, R. (1935) *Libyan Sands*. London: Immel Publishing.
Davies, B. (2001) *SAS Desert Survival*. London: Virgin Books.
Dhillon, S. (2002) Desert expeditions. In: Warrell, D. and Anderson, S. (eds), *Expedition Medicine*, 2nd edn. London: Profile Books. Available from www.rgs.org/eacpubs
Johnson, M. (2003) *The Ultimate Desert Handbook: A manual for desert hikers, campers and travellers*. A Ragged Mountain Press Outdoors Paperback. New York: McGraw Hill.
de Saint Exupery, A. (1939) *Wind, Sand and Stars*. London: Pan Books/Heinemann.
Scott, C. (2000) *Sahara Overland: A route and planning guide*. Hindhead: Trailblazer Publications.
Sheppard, T. (1998) *Vehicle-dependent Expedition Guide*. Hitchin: Desert Winds.
Sheppard, T. (1999) *Off-roader Driving*. Hitchin: Desert Winds.
Sheppard, T. (1988) *Desert Expeditions*. London: RGS–IBG Expedition Advisory Centre.

18 TROPICAL FOREST EXPEDITIONS

Clive Jermy and Corrin Adshead

Tropical rain forests cover less than 6 per cent of the Earth's land mass and are defined by their location (between the Tropic of Cancer 23° 27' N and the Tropic of Capricorn 23° 27' S) and by their high rainfall. They are therefore hot and humid environments, although upland forests may often be cold enough to require a blanket or lightweight sleeping bag at night. Some forest floors may be under water for much of the year.

This chapter is intended to provide a basic outline of tropical forest logistics for small expeditions. It covers most of the topics relevant to living, travelling and

Figure 18.1 *The use of rivers is a pleasant and traditional means of travel in the jungle (© Corrin Adshead)*

working in the tropical forest environment. Further information, especially in relation to somewhat larger expeditions, is given in the Royal Geographical Society–Institute of British Geographers (RGS–IBG) Expedition Advisory Centre's *Tropical Forest Expeditions* manual (Chapman et al., 1983 and later editions).

ACCLIMATISATION

Humans may once have been "tropical animals" but for many this is definitely past tense. Those transported abruptly by aircraft from temperate climes are likely to suffer from fatigue, lethargy, poor sleep and reduced exercise tolerance. They should maintain hydration and exercise little in the first few days, and this limit on activity should be considered when formulating the expedition itinerary. Acclimatisation is assisted by slower transit overland or on a boat but otherwise takes about 8–10 days (slightly longer in children). Not surprisingly, air conditioning delays acclimatisation and is best avoided. In preparation, careful exercise in a hot humid environment, as simulated by exercising in warm clothing indoors for an hour a day for at least a week preceding your visit to a tropical environment will aid acclimatisation, although the benefit is lost within a week if not maintained. Care must be taken to maintain hydration and exercise halted if any signs of heat stress develop.

CLOTHING AND EQUIPMENT

Footwear

Jungle boots, specialized calf-length canvas rubber-soled boots, are the traditional footwear for tropical forest expeditions. They may be obtained at various "ex-army" stores. Those that originate in the US army are leather-based and more substantial than commercial copies. On the other hand, baseball or hockey boots (with eyelets above the soles to let out water) are adequate if you are not planning on trekking too far, and are more easily obtained and cheaper. Try to find the thicker-soled variety with treads, because they give better grip on slippery river beds and are more comfortable on roughly cut tracks. Don't assume that you can buy them locally, particularly in larger sizes, although in some countries (e.g. Malaysia) they are available. However, it is best to assume that both baseball boots and local purchases won't last more than 2 months in the field (maybe less, although their lives can be prolonged by drying them whenever possible), and you should have a spare pair, so take as many as you will need. Avoid totally waterproof boots made with "breathable" fabrics because they tend not to dry easily and will certainly become wet inside. The humidity prevents the fabric from "breathing" so you are better off with boots that will allow water in and out easily. The jungle floor can be muddy and wet, so slightly higher-leg boots, with a sewn-in tongue, are preferable.

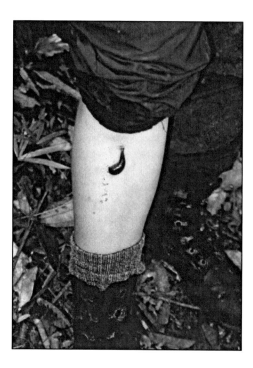

Figure 18.2
Leeches, although a persistent nuisance, are easily removed using insect repellent. Treat the bite with antiseptic (© Corrin Adshead)

Clothing

Wear baggy, lightweight but strong trousers, tucked into socks and boots, and long-sleeved shirts, with plenty of pockets. Underwear, if worn, should be quick drying and should not chafe. Do not take old cast-off clothes; they will not last long. Don't wear obviously military (especially camouflage) clothing, practical though it may be, because this can sometimes be misinterpreted with unfortunate consequences. Socks should be long enough to tuck trousers into for general comfort, but they will not keep out ticks, leeches, etc. If leeches are bad (as they may well be in Asia), it is worth buying "leech socks" locally or making your own. These are over-socks, made from fine-weave cotton, simply styled like a Christmas stocking but with a double seam tightly sewn so that leeches cannot squeeze through. They should come to just under the knee and have a draw cord, which is tightened over the trousers. Leeches that have evaded your defences can be persuaded to leave with a dab of insect repellent, spirit or tobacco soaked in water.

Mosquito (or midge) head-nets, generally designed to be worn over a wide-brimmed hat, can be an enormous asset. Remember that if you are covered up there is less access for insects to bite you. Malaria prevention starts with your clothing. Long sleeves and trousers are far better than shorts and T-shirts.

You should always have at least one set of comfortable clothes for use in camp and for sleeping. Never let them get wet: resist the temptation to keep them on instead of climbing into your clammy field gear on a bleak morning.

Other equipment

Take sunglasses and sunscreen for long boat rides and open-topped transport journeys.

Simple, suede/cotton gardening gloves will save your hands where you are climbing in forest on rough rock, e.g. in limestone areas, and are useful for trail cutting and rope work. Sweat-rags are useful, and a floppy wide-brim hat keeps the sun off during river journeys – and the bugs, especially ants, from going down your neck while cutting trails. Waterproof clothing is generally more trouble than it is worth in a forest – it is simpler to get wet and then to dry again – but a large waterproof poncho (with eyelets) may save a lot of discomfort during river journeys when it can get very cold if you are wet. A poncho can also be rigged as a temporary shelter if required. Small, folding umbrellas for base camp life are worth a thought too.

Your rucksack should be made of synthetic material, and be comfortable and well balanced (as most modern sacs are). Side pockets can be a hindrance in dense vegetation; if you have them but do not want them, then tape or strap them up to keep them out of the way. If you have a "high" pack that protrudes far above your shoulders, this can also be extremely awkward when ducking under branches and logs.

Figure 18.3 *Hammock with flysheet suspended between two trees (© Rupert Grey)*

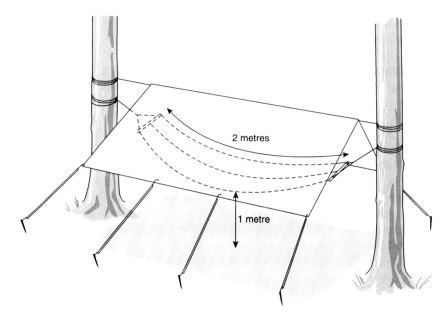

Figure 18.4 *Erecting a hammock and flysheet (© Bob Hartley)*

Attach sleeping mats vertically to the back of the pack. Use waterproof liners, but still pack individual items in polythene bags to prevent them from getting wet. A good selection of self-seal bags is invaluable, and you will probably need many more than you think you are going to.

Cooking and camping

It should generally be possible to cook over wood fires (even in heavy rain you will find dry wood somewhere, or your guides or trackers will know which wood to cut for immediate burning), but take a bicycle inner tube, candle or some other suitable fire-lighter and, if you're not happy about your ability to make fire, take liquid fuel cookers for back-up. If you are working and living in a nature reserve or national park, you may not be able to cut trees at all. If you are in a large park where a small amount of selected cutting is allowed, remember to take them well away from the path that visitors/tourists use. Cut only standing, dead wood, however. Strip off the wet, outer layers of bark if the wood looks soaked. For igniting, carry lighters as well as matches in a waterproof container. Lighters are prone to becoming damp.

Remember that night tends to fall early in the tropics, so if you have to work, read, etc. after nightfall, as most expeditions must, take paraffin or gas lamps. Butane canisters are virtually universal and can normally be bought in most tropical countries, but gas mantles are often more difficult to come by and are best taken from the

UK. If the expedition is frequently on the move you may need quite a few! Hammocks are advisable for camping in the jungle – they can be bought locally but it may be better to purchase (and try out) good-quality hammocks before you depart on your expedition. The nylon mesh hammocks available in Britain tend to be too small for comfort but can be useful for suspending supplies. Several companies in Britain now make lightweight, quick-drying hammocks made from cloth. In Southeast Asia "camping sleeves" (essentially stretchers) are often used, but this involves cutting poles from the forest. Although this is usually no problem unless camping in an area where tree cutting is prohibited, or if a science programme prevents drastic changes in the forest structure, it is better practice to avoid cutting live, young trees at all.

Mosquito nets are worth taking even if there is no risk of malaria in the area – there may be sufficient insects to interfere with your sleep, and nets can keep out other unwanted bedfellows such as scorpions, centipedes, bats and even snakes. In South America you can buy mosquito nets that are specially designed for use with hammocks. Otherwise tuck in the excess material underneath you. Check the inside of your hammock and net before you go to bed. Experiment with all this exotic sleeping gear in the garden before going out into the field! Treat your mosquito net with permethrin to add to its effectiveness in deterring insects.

It can be surprisingly cold at night, especially in a hammock and in a mountainous area, so take a synthetic sleeping bag or warm blanket. If you are planning to be in one place for any amount of time, the best means of camping is to tie a large plastic tarpaulin over a stout wooden framework, lashed together with rope – or your local guides will use lianas or bark ties. Spend a little time making the camp reasonably comfortable – it's very easy to rig up basic tables/work benches and seats from sapling poles, rattan or bamboo, and they make life much more pleasant. For lightweight trekking expeditions it would be better to take "bashas" (single shelter sheets for use with hammocks and mosquito nets), though it is not necessary to get a specially made set. Ordinary hammocks and lightweight tarpaulins, the latter suspended with elastic "bungees" or nylon cord, are fine. Ponchos, although versatile, are often short and tend to leak around the hood seams. A heavy downpour in the jungle will make you wish you had invested in a good-sized basha sheet! On this type of expedition, make sure that you leave yourself sufficient time to set up a decent camp before nightfall.

Camp hygiene is important and latrines should be sited so as not to interfere with your domestic water supply or anyone else's. In a long-stay camp it is necessary to build something substantial, consisting of a seat, shelter and hand-washing facility. If you have a big porter team, remember that they too need these facilities. It is not easy to organise the defecating habits of others, but at least make sure that they are "down river".

Much of the equipment that you should take with you is of course the same as you would take on any other camping trip: compasses, whistles, maps (although these may be of limited use in tropical forests) in plastic map cases, penknives (locking blades if possible – don't include them in your hand luggage on the plane), cigarette lighters, strong nylon cord, insect repellent (and one you know works on your skin – not everybody's body chemistry is the same), waterproof notebooks (e.g. surveyors' notebooks with high wet-strength paper), waterproof watches (on non-rot straps or cords), tin openers, waterproof torches, water bottles, cooking and eating equipment, first-aid kits, etc. Medical supplies should include plenty of fungicidal cream and powder, antimalarials, antihistamine and effective antibiotics (see Chapter 14). Everybody should carry a small, practical survival kit in a belt pouch in case they become separated from their main rucksack. Wear a whistle and a small compass everywhere. It is easy to become disoriented in the jungle.

Machetes

There are very few items of equipment specific to forests, but machetes (or *Parangs* as they are called in Malaysian countries) are one of them. It isn't worth taking them out from Britain – you can obtain them easily when you arrive and you will also avoid problems at border customs points. They are essential in the forest and have a multitude of uses from cutting wood and clearing trails to digging holes. For most general work it is best to get one with a shorter (12-inch/30 cm length) and reasonably heavy blade, and with a sheath. This can be slung over the shoulder on a tape, or worn on a belt. Keep the machete in its sheath when not actually cutting the trail – it is all too easy to fall accidentally on to a razor-sharp machete blade. And it needs to be razor-sharp: take a file or whetstone; the latter may be heavier but it keeps its virtue longer. Attend to the blade every night and do not put the machete on the jungle floor during use, as you will probably lose it.

Cameras and film

You need to be particularly careful to prevent mould from attacking optical lenses (fungus can grow on [and etch] glass) and film. Details of tropical photographic techniques are not provided here, but there are a few primary rules, which will help to prevent disasters. Avoid leaving film in the camera for long periods; process all film as soon as you possibly can. If possible take a camera that is not entirely reliant on electronics, because these can go wrong in the damp. If you are arranging for a runner to take out mail, live plants or animals, post off film to an appropriate processing laboratory at the same time. Keep film, cameras and lenses in sealed containers with desiccant (silica gel with at least a proportion of the colour indicator type) when they are not in use. Dry the silica gel regularly over a fire. Light levels at the forest floor are remarkably low and, although fast film overcomes this problem, it does so at considerable expense of quality. You may be better off using a flash or tripod.

Consider carefully the nature of your expedition and the specialised equipment that will be required. If you are planning aerial supply drops, for example, you will require marker balloons and probably radios, flares and coloured marker panels. Make sure that the radios and other communication systems, for whatever reason you need them, are suitable for the type of terrain in which you will be. If you are going to need to cross a river, you will require appropriate river crossing equipment and know how to use it safely.

Where you are employing porters to carry equipment and supplies, make sure that you know exactly where everything is (and that it is labelled if necessary), and that items that you may need during the day or at short notice are conveniently located. This may mean explaining to porters where they should be in the group, when trekking. Consider how some items need to be carried and whether you should provide rucksacks or frames for this purpose.

Food
Take all the food that you will need along with you. Living off the forest is impractical and very time-consuming and, when in national parks or game reserves, illegal or at least morally wrong. When you are close to, and working with, local people, remember that their economy may not be able to support large numbers of extra people. When taking in substantial numbers of porters, take in sufficient extra food for them too, unless you have arranged that they feed themselves. You should only do the latter when you know that there is a source of food readily available or you may encourage illegal hunting. They, and you, may well be able to supplement your diet with fish, but don't count on it. If you are taking food for your porters, make sure that they are happy with it in advance. This applies particularly to the local staple (rice, manioc or whatever). It's a good idea to get one of them to come to the market when you buy it.

Animals will visit your camp. Food should be suspended under cover as anything left on the ground and not in a tin will probably not be there in the morning. A string hammock is handy for this. Support strings can be soaked in insect repellent if you are worried about ants. Even if you are miles from habitation there may be a substantial endemic rodent population which will gnaw through thin metal screw-tops of food jars – and also through thick polythene bottles regardless of their contents (in this case collecting jars should be suspended too). If you are living out of tins, etc., your diet may need topping up with vitamin pills and fresh food whenever it becomes available.

Surprisingly, water can be extremely scarce in certain tropical forest areas during the dry season. Try to find out about this in advance, and plan accordingly. Water filters (modern pump filters or cotton Millbank filter bags) and/or purifiers (iodine will do) should be taken on all expeditions, but these will be particularly appreciated when the only water available is from a nasty pond. Everybody should carry a bottle

213

of iodine purifier for personal use. Carry two 1-litre-sized bottles each. Collapsible water bags (Ortlieb) are excellent for treks. At the very least, boil all your water for 5–10 minutes. Use large, plastic jerry-cans for holding adequate supplies of water in the base camp. Administer the appropriate quantity of iodine to the water with a large plastic syringe and identify those containers holding purified water.

RELATIONSHIPS WITH AND EMPLOYMENT OF LOCAL PEOPLE

It is essential to establish good relations with local residents both within and around the forest. Make sure that you know which dialects they understand and can communicate in, and have the appropriate interpreters. Even if you have permission to visit the area, remember that the forest may be their spiritual home and therefore respect their ownership. Explain to their leaders (elders) what you are hoping to do in the forest: if you are collecting specimens, make it clear that there is no commercial value and, if appropriate, that when studied the samples will come back to that country. Most local people are nationalistic (even when oppressed by their governments) and feel happier if their plants or animals are being returned to "their" country (even if they will never see them again) than if they go for good.

Figure 18.5 *A good local guide will assist greatly in daily tasks such as fire-lighting, cooking, fishing, choosing routes and campsites (© Corrin Adshead)*

If local people can be hired (as guides, porters, cooks, clothes washers) it can help their local economy and personal wealth. Try to contact the local "headman" when you reach a village or new area. Make sure that you are using the correct rates and spread the cash as far as you can through the community. Always discuss the rates and get an agreement before loads are picked up. Keep your team cheerful. Be sure that they have adequate equipment and bedding for cool nights. Carry contingency cash in case you need to hire extra guides and porters along the way. In many areas cigarettes are a form of "currency" so consider taking a good supply even if you don't smoke. However, cash is a healthier alternative.

If you are travelling in populated areas and going from village to village, discuss your destination with the porters. They will know how long it will take and whether it is feasible within the time that you suggest. There is, however, still enormous scope for confusion on this subject, so be warned! Try to avoid questions that can be answered yes or no. When the destination is unknown, or cannot be defined, make sure that you plan to stop by early afternoon. Respect any local taboo as to where you can walk, travel through or stay. It is usually based on superstition or hunting, but to break it can mean that your whole labour force walks out on you.

By collaborating with local people you will learn infinitely more about the environment, and gain infinitely more from the expedition, than if you try to "go it alone". "Five get lost in the jungle" sounds pretty exciting and romantic, but really it's an utterly barren and unrewarding experience in comparison.

TRAVEL

By river

Most expeditions use river transport in tropical forests as a rapid means of getting from A to B. This generally entails using local transport facilities, which in many cases means longboats (wood or aluminium) with outboard motors. Inflatables are only really suitable for large rivers, and even then they may not be ideal during the dry season. If your expedition logistics hinge on river travel, make sure that you go there at a time of year when the rivers are navigable with relative ease (i.e. not too much water and not too little).

You should hire experienced local boatmen to take you because they will know the vagaries of their craft and the intricacies of their rivers. Although these people may adopt a somewhat cavalier attitude towards river travel, for the sake of safety there are a number of points that should be observed:

- Wear life jackets (it is not an insult to the boatman), and carry paddles and balers. If possible, use outboards without shear pins in rivers with rapids. If you are providing the motors, it is worth considering using those with long

Figure 18.6
Longboats require careful loading but are a useful means of travel in the jungle (© Corrin Adshead)

propeller shafts for negotiating shallow stretches of river. A 25 hp motor is enough to get most boats up most passable rapids, but you may need to haul boats by bow and stern ropes (which should be at least 20 m long) in leap-frog fashion past difficult stretches. You may need to portage (carry) and for that reason fuel and equipment should be distributed in manageable loads. For rivers with rapids you should have two boatmen: one in the bow with a pole and a paddle and one at the stern. The motor should be attached to the boat, as should everything else (except you!). For this purpose, ropes, karabiners and a large net are useful. Sensitive and valuable items should be stored in waterproof (floating) plastic containers. Be sure to have enough tools, spare parts, shear pins (if required) and fuel. A fuel funnel can be handy.

- A 25 hp engine can travel for about 3 hours at 15 knots on a 5-gallon tank, but fuel consumption depends very much on the speed of the river and the number of rapids. Descending rapids is more hazardous than ascending, because once you have committed yourself you cannot stop. Don't wear heavy boots when travelling on dangerous rivers, and remember that you may be burned or dazzled by sunlight reflecting off the surface of the water. Do what the boatman tells you to do.

On land

It is advisable to travel through the forest using the pre-existing trails cut by local people rather than cut your own. Unless you really are completely out in the sticks, these trails will exist. Travel in single file at the pace of the slowest member of the team. It is easy to over-estimate the speed at which you are travelling in forest, so if this is important you may need to count your paces. On a good trail you may cover 3–5 km/hour, whereas on a bad one you may make only 1–3 km.

River crossings can be extremely dangerous, and the current deceptively strong. There will be plenty of submerged hazards in jungle rivers, such as dead tree trunks. Complete a good reconnaissance of the crossing point. Use a rope if you are wading across anything remotely dubious, and always unfasten your rucksack so that it is slung on one (downstream) shoulder. Cut yourself a stout walking stick for stability; this can also be extremely handy on steep slippery slopes. If you are crossing bridges of felled trees over dangerous water, always use an improvised handrail and make sure that you have a safety plan in case anyone falls in. The crossing of larger rivers requires rather more complex procedures. These involve different rope techniques and require practise by the team and clear briefings before the river crossing is carried out. Make sure in advance that you have the necessary equipment with you if your route is likely to take you to such a river. Ideally you should not try to cross difficult jungle rivers, but alter your route instead. Rainfall can cause flash floods, which arrive without warning, so take pains to choose a campsite well away from the river flood-plain. Check the watermarks on trees near the banks and never camp in a dried-out water source.

If your route involves road or air transport, remember that roads and airstrips in certain regions will be closed for climatic reasons at certain times of the year and jungle helipads become overgrown quickly if not maintained.

Navigation in the forest

Navigation in tropical forest is not easy. Even on short journeys from a forest camp in virgin forest it is very easy to get disoriented and lost. Mark saplings and trees clearly on both sides by shaving off a length of bark, so you can see the mark when approaching from the other direction. Regularly used routes can be marked with coloured plastic tape. Follow rivers or ridges where possible and check their direction of flow, and compass bearing with the map. Remember that, in tropical forest areas, it is likely that many streams will not appear on your map.

When cutting trails for regular use, make sure that your machete is sharp and that other people are well clear of the area of activity. You will probably be surprised how slowly you cover the ground. It may be necessary to cut in straight compass-controlled lines for demarcating research areas, and these may be used for general travel. It is relatively easy to cut straight lines in lowland alluvial forest and navigation by compass is possible on such occasions. In mountainous areas this may be much more difficult. If you need to be particularly accurate with your direction finding, use a "leap-frog" system of the type described by Chapman et al. (1983), and back bearings. Similarly, if your distance measuring is important, compare your pace count with a count made over a known distance on identical terrain. Remember that machetes tend to become magnetized during sharpening and will deflect your compass needle if carried too close.

When travelling by boat on larger rivers, maps can be more easily followed, but remember that rivers can change course and meanders may form anew. In areas where you will be exploring by such rivers, recent aerial photographs, if available, will be the best navigational aid to follow. Consider taking a GPS with you. Although this should not replace your guide or basic navigation skills, it is a handy tool for navigation and many models do work under the thick tree canopy. It is very useful for fixing the location of clearings that could be used for helicopter evacuation from the jungle, and for supply drops, etc.

Safety

Tropical forests are not the dangerous places people like to make out, but you do need to maintain a modicum of common sense. Carry out introductory training before you depart, covering emergency procedures, camping techniques, navigation, base-camp routine, etc. Find out in advance what potentially dangerous insects and animals exist in the region, and either avoid them altogether or avoid antagonising them. You are very unlikely to be bitten by a snake if you take sensible precautions such as not wandering around in bare feet and not rummaging with your hands in leaf litter. But find out anyway which snakes are venomous, and where the nearest health centre is that maintains a stock of antivenom. Check your clothes and boots for unpleasant animal or insect surprises in the mornings. Avoid malaria, be cautious with rivers, don't camp under very large or dead trees, and don't eat tempting fruits unless you know them to be edible. Avoid wandering off on your own and don't go too wild with your machete!

FURTHER INFORMATION

Further reading

Allen, B. (1985) *Mad White Giant: A journey to the heart of the Amazon jungle*. London: Macmillan.
Allen, B. (1987) *Into the Crocodile Nest: A journey inside New Guinea*. London: Macmillan.
Allen, B. (1989) *Hunting the Gugu: In search of the ape-men of Sumatra*. London: Macmillan.
Bennett-Jones, N. (2002) Base camp health and hygiene. In: Warrell, D. and Anderson, S. (eds), *Expedition Medicine*, 2nd edn. London: Profile Books. Available from www.rgs.org.eacpubs.
Bradt, G. *South America: River Trips. How to travel by dugout canoe, cargo boat, passenger steamer and raft down 11 rivers*. Chalfont St Peter, Bucks: Bradt Enterprises.
Chapman, R., Jermy, C., Adshead, C. et al. (1983 and subsequent editions) *Tropical Forest Expeditions*. London: RGS–IBG Expedition Advisory Centre. Available at: www.rgs.org/eacpubs

Collins, N.M. (ed.) (1991) *The Last Rain Forests.* London: IUCN and Mitchell Beazley.
Collins, N.M., Sayer, J.A. and Whitmore, T.C. (eds) (1991) *The Conservation Atlas of Tropical Forests: Asia and the Pacific.* London: Macmillan.
Cranbrook, G. and Edwards, D.S. (1994) *Belalong: A Tropical Rainforest.* Singapore: Sun Tree Publishing Ltd.
Davies, B. (2001) *SAS Jungle Survival.* London: Virgin Books.
Hanbury-Tenison, R. (1974) *A Pattern of Peoples* (Indonesia). London: Angus and Robertson.
Hanbury-Tenison, R. (1980) *Mulu: The rain forest.* London: Weidenfeld and Nicolson.
Hanbury-Tenison, R. (1982) *Aborigines of the Amazon Rainforest. The Yanomami.* New York: Time-Life.
Hanbury-Tenison, R. (1984) *Worlds Apart. An explorer's life.* London: Granada.
Harrison, J. (1986) *Up the Creek: An Amazon adventure.* Chalfont St Peter, Bucks: Bradt Enterprises.
Harrison, J. (2001) *Off the Map: The call of the Amazonian wild.* Chichester: Summersdale Publishers.
Hatt, J. (1985) *The Tropical Traveller.* London: Pan.
Hemming, J. (1978) *Red Gold: The conquest of the Brazilian Indians.* Macmillan, London.
Hemming, J. (ed.) (1985) *Change in the Amazon Basin,* Vol I: *Man's Impact on Forests and Rivers.* Vol II: *The Frontier after a Decade of Colonisation.* Manchester: Manchester University Press.
Hemming, J. (1987) *Amazon Frontier: The defeat of the Brazilian Indians.* London: Macmillan.
Hemming, J. (1993) *Maracá: Rain forest island.* London: Macmillan.
Hemming, J. (2003) *Die If You Must: The Brazilian Indians in the twentieth century.* London: Macmillan.
Insight Guides (2002) *Amazon Wildlife Insight Guide.* London: Insight Guides.
Jordan, T. and Jordan, M. (1982) *South America: River trips.* Vol II. *Travelling Using Your Own Boat in Venezuela, Suriname, Peru and Brazil.* Chalfont St Peter, Bucks: Bradt Enterprises.
O'Hanlon, R. (1984) *Into the Heart of Borneo: An account of a journey made in 1983 to the mountains of Batu Tiban with James Fenton.* London: Penguin.
O'Hanlon, R. (1988) *In Trouble Again.* London: Penguin.
Ratter, J.A and Milliken, W. (eds) (1998) *Maracá: The biodiversity and environment of an Amazonian Rainforest.* London: John Wiley & Sons Ltd,
Richards, P. (2002) Tropical forest expeditions. In: Warrell, D. and Anderson, S. (eds), *Expedition Medicine*, 2nd edn. London: Profile Books. Available at: www.rgs.org/eacpubs
Whimore, TC. (1998) *An Introduction to Tropical Rain Forests.* Oxford: Oxford University Press.

Useful addresses

BCB International Ltd, Clydesmuir Road, Cardiff CF24 2QS. Tel: + 44 2920 433700
  Helium-filled air marker balloons.
J.E.T Asia: Batam, Indonesia. Website: www.jet-asia.com
  Expedition training company in remote tropical locations. Courses cover leadership, safety and jungle survival.
Liverpool School of Tropical Medicine, Pembroke Place, Liverpool L3 5QA. Tel: +44 151 708 9393, fax: +44 151 708 8733, website: www.liv.ac.uk/lstm/lstm.html
London School of Hygiene and Tropical Medicine, Keppel Street, London WCIE 7HT. Tel: +44 20 7636 8636, website: www.lshtm.ac.uk
Nomad Travellers Store, 3–4 Wellington Terrace, Turnpike Lane, London N8 0PX. Tel: +44 20 8889 7014, fax: +44 20 8889 9529, email: sales@nomadtravel.co.uk, website: www.nomadtravel.co.uk
  Suppliers of tropical equipment and clothing.
Oxford Tropical Medicine Unit, The Nuffield Department of Clinical Medicine, John Radcliffe Hospital, Headington, Oxford OX3 9DU. Tel: +44 1865 220968, fax: +44 1865 220984, email: david.warrell@ndm.ox.ac.uk
Trekforce Expeditions, 34 Buckingham Palace Road, London SW1W 0RE. Tel: +44 20 7828 2275, fax: +44 20 7828 2276, website: www.trekforce.org.uk

19 POLAR EXPEDITIONS

David Rootes

The polar regions are becoming ever more accessible and for good reason: they offer fabulous opportunities for travel and expeditions. Remote locations, breath-taking scenery, unusual wildlife – not at all as depicted on television and in the movies. Certainly, there can be bad weather and blizzards but there will also be tranquil moments during the endless twilight or, if you travel far enough north or south, midnight sun and 24-hour daylight.

The polar regions contain a bit of most environments: much of the Arctic is surrounded by deep and fascinating forests; tundra is often braided by rivers that vary enormously in flow during freeze-up or thaw; ponds and lakes are common; mountains and glaciers abound. There are a burgeoning number of polar outfitters, and cruise or travel companies to guide your research.

The polar environment

Much of the North is ocean and there is no single geographical definition to the southern boundary of the Arctic: it is variously described as contained within the Arctic Circle, the extent of continuous permafrost, the tree-line. The land and ice mass of Antarctica are better defined, surrounded by the cold Southern Ocean. At either end of the world, seasonal expansion and break-up of sea ice swells and deflates the extent of the polar regions like some monstrous breathing organism.

This chapter is concerned with environments that have "polar"-like conditions. Mostly, these are at high latitudes but there are ice caps at more temperate zones and even glaciers at high altitudes near the equator. Polar conditions are found in the land and oceans adjacent to the Poles but include Greenland, Iceland, Svalbard and the Patagonian ice cap. In the British Isles, Welsh or Scottish highlands are excellent training grounds for polar travellers and the Cairngorms in the depths of winter have a tundra-like quality.

Figure 19.1 *Polar regions offer fabulous opportunities for wilderness travel and field research (© Dave Rootes)*

Polar people and governments

The Arctic is peopled by many native groups, well adapted to the conditions. It is their home, spiritually as well as physically, and deserves our respect. In addition, many southerners have moved north to work and some have stayed. Antarctica has no permanent residents.

Style of government varies: Greenland, despite home rule, has strong ties with Denmark and expeditions are administered by the Danish Polar Centre; Norway administers Svalbard under an international treaty. Expeditions to Antarctica require a permit or authority to travel from your "home" government.

Many agencies now require some level of environmental impact assessment before authorising access. It makes good sense to prepare a risk assessment of your plans to satisfy yourself and your backers that you have suitable contingency planning.

Make sure that you have the right permission, you know the rules and, for the sake of local inhabitants or those that follow, leave as little evidence of your visit as possible.

Key information sources

Your starting point should be the huge volume of polar information on the web. Contact one of the polar libraries for more depth or detail. The Scott Polar Research

Figure 19.2 *The weather in polar environments is characterised by strong winds (© David Rootes)*

Institute, part of Cambridge University, has one of the best polar libraries in the world. Call to discuss your research needs. For a small fee they will search their accessions database and send you an annotated bibliography, saving hours of work at the library.

Weather

Polar weather is characterised by its unpredictability. An idyllic moment can rapidly transform into one of utmost severity as a katabatic wind tumbles off a nearby glacier; 0 to 60 mph in under half an hour is not uncommon. At very high latitudes, and as winter deepens, the likelihood of rain may be small but during summer in much of the Arctic, coastal Antarctica and Greenland, for example, temperatures hover around 0°C. In these conditions, it could rain, snow and be sunny during the compass of a single day.

Extreme daily temperature ranges are uncommon. The difference between summer and winter conditions may be wide and, of course, greater daily temperature ranges can be expected where continental weather prevails, away from the levelling influence of the oceans.

Fortunately, there are good websites listing current and historical weather conditions, although information is limited to reporting stations, e.g. towns, airfields, military or other significant installations. Some country weather bureaux are exceptionally helpful.

HAZARDS OF POLAR TRAVEL

Blowing snow and white-out

Travelling on snowfields and sea ice can be frustrated by blowing snow. It may be only ankle or waist high but surface detail is lost and crevasses, sastrugi or leads become impossible to detect.

White-out conditions form as light reflects back and forth between overcast cloud cover and the snow surface, creating a diffuse and shadow-less effect. This is very hazardous and travel should be avoided unless in an emergency. In a white-out, the horizon is lost and there is no contrast or shadow to define the surface. In certain types of white-out it is impossible to judge whether you are on a flat or sloping surface or how fast you are moving on ski or sleds. It is very easy to become disoriented and lost.

Tents and equipment at camps should be linked by flags or lines to save losing people in blizzard or white-out conditions. If you must travel, the chances of finding camp again can be vastly improved by laying out flag lines around it.

Wind chill

The combination of wind and low temperatures increases the loss of body heat alarmingly. The effect is known as wind chill (Figure 19.3). Put simply, the stronger

Wind speed		Ambient temperature (°C)							
mph	kph	–40	–30	–20	–10	–5	0	+5	+10
		Equivalent temperatures (°C) and danger of hypothermia for a fully clothed person							
		GREAT (exposed flesh may freeze)			INCREASING			SMALL	
46	74	–87	–71	–54	–38	–29	–21	–13	–4
35	56	–84	–68	–52	–36	–28	–20	–12	–3
23	38	–77	–62	–49	–31	–24	–16	–9	–1
12	20	–62	–49	–36	–23	–16	–10	–3	+2
6	10	–48	–37	–26	–15	–9	–3	+1	+7
0	0	–40	–30	–20	–10	–5	0	+5	+10

Figure 19.3 *Wind chill index*

the wind, the colder the "apparent" temperature on exposed flesh: a 10 mph wind at -10°C "actual" temperature has the effect on your skin of about −22°C "apparent" temperature.

A drop in ambient temperature, a rise in wind speed with no change in actual temperature or any freezing temperature in a wind above 15 mph can greatly increase the danger of cold injury (frostbite) and hypothermia.

Wildlife

Many polar species are protected and some are top carnivores. Only the foolhardy travel without firearms in polar bear country. If unsure of what you are doing or inexperienced with firearms, seek a local guide.

Generally, Arctic fauna are wary of humans as a result of being hunted. In Antarctica, animals show much less fear but any animal will attack if startled. Seal bites are particularly nasty because of their bacterial load. Check that you have suitable medication if you are likely to come across them.

Arctic insects (blissfully absent in Antarctica) can, and do, drive humans and other wildlife to distraction. Fortunately, there are few transferable diseases but bites will become infected if not well managed.

MEDICAL

Conventional medical support may be some considerable distance away and good medical back-up must be included in your planning. Fortunately, there are few diseases about and your main concerns will be broken limbs, infected cuts, scratches or bites (insect or mammalian), and cold injury.

Potentially serious are dehydration, exhaustion, cold injury and hypothermia. Note the order – it all starts to go wrong when you get thirsty – and don't forget the importance of adequate nutrition. Good preparation and equipment will reduce the chances of dehydration and exhaustion and lessen the likelihood of cold injury or hypothermia. Nevertheless, suitable provision must be made in case of accident. The key points to note are:

- Dehydration, inadequate nutrition, exhaustion and hypothermia (and altitude) are all interlinked.
- Minor deficiencies and problems quickly escalate, and have a knock-on and vicious-circle effect.

Ozone and sunburn

Damage to the protective ozone layer is particularly marked over the polar regions where the level of ozone is greatly reduced during the transition from winter to summer. Exceptionally clear air, high levels of solar radiation and the reflective effect of snow and cloud combine to "burn" unprotected skin in short order. The likelihood of skin damage leading to cancer is very high. Never go out without first applying high-factor sun cream, even on cloudy days. Especially vulnerable are earlobes and the bottom of the nose, which burn by UV reflection off snow- or ice-covered surfaces.

Snow blindness

The cloud-penetrating ultraviolet (UV) radiation that causes sunburn can also damage your eyes. Snow blindness can be excruciatingly painful and may disable a victim for several days. Even a light touch will have you running for a darkened tent feeling like sand has been thrown into your eyes.

Wear sunglasses certified to guard against UV light and with side pieces – simple Polaroids do not give sufficient protection. Suitably rated goggles are just as good and may be the only option for people who normally wear glasses. Check that you have relevant medications to deal with snow blindness, take spare sunglasses and learn how to improvise some.

Frostbite

Nobody on a well-run expedition should suffer frostbite (deep freezing of tissue). Mild frostbite or superficial freezing, sometimes called frostnip, is more common, especially on the face, hands and feet. Damage to hands and feet can be the most insidious simply because they are usually covered and deep freezing may occur without being noticed.

A blanching or waxy appearance to the skin is a sign of superficial freezing and, if caught quickly, can be reversed with little likelihood of permanent damage. As the face, hands or feet become colder they may start to tingle or sting, followed by numbness. Any loss of sensation is a danger signal not to be ignored.

Check yourself and each other frequently and stop to warm affected areas immediately. Cheeks, earlobes or nose can be thawed by covering with a warm hand. Superficially frozen hands and feet can be warmed under warm clothes. Anything more serious will require shelter or, possibly, evacuation.

Water, fuels and wind greatly magnify the dangers of frostbite because evaporating liquids strip heat from your skin. Bare skin will freeze to very cold metal. Don't learn the hard way by using your mouth as an extra hand and freezing a karabiner to your lips – it's very painful.

Hypothermia

Fumbling with gear and shivering are surprisingly common symptoms that are often ignored. If you become apathetic or uncooperative and your judgement and decision-making abilities deteriorate, you are likely to be hypothermic.

Hypothermia occurs when you lose heat faster than you can generate it.

Hypothermia is defined as a core temperature less than 35°C (normal is 37°C). There are two types: acute, which is sudden, e.g. immersion in icy water, and chronic caused by gradual cooling, e.g. stranded on an icecap. Medical professionals divide either type into several levels based on core temperature.

Measuring core temperature during an expedition is difficult, so a useful field classification is mild – fully conscious and shivering (core temperature 35–32°C) – or severe – not shivering or with decreased consciousness (core temperature < 32°C). Management depends on which level you have reached.

All polar travellers (and mountaineers for that matter) should recognise the symptoms of hypothermia and know how to manage it. The principles are:

- prevent further heat loss
- restore body temperature to normal.

In brief, the following gives the management of hypothermia:

- For *mild hypothermia*: remove from cold environment; protect from wind, wet, cold; add dry clothes, and especially keep head warm; insulate from ground; give hot drinks with sugar; apply external heat by getting into bag with victim.
- For *severe hypothermia*: victim may have no obvious pulse or breathing. Administer oxygen; do not remove wet clothes; add thermal clothing; get into shelter, preferably a casualty bag; gently evacuate for slow re-warming in hospital. Do not give cardiopulmonary resuscitation (CPR) unless an ECG is available and it can be continued until hospitalisation.

Severe hypothermia is extremely dangerous but treatable. Prevention is best: STOP when you feel thirsty or hungry and sort out the problem before it escalates.

"Space blanket"-type products have improved immeasurably beyond the wrap-around sheet. Take at least one Blizzard Pack (www.blizzard.co.uk). These have been shown to be much better than traditional space blankets for warming cold patients. At a push they can be used for a bivvy, matching a two- or three-season bag for warmth.

Carbon monoxide poisoning

It's cold and blowing a gale. You close the tent tightly to keep out the wind and snow and make a brew. Your eyes sting and you get a headache. You are probably suffering from carbon monoxide poisoning. Carbon monoxide is an odourless and colourless gas produced by incomplete combustion of fuels – by stoves, lamps or vehicles.

The treatment and solution are the same and simple: always make sure that you have adequate ventilation, however nasty the weather. Snow holes, in particular, can easily become airtight as the inside glazes over. An axe or ski-stick left poking through to the outside, and frequently rattled, will keep the airflow going.

Personal hygiene
Hygiene soon slips in field camps and cold is a great disincentive to washing. Dirty clothes, particularly socks, lose their insulating properties. Make some attempt each day to keep yourself clean and try to rinse clothing once a week or when you can.

Make sure that snow/ice/water gathering points are well away from washing/toilet areas, and make sure *everybody* knows the camp rules and layout.

Food and drink
Regular and frequent hot meals and plenty to drink are the first steps in keeping warm, not to say having a comfortable time.

It's obvious to say but it is surprisingly easy to be put off eating or drinking when travelling, especially when it is cold or at altitude. All those layers worn to keep warm are a right nuisance when it comes to having a pee or bowel movement. Provision must be made so that people don't avoid drinking just to save the hassle of urinating. The chances of suffering hypothermia are increased by dehydration and lack of food.

Dehydration
If you feel thirsty you are partially dehydrated. Our bodies' response to lack of fluid is a behavioural one and we can consciously refuse fluids all day, so becoming very dehydrated.

Various things contribute directly or indirectly to dehydration but there are problems that are peculiar to polar regions. Cold affects your blood circulation. Chilly hands or feet will drive blood into the core, resulting in the desire to urinate. Cold decreases the ability of the kidneys to concentrate the urine and so conserve water. Cold air is dry and water loss during respiration increases dramatically, as it does at altitude.

Dehydration has a number of insidious effects that should not be underestimated. A decrease in exercise tolerance will occur even before you feel thirsty and may have a dramatic effect on performance. Headaches, dizziness, fatigue, nausea and loss of concentration are all effects of dehydration. Dehydration also reduces the desire to eat.

How much should you drink to keep hydrated? There is no such thing as a "normal" fluid intake because it depends on the rate of water loss:

- An *absolute minimum* for a 70-kg man doing nothing would be about 3 litres/day.
- Plan for 5 litres/day when hanging around camp.
- ADD a litre for every hour of exercise.
- Hard exercise will bump up your requirement to 10–15 litres/day.

EQUIPMENT

The polar environment encompasses cold–wet to cold–dry, bare ice to deep soft snow, and it is often windy. It is best to use tried and tested equipment and beware of spending the whole expedition budget on the latest style of clothing, tent or sleeping gear. Study expedition reports for lists of equipment that others have taken and seek advice from reputable cold weather equipment suppliers.

Clothing

Keep clothing versatile so that varying weather conditions can be met by adding or omitting layers. Ideally, wear just sufficient clothing to keep warm but prevent over-heating. Sweat, generated while exercising, will rapidly chill you when you stop. Try to pace yourself or remove layers to avoid perspiring excessively.

First-time polar travellers tend to wear all the clothing that they have and fail to adjust to the prevailing conditions. Experienced travellers will wear the minimum, keeping items in reserve for when temperatures drop or wind strength increases. Thick, bulky garments should be avoided unless you expect a lot of standing around or travelling on open vehicles. The following are a few key pointers:

- Don't wear tight boots or gloves – this restricts circulation and your feet/hands chill more quickly.
- Don't sweat too much – adjust layers and carry enough to keep you warm when you stop.
- About 10 per cent of heat loss is through the head – carry spare head gear.
- Anchor gloves so that they do not blow away.
- Dry your gear at night – including headgear, gloves, socks, boot insoles.
- Always wear goggles or glasses when travelling over snow, especially if overcast.

Navigation and communication

The ability to navigate is crucial in the polar regions where conditions can deteriorate so quickly. Not all countries are as well mapped as the UK. Aerial photographs, satel-lite images, map and chart sources can all be found on the web. Seek local knowledge to help fill the gap.

Proximity to the magnetic poles makes compasses unreliable at high latitudes. Global positioning systems (GPSs) are a good alternative but they use an internal grid. Check that you are using the best grid for your locality and check its accuracy in the field.

For emergency communication, take an EPIRB (Emergency Position-Indicating Radio Beacon) and, for more general communications, a satellite phone. Many sat-phone systems do not work at high latitudes because the satellite (lying over the Equator) is "invisible" as a result of the Earth's curvature. Only polar-orbiting satellite communication systems will work, e.g. Iridium, at latitudes greater than about 70°.

But with all this techno-kit, don't forget the basics: a signal mirror is still the best way to show your position, especially to aircraft; standard and mini-flares are invaluable for searches (and for keeping inquisitive bears at bay).

Transport

Dog teams, all terrain vehicles (ATVs, usually six-wheelers; unstable four-wheelers – quads – should be left for sports drivers), snowmobiles and boats are all readily available in the Arctic and, with the addition of light aircraft, are invaluable to reach your start point, if not for the expedition itself. All require a degree of skill to drive and experience to deal with associated hazards.

Check local limitations on use of motor transport at your destination, usually imposed to protect vegetation. In some Arctic countries helmets must be worn on ATVs or snowmobiles. You may think you look daft but more people are injured in the Arctic falling off snow vehicles than in any other way. ALWAYS wear a helmet whether you are legally required to or not.

Skis and sleds

Skis, snowshoes and crampons assist travel over otherwise impossible surfaces and each has its techniques for use. Like all techniques, some people are good, some not so, and all need practice. Research the type of surface that you will travel over and seek advice from suppliers for the best footwear and accessories.

Sledging is an exceptionally valuable technique, allowing heavier and more bulky loads to be taken than can be carried. Sledges range from small high-impact plastic trays (kids' sleds), to 2-m Pulks, to wooden Nansen sledges. Your research will indicate the one best suited to your expedition.

Boating

Many polar expeditions take to boats as an excellent way of covering long distances up-river or along coasts. Ice and water temperatures around freezing make boating more hazardous than in temperate climates. A normally clothed body lasts only a few minutes before becoming paralysed. In many regions coast guard cover is patchy or absent.

Plan for worse conditions than the North Sea in winter, avoid ice fields unless your vessel is designed to work in them and use survival suits (see "Supplier details").

Energy and power sources

Availability of fuels varies considerably throughout the Arctic; there are no suppliers in Antarctica. White gas (or naphtha, "bencina blanca" in Latin America, a form of gasoline) is widely used in the North American Arctic; kerosene can be harder to

Figure 19.4 *Inflatable boats are an excellent way to travel along the coast, but try to get some training in boat handling techniques before you go (© David Rootes)*

find; wood is still the fuel of choice on the edge of the Russian forests.

Fuel consumption increases markedly over use of melting snow (uses more fuel) or ice (uses less fuel) for water. As a guide, you can get by with an MSR-type stove and about 0.33 litre of white gas per person per day. Allow more for base camps and lanterns.

Solar panels work well in polar areas and can be taken for charging batteries for all the electronic kit such as radios, phones or CD player. Make sure that it has a good protection circuit to prevent overcharging.

FURTHER INFORMATION

Suggested reading/research

Alexander, B. and Alexander, C. (1996) *Evocation of Arctic Native Peoples*. London: Cassell.

BBC Enterprises. *Kingdom of the Ice Bear*. Arctic wildlife video.

BBC Enterprises. *Life in the Freezer*. Antarctic wildlife video.

Beletsky, L. and Paulson, D. (2001): *Alaska, The Ecotravellers' Wildlife Guide*. London: Academic Press. Field guide to the most commonly encountered Alaskan wildlife. Includes short overview of conservation and parks.

Cherry-Garrard, A. (1997) *The Worst Journey in the World*. Reprint. New York: Carroll & Graf Publishers. Scott's doomed last expedition; possibly the best adventure tale ever written.

Churchill. *Birds of the Canadian Arctic.* RSPB. Ornithological video.
Davies, B. (2001) *SAS Mountain and Arctic Survival.* London: Virgin Books.
Duncan, R. (ed.) (2003) *Polar Expeditions.* London: RGS–IBG Expedition Advisory Centre. www.rgs.org/eacpubs
Johnson, C. (2002) Polar expeditions. In: Warrell, D. and Anderson, S. (eds), *Expedition Medicine*, 2nd edn. London: Profile Books. Available from www.rgs.org/eacpubs
Leffman, D. *Rough Guide to Iceland.* London: Penguin. Rough Guides.
Lindenmeyer, C. (2003) *Trekking in the Patagonian Andes.* London: Lonely Planet.
    Descriptions of walks throughout Patagonia.
Lonely Planet guide. *Iceland and Greenland.* London: Lonely Planet Guides. Travel video.
Lopez, B. (1986): *Arctic Dreams, Imagination and Desire in a Northern Landscape.* London: Macmillan.
    Celebrated meditation on Arctic travels throughout the North.
Poles Apart. *Polar Updates: Antarctica 2004.* Cambridge: Poles Apart.
    Details legislation and environmental requirements to visit Antarctica.
Rubin, J. (1996) *Lonely Planet Guide to Antarctica.* London: Lonely Planet Guides.
    Comprehensive guide to Antarctic wildlife and history.
Sage, B. (1986) *The Arctic and its Wildlife.* Oxford: Facts on File.
    Excellent review of Arctic wildlife.
Soper, T. and Powell, D. (2001) *The Arctic, A Guide to Coastal Wildlife.* Chalfont St Peter, Bucks: Bradt Travel Guides.
    Illustrated guide to coastal marine mammals and seabirds of the circumpolar north.
Soper, T. & Scott, D. (2000) *Antarctica, A Guide to the Wildlife.* Chalfont St Peter, Bucks: Bradt Travel Guides.
    Field guide to Antarctic wildlife.
Soublière, M. (ed.) (2003) *Nunavut Handbook.* Iqaluit: Nortext Multimedia Ltd.
    Comprehensive guide to Nunavut (Inuit part of North West Territories), Canada.
Steadman, R.G. (1971) Indices of wind chill of clothed persons. *Journal of Applied Meteorology* **10**: 674–83.
    Detail about wind chill.
Swaney, D. (2001) *Lonely Planet Guide to Iceland, Greenland and the Faroe Islands.* London: Lonely Planet Guides.
Swaney, D. (1999) *Lonely Planet Guide to the Arctic.* London: Lonely Planet Guides.
    Comprehensive guide to the circumpolar north.
Umbreit, A. (1997) *Guide to Spitsbergen.* Chalfont St Peter, Bucks: Bradt Travel Guides.
    History, wildlife and natural history of Spitsbergen.

Supplier details

Acton International Inc., 881 Laundry Street, Acton Vale, Quebec, Canada, J0H 1A0. Tel: +1 450 546 3735.
    Cold weather boots.
Baffin. Website: www.baffin.com
    Cold weather boots.
Blizzard. Website: www.blizzardpack.com
    Blizzard pack website.
Cotswold. Website: www.cotswoldoutdoor.com
    Contract outdoor clothing supplier.
Expedition Kit Ltd. Website: www.expeditionkit.co.uk
    Expedition equipment and clothing.
Montane, Unit 7 Jubilee Industrial Estate, Ashington, Northumberland NE63 8UA
    Clothing supplier.

Multifabs Survival Ltd: Multifabs Survival Ltd, Kirkhill Place, Kirkhill Industrial Estate, Dyce, Aberdeen AB21 0GU
  Survival suits for hire or purchase.
Open Air, 11 Green Street, Cambridge CB2 3JU
  Knowledgeable polar equipment suppliers.
Snowsled Ltd. Website: www.snowsled.com
  Designs and manufactures sleds, pulks and tents.

Useful websites

British Antarctic Survey. Website: www.bas.ac.uk
  UK Antarctic research centre.
Central Office of Intelligence, USA. Website: www.cia.gov/cia/publications/factbook/
  World fact book – good starting point for research.
Cold Regions Research and Engineering Laboratory. Website:
  www.crrel.usace.army.mil/library/crrel_library.html
  Vast collection of scientific and technical literature on cold regions.
Conservation of Arctic Flora and Fauna. Website: www.caff.is
  Arctic conservation site.
Danish Polar Centre. Website: www.dpc.dk
  Administers expeditions to Greenland.
European Centre for Medium-Range Weather Forecasts. Website: www.ecmwf.int
  Links to national weather centres.
GRID-AMAP. Website: www.grida.no/amap
  Arctic monitoring and assessment.
NASA. Website: http://science.nasa.gov/
  Provides ozone details.
National Ice Centre, USA. Website: www.natice.noaa.gov
  Sea ice and iceberg charts and data. Current and historical data.
Norsk Polar Institute. Website: http://npolar.no
  Administers expeditions to Svalbard.
Polar Photographers. Website: www.arcticphoto.co.uk/
  Great site for polar images.
Polar Web. Website: www.urova.fi/home/arktinen/polarweb/polarweb.htm
  Guide to Arctic and Antarctic internet resources.
Scott Polar Research Institute (SPRI). Website: www.spri.cam.ac.uk
  World's most comprehensive polar library and archives.
UK Meteorological Office. Website: www.metoffice.gov.uk
  Useful surface weather charts and links to other bureaux.
UNEP World Conservation Monitoring Centre. Website: www.unep-wcmc.org
  Conservation of species and protected areas.
World weather from reporting stations. Website: www.wunderground.com
  Gives current and historical weather; latter rather tedious to download.
World Weather Links: www.landings.com/evird.acgi$pass*55526469!_h-www.landings.com/
  _landings/pages/weather.html
  Links to detailed weather sites.

20 UNDERWATER EXPEDITIONS

Juliet Burnett

Marine and freshwater projects open up exciting opportunities for expeditions. With more than 70 per cent of the surface of the planet under water, the world is literally awash with opportunities for an interesting project. There are several additional risks associated with working near, on or under the water. However, careful planning, appropriate equipment and the correct skills and training, will keep your group safe, and you will gain a life experience that might not be achieved on a less ambitious terrestrial project. This chapter highlights some of the main points that you should consider, but should be read in conjunction with the other more detailed and specialist literature available.

Figure 20.1 *Diving in a marine reserve (© RGS–IBG/Paul Kay)*

HEALTH AND SAFETY

Working near, on or under water, particularly if you intend to use boats and/or add SCUBA diving to your itinerary, is potentially hazardous and must not be undertaken lightly. It is vital that your expedition team includes personnel with appropriate qualifications and skills. All members of your team must be competent swimmers, sit a first-aid course and know how to rescue and resuscitate a drowning victim.

Common ailments

Common ailments that you might face will include all of those of land-based expeditions, including cuts, burns and grazes. Bear in mind that such minor injuries are less likely to heal quickly in humid or wet environments, and additional care should be taken to avoid them. In particular your feet are valuable and vulnerable – it is all too easy to think that walking barefoot down a beach is pleasant, but a nasty shell or coral cut to a foot can prevent you from diving for several days, and is painful to try to squeeze into a diving bootee. Take a good supply of an iodine-based solution (such as Betadine) which can be painted on to wounds on a regular basis – the paints are better than creams because they dry out a wet wound.

You should be particularly aware of sunburn and dehydration, especially when working from boats when a breeze might mask the heat of the sun. Extra care should be taken to drink enough water, particularly when diving. If you make contact with the coral on a dive, the resulting coral scrapes can take a long time to heal, but, again, prevention is the key – if your buoyancy is good you should not make contact with the coral anyway. Ear infections can be a real nuisance on diving projects, and ears should be dried carefully each time you get out of the water. It is also a good idea to speak to your GP about an appropriate ear-wash solution of alcohol and glycerol which some people like to use before and after dives, and a good supply of antifungal ear drops and topical creams. Good personal hygiene is the simplest way to stay healthy – and keeping your skin rinsed in fresh water where possible will help prevent problems.

There are some dangerous marine creatures out there, but you'll be lucky even to see one and are unlikely to be bothered by them. Sessile creatures, such as hydroids and corals, do sting if you make contact with them, as do jellyfish, and it is worth having small bottles of vinegar with you, especially on boats, to treat stings. Gloves should not really be worn because this encourages divers to touch things under water – it is far better not to touch the bottom at all. The best precaution against the stings of benthic fish is to wear shoes in the shallows. Do find out what creatures are in your area, and make sure that all members of the team are aware how to recognise them and treat any resulting injury.

Use of boats

If you are working in a remote area and have to take your own boats with you, ensure that a suitable number of your expedition team are very experienced in using and handling them. They should hold a current qualification (e.g. Royal Yachting Association). Bear in mind that if you have only one or two members who are qualified, your work and indeed safety could be in jeopardy if they are both unwell and unable to drive the boats. Shipping inflatables is costly and you will need to take all spares with you. It might be more appropriate to hire local craft on arrival if you can confirm their availability in advance.

An alternative to driving the boats yourselves is to try to hire boats with local boatmen. This carries the massive benefits of local knowledge of the waters, weather patterns, tides and other factors, as well as supporting the local economy and potentially integrating yourself more into the local community. The potential downsides are that you never quite know what craft you will get until it turns up, and how safe these vessels will be, so ensure that you bring all additional safety equipment, including life jackets, with you. You should also be aware that local boatmen may never have worked with divers before, and may not know how to approach them in the water safely. Spend time in the first few days running through drills, such as person-overboard practices and radio communications.

You should always take a well-stocked "boat box" with you on every trip – use a waterproof case and check and replenish it after every trip. This should include fresh water and snacks, radio, your standard first-aid kit plus vinegar, sun creams, flares, boat and diving spares, diver-recall system (or agree an alternative signal such as knocking on the bottom of the boat) and any medication required by any member of your group.

Outboard engines can also be a problem – if shipping your own, ensure that you know how to maintain the engine and that you've got all spares with you, including propellers. If using local engines, try to find out what make they are before you depart and whether spares are available locally – if not, again take them with you, because they will always make good thank-you presents at your departure if you have been fortunate enough not to need them. Fuel and oil can be a issue in remote areas – do check out the supply before you leave and, when in-country, allocate one responsible member of your team to look after supply, storage and mixing of fuel.

Ensure that you know the correct anchoring and mooring systems for small boats, and that all members of the team know how to secure boats. Local anchors are often not the most environmentally friendly, so do bear this in mind in reef environments, and re-lay the anchor at the beginning of dives if it is resting on a fragile environment.

Finally, if all this talk of shipping or hiring gear is too much, you could consider using a hired-boat or charter vessel, where accommodation, storage, compressors, security, etc. may all be available. This may be a good option, but it is important that you get a written agreement that the skipper and crew understand the aims of your expedition, and your requirements for their boat. They might be unexcited by the

Figure 20.2 *Ensure that all members of your team have the appropriate qualifications to dive (© RGS–IBG/John Nortcliffe)*

locations that you choose for research purposes, and put pressure on you to change your plans. Do also check whether there will be a smaller boat on board to take you into shallower locations, and pick up divers or kit if necessary.

SCUBA diving

This has its own associated risks, in addition to just working near the water. The most important factor when planning an expedition that includes diving, especially research diving, is that all members of your group are appropriately qualified and at least some of the team have had experience of diving in similar locations.

Not all divers are appropriately qualified to dive with each other, and the lowest level of qualification will not be suitable for an expedition environment. It is important that every member of the expedition gains a suitable level of qualification to ensure that they can dive with any other member of the team – when planning dive rotas the last thing you want to be trying to factor in is who is qualified to dive with whom, particularly if members of the team are unwell or unable to dive. There are a number of certifying agencies worldwide, and it can be confusing to work out the equivalent grades between these systems. The most common agencies that you will come across are the British Sub Aqua Club (BS-AC) and the Professional Association of Diving Instructors (PADI). Either of these organisations can give you advice on equivalent diving standards, as can the World Underwater Federation (CMAS).

Before you depart, and when in the field, it is a good idea to appoint one member of your team as a diving officer (DO) who takes responsibility for running all diving activities. He or she should be a well-qualified diver (ideally PADI Dive Master/BS-AC Advanced Diver) and should have had experience in remote diving and over-seeing group diving activities. The DO should also take responsibility for recording all dive information, and a logbook including names of divers, dive details (time, depth, etc.) and any other additional information such as weather or diver illness. Such a log can be an invaluable history of a diver's activities should an incident occur that needs treatment or reporting to the authorities.

It is important to find out details of the local recompression facilities/chamber before you embark on a diving project, and decide whether the location is safe. All divers should be well aware of the very real dangers of nitrogen narcosis and decom-pression sickness, but even if you implement all safety measures it is still possible that you will need the use of a chamber. Ensure that you have a written and agreed plan for a casualty evacuation to a chamber if necessary, and that this has been agreed with the relevant local authorities (such as the chamber operators, coastguard if present, etc.) and your insurance company.

Do remember that, when you have additional tasks to complete underwater, such as collection of survey data, ordinary sport diving safety rules, such as monitoring your buddy, can get forgotten and, as such, research diving carries additional risks. All steps to prevent a diving incident should be taken, and specialist advice sought from experienced divers on how to dive more conservatively to reduce your risks. The following are key suggestions for additional safety precautions:

- Ideally no decompression diving and all dives limited to 30 m maximum. If dives outside these limits are required, they should be planned on an individual basis with additional risk assessments and safety precautions.
- All dives should be within tables if possible – computers should be used as a dive record but not used to calculate bottom time
- A minimum of one day in every seven should remain dive free – ideally one day in every four.
- Minimum of BS-AC Dive Leader/PADI Dive Master for all expedition divers – if the team has to be made up of less experienced divers then they should be limited to shallower, shorter dives.

If you are diving, it is important that you have a supply of oxygen for treatment of casualties, and at least two or more of your team are trained in its administration. Taking full oxygen cylinders with you on a flight can cause difficulties, and will need to be set up in advance. You need to check whether oxygen cylinders and fills are available from a reputable source before you depart – if so it might still be a good idea to take your own regulator and administration kit so that you can be confident that these are functioning correctly.

237

Disability awareness

If one of your potential team has a physical disability, don't assume that you or they cannot join the team, because all members of teams on integrated expeditions tend to gain from the range of skills in the group. Many disabilities do not prevent diving safely and well and, indeed, many disabled people come into their own underwater. If you are concerned about any safety or logistical considerations, discuss this with the team member him- or herself and with his or her GP, or seek expert advice from BS-AC, PADI or one of the disabled diving organisations listed at the end of the chapter.

EQUIPMENT AND CLOTHING

It is important that you plan your equipment and clothing carefully to match your planned environment, particularly if you are unlikely to be able to purchase spares during your expedition. A reconnaissance visit to your planned location, or at least to talk to people with first-hand knowledge of the area, will help ensure that you take equipment that is suitable for the temperature and other conditions that you will encounter. Sun, sand and saltwater all take their toll on kit, as well as on you, so don't assume that the gear that has served you well in the UK is the best for your expedition.

Figure 20.3 *It is vital to ensure that all your diving equipment is properly maintained*
(© RGS–IBG/Tom Hooper)

Diving kit

Your diving kit is your lifeline underwater, and it is vital that you purchase the best you can afford, that it is properly serviced and that you look after it in the field. Don't think that your kit is the best place to cut corners on your budget – remember that it is your life that you are dealing with. However, do bear in mind that high quality doesn't mean with all the bells and whistles – simple, well-constructed kit that is strong and easy to service is what you are after. Ensure that at least one member, preferably more, of your team is competent to service the kit that you have, and that you have tools and ample spares to do this. Consider sending several members of your team on training courses, specific to the brands of kit that you have. This goes for everything, from regulators, tanks, torches, compressors and boat engines right down to your basic kit. Where possible, try to have all members of the team with the same brand of kit, particularly your regulators, and that they are not too complex to be serviced easily. Take more "O" rings than you think you need, and carry them with you, along with spare mouthpieces, mask and fin straps, and lots of silicon jelly.

If you are intending to hire cylinders and weights in country, do check that the source is reputable, and that you have the booking in writing – you don't want to arrive and find someone else is using the tanks that you need. Do check the inspection test date and fill capacity on the neck of all the bottles. Again, this is better done on a recce visit to check that the tanks are well serviced.

If you are camping, take your kit out in strong bags or even pelican cases, and take padlocks to keep them secure. Waterproof boxes and/or waterproof kit bags (the best are the ones that roll over and seal at the top) are great for taking your gear on small boats, and may even float for a while in the unfortunate event of a capsize or loss overboard. Similarly, take a couple of cork key-savers to use on the boat, and waterproof bags for GPS or camera gear. If you have space, also take some *large* plastic bins that can be filled with fresh water to wash your kit, some big rolls of paper towels (like those used in laboratories) and a good store of cotton buds for cleaning out your gear.

A good tool kit is a vital part of your equipment, and should include metric and imperial spanners, Allen keys and sockets, plus any tools specific to your diving gear. Fill up any other space with lots and lots of plastic cable ties (various sizes), duct tape, self-amalgamating tape, Super-Glue, strip-seal and Ziplok plastic bags and some condoms (yes, seriously), because they're really useful for waterproofing kit or cuts on fingers. Do check out the local hardware stores and outboard dealers when you arrive in your host country, so that you know what you might be able to get if the situation arises.

In terms of diving suits, take what is appropriate for the water temperature – investing in the right dry/wet or Lycra suit for the conditions will make your trip a much more pleasant experience.

Once you have invested so much of your budget in your kit, you have an added incentive to really look after it. Establish an equipment officer, and/or have rotas for use and servicing of compressors, etc. Security will also be an issue, so make sure that things are as secure as possible in your base camp. Remember, your Mr or Mrs "Fix-it" may be the most valuable member of any team, but you cannot rely on them always being there, or indeed being well enough to fix something. It is better for you all to take responsibility and learn what you need to know before you go.

LOGISTICS

In addition to all the other logistical considerations of any terrestrial expedition that are included in earlier chapters of this manual, there are certain additional consider-ations for water-based expeditions.

Flights

Do check that your airline allows you to transport your diving kit (if necessary) and oxygen sets, and it is worth checking whether the airline gives an extra baggage allowance to divers.

Base camp

Your primary concern in planning your expedition base is the safety of you and your team members and your equipment and, with diving equipment, good storage and maintenance are vital to ensure this personal safety. Remember that sand, salt and kit do not mix well, and so you must try to create an environment that keeps these sepa-rate. If possible, use an existing field station (check out the RGS–IBG World Register of Field Centres on www.rgs.org/fieldcentres for a list that caters for marine research) or a building because these are much easier to keep clean. If you are intending to use tents, take at least one large tent with you, in addition to your mess tent, and have it dedicated solely to the storage of your equipment. Use plastic tubs outside the doors to rinse sand off feet, and ensure that shoes are left at the door. Security of your kit will be a problem in tents and, if you are really concerned, never leave your camp unattended.

Ensure that your cooking facilities and compressor are well separated, and that the compressor is up-wind of the rest of the camp and has a good supply of fresh air.

Your fresh water supply is vital, and a main concern in planning an expedition location. Do not underestimate the amount of water that you will need to drink, or wash both yourselves and your kit to ensure that things stay functioning well. If you need to transport the water, remember how heavy it is and the logistical considera-tions of this.

A secure anchorage for your boats is vital – seek local advice on this if possible.

Communications

These are vital – within parties in the group and to the outside world to set in place your agreed emergency evacuation procedure. Do ensure that all your planned systems – radios, mobile phone, satellite phone, etc. – actually work before you rely on them. If using VHF radios, staff must be properly trained, professional and keep "chat" to a minimum. You will be very unpopular, and look amateurish, if you block up emergency channels with trivia. Also think about power supplies for batteries, and remember that batteries last a very short time in tropical, humid environments. It is good practice to have agreed report-back times if a team are off in a vehicle or boat. Make sure that such parties sign out before they depart, stating who has gone, where they have gone and when they expect to return. This is important so that you know when to start a search and rescue party if necessary.

Insurance

Finally, do ensure that your insurance company knows precisely what you are doing, and that your policy covers you and any third party for water-related activities. Many policies exclude not just diving but also use of small craft (remember, if your small boat causes a large yacht to run aground, you could be looking at a very large suit against you), and evacuation of a diving-related incident at low altitude, requiring a specific flight. You might consider taking out a separate policy specific to your diving and boating needs.

Permissions

Many countries require you to have permits to dive, drive boats or conduct research in their waters, so do check this out before you go to prevent a major upset on arrival in the country.

RESEARCH AIMS

If you are interested in running an underwater project, you have many options for a research aim – there is still so much to be discovered about our underwater world. With even fairly modest research aims, you can conduct a study that is genuinely useful to yourselves and your host nation, and this must be your starting objective.

Start by contacting the relevant parties in your host country – ask them whether you can run an expedition, what work they would like to undertake and what expertise they can provide. You may well find that there is a project on a plate just waiting for the people to run it. Most countries have research organisations and/or government departments responsible for their wetlands or coastal zone, and they will be able to put you in touch with the right people. If a web search doesn't raise an appropriate institution, try the British Council office in country. Most countries also have

Figure 20.4 *Collecting algae on a rocky shore in Rodriques, home to one of many marine research stations around the world that welcome visiting scientists (© RGS–IBG/Jeremy Neech)*

non-governmental organisations (NGOs) that often conduct environmental work. Finally, check out what other research work has been conducted in the area, perhaps by groups such as your own – via the RGS–IBG Expedition Advisory Centre. You may find that you are able to repeat field research conducted perhaps a few years before, and compare data to see how things have changed.

Your next task is to research your host country thoroughly, including any relevant background to your research area and especially the current environmental issues facing the area that you want to study. It is important that you don't just appear to be a tourist looking for a jolly – countries rich in natural resources and habitats are plagued by requests from individuals, well meaning or otherwise, and you need to demonstrate that you know what you are talking about.

Once you have identified a research topic, you need to work out the best way to conduct the research. Given the logistical, safety and cost implications of diving that have already been raised, even if you came into this wanting to run a piece of diving research, it is important to ask yourself whether this is the most efficient way of collecting data. Remember that a lot of your research will take place in very shallow water, which is often easier and safer to work in through snorkelling than diving. Don't feel you've failed if you choose to snorkel – remember you can always add a week or more recreational diving at the end, perhaps through a registered diving

Figure 20.5
Snorkelling is a simple way to investigate the subtidal environment
(© RGS–IBG/Jeremy Neech)

school, once your work is complete, without all the hassle and responsibility of organising research diving.

The best rule of thumb for any research project is to keep your research objective and techniques as clear and as simple as possible. You want to be able to collect enough data to validate your statistical studies, and so that they can be repeated. Do conduct at least some data analysis on site, so that you can pick up inadequate or suspect data-sets. Try to avoid taking samples for analysis in UK – if this is a must then ensure that you know how to store and transport them, and that you have written permission to take samples out of the country and back into your home country. Make sure that you photograph species you want to identify when they are still fresh. Simple water quality analysis can be conducted on site, and training in the use of reagents or electronic sampling equipment can be carried out before you leave. Make sure that you have permission for transporting any chemicals or expensive-looking kit in and out of the country.

There are several bodies who can give you further advice; some of these are listed at the end of the chapter. Also use your own research institutions or universities – those with active marine science departments include Heriot-Watt, Liverpool, Newcastle, St Andrews, Stirling, Southampton, Portsmouth, Plymouth, University of Wales (Bangor and Swansea) and York.

Finally, take seriously your responsibility to thank those who helped you and to report back to your host nations and sponsors on your return. Who knows when you, or others, will need their help again!

FURTHER INFORMATION

Further reading

Anon (1990) *Cave Diving: The cave diving group manual.* Castle Cary, Somerset: Mendip Press.
British Sub-Aqua Club (1988) *BS-AC Sport Divers Manual.* London: Stanley Paul.
English, S., Wilkinson, C. and Baker, V. (eds) (1997). *Survey Manual for Tropical Marine Resources.* Townsville, Australia: Australian Institute of Marine Sciences. www.aims.gov.au.
Farley, M. and Royer, C. (1980) *SCUBA Equipment: Care and Maintenance.* Port Hueneme, CA: Marcor Publishing.
Flemming, N.C. and Max, M. (eds) (1996) *Scientific Diving: A general code of practice.* Paris: UNESCO.
Palmer, R. (1990) *Underwater Expeditions.* London: RGS–IBG Expedition Advisory Centre.
Pitkin, A. (2002) Underwater expeditions. In: Warrell, D. and Anderson, S. (eds), *Expedition Medicine* 2nd edn. London: Profile Books. Available from www.rgs.org/eacpubs
Rose, P. (2002) *Diving Regulations for the RGS-IBG Shoals of Capricorn Programme.* London: RGS–IBG Expedition Advisory Centre.
Rowlands, P. (1983) *The Underwater Photographer's Manual.* London: Macdonald & Co.
Sisman, D. (1982) *Professional Divers Handbook.* London: Submex.
SPC/UNEP (1984) *Coral Reef Monitoring Handbook. Reference methods for marine pollution studies 25,* UNEP.
Veron, J.E.N. (2000) *Corals of the World,* 3 vols. Townsville, Australia: Australian Institute of Marine Science.
Wells, S. (ed.) (1988) *Coral Reefs of the World,* 3 vols. Cambridge: UNEP/IUCN.
White, P. (1996) *Outboard Troubleshooter.* Arundel, W. Sussex: Fernhurst Books.

Useful addresses and websites

Note that most of these sites have links to others, and some good web searching will help you find out more specific information for your expedition.

Australian Institute of Marine Sciences. www.aims.gov.au
British Society of Underwater Photographers, 12 Coningsby Road, South Croydon, Surrey CR2 6QP. Tel/fax: +44 20 8688 8168, email: b.pitkin@nhm.ac.uk, website: www.bsoup.org
British Sub-Aqua Club, Telford's Quay, Ellesmere Port CH65 4FL. Tel: +44 151 350 6200, fax: +44 151 350 6215, email : postmaster@bsac.com, website: www.bsac.com
Confederation Mondial des Activites Sub-Aquatique (CMAS), The World Underwater Federation HQ (Rome). Tel: +396 3751 7478, website: www.cmas2000.org
Coral Cay Conservation Ltd, The Tower, 125 High Street, Colliers Wood, London SW19 2JG. Tel: +44 870 750 0688, email: info@coralcay.org, website: www.coralcay.org
Coral Reef Alliance, 2014 Shattuck Avenue, Berkeley, CA 94704 1117, USA. Email: info@coral.org, website: www.coral.org
Cousteau Society Inc, 870 Greenbrier Circle, Suite 402, Chesapeake, Virginia 23320, USA. Email: cousteau@cousteasociety.org, website: www.cousteau.org/en
Dive Rescue International, 201 North Link Lane, Fort Collins, CO 80524-2712, USA. Tel: +1 970 482 0887, fax: +1 970 482 0893, email: Training@DiveRescueIntl.com, website: www.diverescueintl.com/
Divers Alert Network (DAN Europe) PO Box DAN, 64026 Roseto (Te), Italy. Tel: +39 085 893 0333, fax:

+39 085 893 0050, email: mail@daneurope.org, website: www.diversalertnetwork.org

Diving Diseases Research Centre, The Hyperbaric Medical Centre, Tamar Science Park, Research Way, Plymouth, Devon PL6 8BU. Tel: +44 1752 209999, fax: +44 1752 209115, email: enquiries@ddrc.org, website www.ddrc.org

Global Coral Reef Monitoring Network, c/o Australian Institute of Marine, Science, PMB No 3, Townsville, MC 4810, Australia. Website: www.coral.aoml.noaa.gov/gcrmn

HSA (Handicapped Scuba Association) International, 1104 El Prado, San Clemente, CA 92672-4637. Tel: +1 949 498 4540, fax: +1 949 498 6128, website: www.hsascuba.com/

International Association for Handicapped Divers (IAHD). Website: www.iahdeurope.org/iahdmain.html

International Association of Nitrox and Technical Divers (IANTD United Kingdom, Ltd), 11 Telford Road, Ferdown Industrial Estates, Wimborne, Dorset BH21 7QP. Tel. +44 1202 632211, fax: + 44 1202 632319, email:iantduk@cs.com, website: http://www.iantd.com

International Coral Reef Action Network (ICRAN), c/o UNEP World Conservation Monitoring Centre, 219 Huntingdon Road, Cambridge CB3 0DL. Tel: +44 1223 277314, email: icran@icran.org, website: www.icran.org

Irish Underwater Council, 78a Patrick Street, Dun Laoghaire, Co Dublin, Eire. Website: www.scubaireland.com

Marine Conservation Society, 9 Gloucester Road, Ross-on-Wye, Herefordshire HR9 5BU. Tel: +44 1989 566017, fax: +44 1989 567815, email info@mcsuk.org, website www.mcsuk.org

National Maritime Museum, Romney Road, Greenwich, London SE10 9NF. Tel: +44 20 8858 4422, website: www.nmm.ac.uk

Nautical Archaeological Society, Fort Cumberland, Fort Cumberland Road, Eastney, Portsmouth, Hampshire PO4 9LD. Tel: +44 123 9281 8419, email: NAS@nasportsmouth.org.uk website: www.nasportsmouth.org.uk

NOAA Coral Health and Monitoring Program (CHAMP). Website: www.coral.noaa.gov/

PADI International Ltd, Unit 7, St Philips Central, Albert Road, Bristol BS2 0PD. Tel: +44 117 300 7234, fax: +44 117 971 0400, email: general@padi.co.uk, website: www.padi.com

Reef Check Foundation, 1362 Hershey Hall 149607, University of California at Los Angeles, Los Angeles, CA 90095-1496, USA. Tel: +1 310 794 4985, fax: +1 310 825 0758, email: Rcheck@UCLA.edu, website: www.reefcheck.org

Reef Conservation UK. Website: www.rcuk.org.uk

RYA – Royal Yachting Association. Tel: +44 845 345 0400, website: www.rya.org.uk

Scottish Sub-Aqua Club, Cockburn Centre, 40 Bogmoor Place, Glasgow, Lanarkshire G51 4TQ. Tel: +44 141 425 1021, fax: +44 141 425 1021, email: ab@hqssac.demon.co.uk, website: www.scotsac.com

Society for Underwater Technology, 80 Coleman Street, London EC2R 5BJ. Tel: +44 20 7 382 2601, fax: +44 20 7 382 2684, website www.sut.org.uk

Southampton Oceanography Centre, Waterfront Campus, University of Southampton, European Way, Southampton SO14 3ZH. Tel: +44 23 8059 6666, fax: +44 23 8059 6667 email: external-affairs@soc.soton.ac.uk, website: www.soc.soton.ac.uk

Sub-Aqua Association, 26 Breckfield Road North, Liverpool, Merseyside L5 4NH. Tel: +44 151 287 1001, fax: +44 151 287 1026, email: admin@saa.org.uk, website: www.saa.org.uk

UK Diving: the UK Divers Internet resource. Website: www.ukdiving.co.uk/ukdiving.htm

UK Sports Diving Medical Committee. Website: www.uksdmc.co.uk

UNEP World Conservation Monitoring, Centre, 219 Huntingdon Road, Cambridge CB3 0DL. Tel: +44 1223 277314/info tel: +44 1223 277722, email: infor@unep-wcmc.org, website: www.unep-wcmc.org

US National Association for Cave Diving. Email: gm@safecavediving.org, website: www.safecavediving.com

US National Association of Underwater Instructors (NAUI). Website: www.naui.org

21 CAVING EXPEDITIONS

Dick Willis

Of the various possible types of expedition, cave exploration is one that requires specialist skills that can be developed only through experience. This experience should be gained by caving in the UK before launching an expedition overseas. It is therefore assumed, in the preparation of these notes, that at least some of the team members will have appropriate caving experience.

Other individuals who are not practised cavers should take all reasonable steps to perfect their basic skills before caving overseas. In most cases, membership of a caving club will provide the opportunity to gain a range of experiences, access to shared equipment and, often, a platform for expedition planning. Most experienced cavers interested in overseas activities will be members of both clubs and the British Cave Research Association (BCRA), the major body in Britain providing information and services for the overseas caver.

OBJECTIVES

The best objective for most small caving expeditions would be a fairly remote limestone area not already explored by cavers, where a few caves or known entrances give a hint of discoveries yet to be made. Although such areas become fewer each year, there are still many opportunities within the scope of a small expedition. The increasing availability of cheap flights puts ever more remote countries within reach. Normally an expedition would be instigated in response to some research or the comments of an individual who has already visited the area. The BCRA Foreign Secretary is then an invaluable source of information about previous explorations.

A second type of objective may be the extension or further exploration of a known cave system. Many of the larger caves in France and Italy have not been visited enough to count as being fully explored, but it is often difficult to know which caves are suitable for an expedition, and previous knowledge and local contacts are thus essential.

Figure 21.1 *Stream Zero, part of the 2000 extensions to Cobweb Cave, Mulu*
(© Andy Eavis)

There is no substitute for detailed preliminary research.

Once the area is selected, the expedition should aim to carry out appropriate underground and surface exploration, together with some scientific observation if possible. Biological collecting, geological mapping and water tracing are possibilities depending on the location and the skills of team members. In many developing countries local government agencies may be willing to collaborate to carry out basic research towards water supply or tourism developments. Such local links are important for continued good relations and also contribute to making the expedition more satisfying.

247

As a bare minimum, all discoveries must be surveyed. Make sure that the team can survey properly before leaving the UK – practise in a known cave where it is possible to compare your survey with the published version. Remember that the cave survey is the basic tool for all research and further exploration, so it must be accurate. If possible, get advice from an experienced expedition cave surveyor – essential to augment book learning in such a practical subject. Cavers are making increasing use of computers to record, plot and disseminate surveys and BCRA has a cave surveying special interest group to carry out research and provide advice on this issue.

For further ideas, and to make sure that your plans are realistic, go to the annual BCRA National Caving Conference (usually held in September; details are given on the BCRA website) to hear the lectures and check the displays. This will also give you the opportunity to network with other explorers. You should also read the reports of expeditions that appear in the UK caving press and foreign journals.

PERMISSION

Be aware that in some countries specific permits are required to carry out a caving expedition. In most cases the permits are easily obtainable but without them you will be ejected by the police, or worse! If you intend to carry out scientific research, particularly biological collection, it may be necessary to obtain additional permits from the relevant host government agencies. Advice on such matters can be obtained from the BCRA Foreign Secretary.

Cavers are notoriously touchy about outsiders trespassing on "their patch" and you should make sure that no other group, British or overseas, is or has been working in the area that you intend to visit. If this is the case, it is essential that you liaise closely with them before setting out. Magazines, reports and the BCRA Foreign Secretary should be able to help clarify such information.

EQUIPMENT

Your equipment will need to fit your objectives. Factors to be considered will include your transport/freight arrangements, likely depth of exploration, the type and amount of potential rope use, the surface and underground temperatures in your area, whether or not the caves are wet, rescue contingencies, etc.

Unlike British caves, large European caves may contain long lakes or canals, which may require specialist clothing or dinghies. In the latter case, a cheap beach-type dinghy will probably suffice but don't forget a repair kit and a long thin line to pull it back for the next person. Underground camping is normally only necessary on extended deep explorations or in very long cave systems, rarely on a small first-time expedition. However, if it's a likelihood, make sure that all your team members have

Figure 21.2 *Camp in Cobweb Cave, Mulu (© Andy Eavis)*

practised in a wet UK cave beforehand – so that they learn how to live and sleep in the most uncomfortable conditions!

Variations in temperature and water conditions in caves will determine your personal equipment needs. Alpine-style caving will require good thermal and water-proof protection and, if the cave contains long swimming sections, wetsuits may be necessary. In warmer climates and closer to sea level, wetsuits may not be required unless the trips underground are very long.

Once again, preliminary research is essential and nothing can beat the advice that you will get from other cavers who have visited your area or similar areas. Never hesitate to ask – most cavers are only too happy to talk for hours about their experiences and the advice that they give may make the difference between success and failure.

It is possible to hire specialist items of equipment for expedition use and advice on such matters can be obtained from BCRA.

SAFETY

Cavers in the UK are cushioned both by the best cave rescue organisation in the world and by the relative proximity of our caves to easy means of transport. On an expedition you are much more on you own, so take care – a lot more care than you would normally. In some countries there will be no possibility of outside assistance and one broken leg in a small expedition team can require a monumental effort to evacuate the casualty from even quite a short cave: a relatively minor injury could easily prove fatal.

Your team members must be familiar with self-rescue techniques and systems to rescue another person who is hanging injured on a rope. This is not a matter for book learning and must be learned through hands-on experience. There are a number of organisations and individuals now offering expert training in such techniques.

It is essential that the expedition is adequately equipped with medical supplies and first-aid skills. The medical kit should contain not only treatments for the normal range of disabling infections, such as diarrhoea, but also strong painkillers for use in the event of a major injury. One word of caution though – always check on the customs requirements for the import and possession of such drugs; failure to do so could result in arrest or imprisonment.

In many western European countries there are efficient cave rescue services, but they can take a long time to reach a remote area. Their services are not free, as in the UK, and you may get a massive bill for a rescue, particularly in France where heli-copters are routinely used. Check to see whether any other expeditions are going to be working near you. If so, you may be able to make contingency plans for a rescue and share the costs of specialist items such as an underground stretcher.

Insurance against medical and rescue expenses is essential for any group leaving the UK. A number of brokers can negotiate appropriate insurance but the BCRA operates an expedition insurance scheme for cavers.

FUNDING AND SPONSORSHIP

Besides the normal sources of finance that are available to all expeditions, grants are available specifically for caving expeditions. The Ghar Parau Foundation gives grants each year, with closing dates of 31 January and 31 August. The Foundation is linked to the BCRA and details of its grants are published in *Speleology* and on the BCRA website, together with other relevant sources of grants.

Commercial sponsorship represents an invaluable source of support for expedi-tions. However, caving expeditions lack the glamour of many competing activities and sponsorship is increasingly hard to obtain. Be clear about what you need and target your approaches accordingly. Be clear, also, about what you can offer in return (reports, photos, gear evaluations, etc.); sponsorship should be of mutual benefit and it is unlikely that you will be given something for nothing. Remember, if you promise something to a sponsor, make sure that you deliver or give a very good reason for not doing so. If you fail to deliver on your promise it will make it less likely that that organisation will sponsor any other caving trips in the future.

FURTHER INFORMATION

Further reading
The specialist magazines, such as *Speleology*, *Cave Science* and *Descent* are the best source of expedition

reports and news. *Speleology* also reviews many significant independently produced expedition reports that should be available from the BCRA Library. The RGS–IBG Library, Map Room and Expedition Advisory Centre are also invaluable research facilities as are the unpublished notes of other expedition leaders – never hesitate to ask.

One significant publication is *Caving Expeditions*, published jointly by the RGS and the BCRA. It contains a series of detailed papers prepared by caving specialists on all aspects of expeditions from finance, through transport, science and photography to reports and medicine.

Pitkin, A. (2002) Caving expeditions. In: Warrell, D. and Anderson, S. (eds), *Expedition Medicine*, 2nd edn. London: Profile Books. Available from www.rgs.org/eacpubs

Willis, D. (1993) *Caving Expeditions*. Joint EAC/BCRA publication. Available from www.rgs.org/eacpubs

Useful addresses

British Cave Rescue Council. Tel: +44 1539 625412, email: PeteAllwright@compuserve.com, website: www.caverescue.org.uk

British Cave Research Association, c/o The Old Methodist Chapel, Great Hucklow, Buxton, Derbyshire SK17 8RG. Email: enquiries@bcra.org.uk
Current officers' addresses are at www.bcra.org.uk
BCRA Library: Local Studies Library, County Office, Matlock, Derbyshire DE4 3AG. Tel. +44 1629 580000, ext 6580
BCRA Librarian. Email: librarian@bcra.org.uk
BCRA Foreign Secretary. Email: foreign-secretary@bcra.org.uk
BCRA Travel Insurance. Email: travel-insurance@bcra.org.uk

Cave Diving Group. Current officers' addresses are at www.cavedivinggroup.org.uk/

Ghar Parau Foundation. Secretary. David Judson. Hurst Barn, Castlemorton, Malvern, Worcs WR13 6LS. Tel/fax: +44 1684 311057, website: www.bcra.org.uk

International Union of Speleology, UIS General Secretary, Dr Pavel Bosák, c/o Czech Speleological Society, Kalisnická 4–6, CZ-130 00 Praha 3, Czech Republic. Email: bosak@gli.cas.cz, website: http://clik.to/speleo/

National Caving Association. Current officers' addresses are at www.nca.org.uk

William Pengelly Cave Studies Trust. Website: www.pengelly.org

Speleobooks, PO Box 10, Schoharie NY 12157-0010, USA. Email: speleobooks@speleobooks.com, website: www.speleobooks.com

SpeleoLinks. Website: http://clik.to/speleo/

Spéléo-Secours Français (French cave rescue). Website: www.speleo-secours-francais.com/

Training organisations

Dave Edwards & Associates, Buxton. Tel/fax: +44 1298 85375, email: dave-edwards@lineone.net, website: www.dave-edwardsandassociates.net

Farrworld, Crickhowell. Tel: +44 1873 811085, email: martyn@farrworld.co.uk website: www.farrworld.co.uk

Ingleborough Hall, Clapham. Tel: +44 1524 251265, email: cave@ingleboro.co.uk, website: www.ingleboro.co.uk

NCA Training Officer. The current address is at http://www.nca.org.uk

Andy Sparrow, Wells. Tel: +44 1934 741427, email: andy@mendipnet.co.uk, website: www.ascaveservices.demon.co.uk

SRT Indoor Practice Area: Inglesport, Ingleton. Tel: +44 1524 241146, email: info@inglesport.co.uk, website: www.inglesport.com

SECTION 5

FIELD RESEARCH

22 CHOOSING AN EXPEDITION RESEARCH PROJECT AND PUTTING IT INTO PRACTICE

Rita Gardner

CHOOSING YOUR RESEARCH PROJECT

Choosing your research project is certainly difficult and challenging but it is not the Everest of a task that it often appears at the beginning. The advice I often give to undergraduates thinking of planning their research is fourfold:

1. Think small rather than big – a well thought-out and properly executed small contribution is of immensely greater value than an over-ambitious, poor-quality flop.
2. Projects do not come out of thin air. All research builds on earlier studies and it is most important that any research is firmly rooted in the context of previous studies. This context not only prevents needless repetition, but also provides a background for comparison and evaluation of your work, and allows you to slot your work into a research niche. Most ideas for future studies stem from a good understanding of earlier work.
3. Help is usually at hand from a tutor, lecturer or someone recommended by them. But help will be much more forthcoming if you have some ideas about what you want to study, where you will be going, and if you can demonstrate that you have done some reading before seeking the advice of a tutor. Turning up one month before departure asking plaintively "What can I do in X?" will rarely elicit a favourable response.
4. Leave yourself plenty of time to develop your project.

FOCUSING IN

Why am I doing a scientific research project?

Understanding why you are doing a scientific research project, and its importance in terms of the expedition as a whole, can help when choosing the type of project because it gives a guide to levels of interest and commitment. There are several reasons why you may be involved in a scientific project:

1. Science is a passion and you are fascinated by the idea of undertaking original research in order to further our knowledge about the world in which we live.
2. Science is interesting, but more to the point you have a dissertation to do for your degree course. Dissertations are there to give students experience in the skill of designing and undertaking research; they are not intended to be scientific masterpieces.
3. You want to participate in an expedition for a number of different reasons (e.g. travel, leadership skills, etc.), and you are taking the opportunity to do your dissertation at the same time.
4. You want to participate in an expedition for personal and social reasons, but you feel that you would like to spend some of your spare time doing something useful.
5. The expedition is primarily for adventure, but you feel that a scientific component will help you to attain funding.

If you fall into the first category then the whole expedition will revolve around the research project, and you may well chose your expedition area to fit around the research aims. For the other categories the research component becomes of decreasing importance, and the scientific aims are less and less likely to exert much influence on the choice of expedition area. The question changes from "Where shall I go to do X research?" to "I really want to go to Y. What research can I do there?" In short, the first two categories have research as a main aim of the expedition, whereas the last three have research as the supporting objective. In the fifth category it is worth thinking seriously about whether research is sensible, because a grant-awarding body can generally see when research has been tacked on to the expedition to give it respectability.

Clearly, these statements are over-simplified, and different members of an expedition team may have different research expectations and motivations. However, it is worth clarifying the standing of the research within the overall expedition plan because this will affect the scale of your research project and the type of research project(s) that is most appropriate to your circumstances. All levels can, of course, contribute valuable information.

What type of research project?

How motivated you are to undertake original research is just one of several factors that help to determine the most appropriate type of field research project(s) that you might undertake. Before considering these other factors it is probably worth describing some different types of project.

Classification and mapping
Classification and mapping projects are often the most straightforward to plan and undertake. These involve identifying the existence, and sometimes mapping the distribution, of objects or features. They can be applied as an overview of areas not previously studied, as a detailed investigation of one particular feature/object not previously studied in that area, or as a follow-on study to evaluate change over the period of time since a former study. Every discipline involved in field science has examples of this type of project, including geological and geomorphological mapping, distribution of fauna and flora, and ground "truthing" for remotely sensed imagery. They can be ideally suited, at a simple level, to those for whom research is a secondary aim of the expedition.

Adoption and adaptation
Adoption and adaptation make direct use of earlier research through the application of the same research aims and hypotheses to a different area, or the same area at a later time (adoption), or through adaptation of one or more parts of an earlier study to a new setting. This group of projects includes the applications and testing of models in new field settings. The same field and laboratory methodology can often be followed as in the earlier research, and comparisons of results with earlier research can provide a good basis for the discussion. This type of project is often employed effectively in dissertations.

Impact evaluation
This involves evaluation of the effect of a change in one parameter – often a human factor – on other aspects of the environment and/or social system. Although such projects may seem conceptually easy, they can be very difficult to undertake in the field, and at worst they lead to vague and inconclusive generalisations. In particular, there may be problems in establishing and proving the causal links, and in obtaining reliable information on the conditions before the change.

Research frontiers
This is the most challenging type of project, best suited to those with a passion for research because it usually requires more effort, dedication and insight than other types of projects. Earlier work forms the building blocks on which this new research adds the small next step. Clues about useful avenues to follow may be found at the end of some research papers and monographs.

The type of research project and, indeed, the specific aims of the research depend not only on the motivation of the expedition members. There are two other areas of consideration. First, there are the wider needs of the research communities, which can fall into any of the project type categories just mentioned, but are more likely to be pushing forward research frontiers. These needs are examined in more detail for specific environments later in this section. Second, and equally important, are a number of personal and logistical constraints, which will affect the particular research projects that you can do within each of the project type groups. The following are the most important constraints:

- research skills of the members: including training in field monitoring, description and measurement techniques, and laboratory analysis and data handling
- availability of specific field equipment and laboratory facilities, and where these facilities are located
- ability and ease with which samples can be transported, exported and imported
- levels of practical and intellectual ability of members of the group
- compatibility of research interests within the expedition members; and whether each individual is engaged in one project; if so, whether the projects are designed to dovetail
- time and manpower in the field
- environmental constraints: access, altitude, weather conditions
- timing of the expedition: particularly important for projects that require specific weather conditions
- communication skills (language) if your research project involves local people

After developing a clearer idea of the personal and logistical constraints that face your group – the levels of motivation, the types of project that can be done, the possible research area and important research needs in that area (see below) – you then have the challenge of defining your precise research objectives or hypotheses. At this stage reading and secondary sources of information are vital components. Examine maps, aerial photos and satellite imagery if available, and seek out reports of past expeditions, published articles in journals and people who have been to the area before. Read about the chosen area, the research field that interests you (including research methodology) and the research that has been done in the area. It is then time to see your tutor and get some feedback on what you intend to do.

To sum up, a good research project is one that lies within the constraints imposed by personal characteristics and logistical problems, is appropriate to the levels of motivation within the group, has a clear set of objectives, can relate those clear objectives to the wider research context, and does not attempt to do too much!

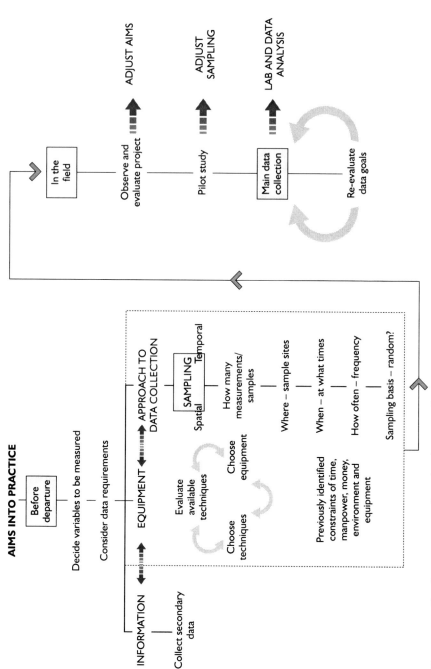

Figure 22.1 *Putting your research project into practice*

PUTTING YOUR PROJECT INTO PRACTICE

After putting a great deal of effort and thought into the logistics, fund-raising and formulation of the scientific objectives of their expedition, many small scientific expeditions forget that putting these ideas into practice actually starts before you go into the field. Then 2 weeks before departure, often just after the end of exams, there is a frantic rush to assemble the necessary equipment; photocopiers are worked day and night to assemble articles on use of equipment and reports of earlier studies; and any polite enquiry such as "Have you thought about your sampling design?" is guaranteed to elicit a response close to hysteria.

Reference to the flow chart in Figure 22.1 will show that putting your research project into practice can be divided into two parts: before departure and in the field.

Before departure

You should already have a good idea of what variables (characteristics) you will want to measure in the field, because these lead directly from the aims of the project. However, there are a number of decisions to be made, or at least considered, that relate to these variables.

The first point to be considered is the likely nature of the data that you will be collecting and your sources of data about the variables. Are the data being derived solely from field measurements, monitoring or questionnaires? Do they involve the collection of samples for subsequent laboratory analysis and, if so, what analyses? In what form will the data be collected, e.g. as frequencies of occurrence within predefined categories (i.e. nominal data scale), as actual measured values on an interval scale or as open-ended questions? What methods of data presentation and analysis do you wish to use? How many data do you need to test your hypotheses? All of these issues relating to the likely data will partly affect the two key areas about which decisions have to be made – namely field equipment and field-sampling strategy. It must be remembered that equipment needs, sampling strategy and data characteristics are interlinked, and decisions in one area will affect the others.

If you plan to collect samples for analysis in the UK, it is essential that the regulations for export of different types of samples from the research area and their import into the UK be investigated at an early stage. The restrictions on importing soil, for example, may well necessitate reconsideration of the aims of the project, or of the logistics, to enable analyses to be undertaken in the host country.

Sampling strategy

The first decision concerns sampling strategy. You know what you want to measure and what information and samples to collect, but of course you cannot measure or collect continuously from all areas and at all times. Thus, you need to consider your strategy for sampling in both space and time.

In simple terms this is the "How many, where, when and how often?" question.

How many measurements/collections/interviews do you need to make to test your hypothesis? Over what area? At what time intervals? From exactly where? How often? And at what specific times of day? For most scientific projects some of these questions, if not all, will need to be answered.

Although you will not be able to answer them definitively before seeing the field area, careful thought about the needs of the project, the sampling options and the development of a likely strategy is strongly recommended, not least because it can affect your decision about equipment and, in the case of "human projects", the formulation of the questionnaire design. The simplest way to approach this mine-field is first to consult a good textbook on sampling for field studies, and then to evaluate what other people have done for similar research projects. This is usually quite easy if you have chosen a project that is adopting or adapting earlier research. However, this does not relieve you of the responsibility of still assessing the feasibility of the approach under your (different) circumstances.

Sampling framework

The role of the sampling framework is to gain as representative, unbiased and accurate a set of measurements as possible of the variable being studied, while at the same time keeping within cost and time constraints. The sampling strategy will depend not only on the distribution but also on the spatial and temporal variability of what is being measured. In general, the greater the level of (spatial or temporal) variability, the larger the number of samples required to gain an accurate record of that variability; the shorter the time over which the variability occurs, the more closely spaced in time the sample collection should be, e.g. measuring discharge and sediment yield from flashy and highly responsive gully systems during and after storms will need measurements of water level and collection of water samples every few minutes, and several gullies would need to be studied in order to determine the typical behaviour. On the other hand, a perennial stream would tend to respond more slowly and over a longer period to rainfall, and thus measurements could be spaced out more widely over time.

Ideally a pilot study should be carried out to determine the levels of variability and hence to enable an efficient sampling strategy to be planned. However, this is a luxury that very few small, scientific expeditions can afford or even manage practically. In this respect (and others) the collection of secondary data is most important – data from other studies about the characteristics and responses of similar natural and human systems elsewhere, and as much appropriate information as possible about your precise field area (meteorological information, maps, air photos and even satellite images for an overview). This information is invaluable in putting your aims into practice before departure. First, it helps you to think about the size of area in which you might work. Second, it helps you to determine the likely variability. Third, it will enable you to think about probable sampling locations. Fourth, it will help with the choice of equipment for measuring/collecting your field information. A discussion at this point with researchers who know the area well can be invaluable.

Equipment
The second decision, and one that clearly follows from the first, concerns equipment. Typically there are several different methods for collecting any set of field data, be it physical collection of insects, mapping of geology or measurement of river discharge. The methods will probably vary in their sophistication, accuracy, ease of use and physical equipment characteristics (portability, cost, power needs, availability, etc.). An evaluation of the methods is important, because an understanding of the limitations posed by the methods and the likely practical problems can save a great deal of embarrassment in the field and during the project write-up. You should be able to justify your choice of techniques – a justification that will be based on compromise between availability/cost of equipment and efficiency/accuracy of method, and which will take into account the field conditions, manpower and the level of sophistication of the aims of the project. As a general rule, simple but reliable low-technology equipment has many advantages for small scientific expeditions, especially in relatively remote areas. There are now a number of specialised books that deal with field methods and types of equipment, and it is well worthwhile investing in a good one. It is most important that the equipment and spare parts are checked before departure, and that the research team are trained in using the equipment

To sum up, the successful completion of your project depends in large part on the groundwork that you do before departure. Without this you will probably be faced with an environment that is larger than expected, looks different from that imagined and is infinitely more complex than shown in any textbook, and one where you simply do not know where to start. Even if your deliberations before departure have not come up with any hard decisions about sampling strategy, the simple act of thinking carefully about the issues will have prepared you for coping with them in the field.

In the field
The first task in the field is to observe. Although a week spent walking around the field area, observing the features relating to your project, may seem like a luxury given the time available, it is almost certainly time very well spent. It is at this stage that you have the opportunity to evaluate your project objectively and to confirm that it is fully appropriate for the field setting. In most cases, if the initial formulation of the project aims and the subsequent preparations before departure were undertaken carefully and conscientiously, it is unlikely that you will need to rethink the project. However, no matter how thorough your preparation beforehand, there is always a chance that conditions may have changed since the last report or map was produced, or that unforeseen difficulties – such as roads destroyed by landslides – are present. If you feel that the aims are not achievable, for whatever reason, it takes great

courage to set about changing or modifying them at the last minute, but it is better to try to do this than to plough on regardless with a project that you, deep down, feel will not work.

Far more common is the need to adjust your planned sampling framework once you have examined the field setting in detail. It may not be possible, for example, to gain access at the point where you hoped to collect your samples, or the variability may prove to be much greater than expected, in which case you may decide to sample fewer areas or variables but each in more detail. If you have time and resources for a formal pilot study, this will help greatly in the formulation of an effective strategy. If not, the few days of careful observation, combined with some preliminary measurements if appropriate, is the next best approach, e.g. if you are comparing surface run-off under forest and agricultural land uses, a set of measurements of slope angles, percentage vegetation cover, aspect and simple evaluation of soil types will help you in choosing the locations. You will probably want, in this case, to control for slope angles, so your choice of areas with "typical" slope angles and, of course, similar slope angles is important. The observation and preliminary simple evaluation of the characteristics in your field area relevant to your project will therefore provide you with the context needed to finalise your sampling framework.

At last you reach the stage of collecting the data, which after all the preparation is something of a relief! Data collection should be straightforward. If you are using sophisticated field equipment, a few trial runs are usually worthwhile to build up confidence, and field calibration may be needed. Checks on the reproducibility of results may also be useful for some projects. On a simpler note, take care to collect enough samples for the subsequent analyses, to label sample bags indelibly with name, date, site and sample identifier, and to take detailed field notes on any topic/measurement/sample that may be of relevance. There is much to be said for recording as much detail as possible – in retrospect, aspects that seemed relatively unimportant at the time may prove to be vital later. For some analyses, undisturbed samples and samples with a known orientation may be required. All samples need to be treated carefully to avoid contamination. Unlike projects based in your home country, you often have only one chance with scientific expeditions abroad and the set of observations that you omitted to note down may just prove to be the vital ones.

It is useful to evaluate your progress with data collection as you go along, especially if there is limited time in the field. If it is taking much longer than anticipated, you may need to reassess your "data goals" and even to reassess and shrink the project aims. It is much better for many projects to collect a sufficient number of samples/field measurements to answer one question well, than to collect a few samples from many different locations in order to try to answer several research questions and to find instead that you have insufficient information to solve any of the questions, e.g. five samples from each of ten large moraines will allow very little in the way of comparison of the soils developed on the moraines, and there will be insufficient data for a sensible statistical analysis of the results. On the other hand, 25 samples from each of two moraines would allow a reasonable comparison to be made between them, provided that the sampling design was carefully chosen.

If you are collecting specimens or sediment/water samples, you need to consider their requirements for sample storage and, if necessary, safe transport to the UK. All samples have some special requirements. For water samples collected for chemical analysis, filtration is important to remove possible polluting substances during storage, and the prevention of bacterial growth (or quick analysis before bacterial growth) is essential for samples that are not from pure water sources. Furthermore, gaseous losses may affect the water chemistry. Soil samples can also suffer from problems of bacterial and fungal growth, and they are best air-dried on site if possible (it cuts down weight too!). Biological specimens may need preservation and will certainly need careful packing.

Lastly, just as each sample is unique, so are written data-sets. Field notes, whether habitat descriptions, survey measurements or questionnaire results, are essential for the expedition to produce results, and should be protected from the rigours of the field and transit. Indelible pens, notebooks with tough paper and bindings, and waterproof notebooks can all be used in the field. If possible, make photocopies before leaving the field area, and send one copy back by post or with another person. Likewise, ensure that duplicate copies are kept on return. Electronically stored data are more sensitive, and the same principles apply to them.

FURTHER INFORMATION

RGS–IBG World Register of Field Centres

If you are looking for a location to do field research, check out the growing number of sites on the World Register of Field Centres. Searchable from the website, it provides information on established field centres, in environments from the high Arctic to the Sahara desert. The centres range from small independent field camps to large, long-term international facilities. The only criterion for inclusion is that centres welcome international visitors, be they scientists, students or teachers (www.rgs.org/fieldcentres).

23 EXPEDITION RESEARCH PROJECTS IN TROPICAL FORESTS

Clive Jermy and Ian Douglas

The ecology of tropical forests and their biota presents an enormous range of opportunities for field research. Any biological project carried out in tropical forests is going to bring the expedition members close to a wide range of plants and, to a lesser extent, animals, many of which are likely to be completely new to science. There will rarely be a simple, straightforward textbook that will aid identification, even of the common species. A local scientist who knows his or her flora is indispensable to the success of any expedition that wants to work on the ecology of tropical forests.

The complex environments of tropical forests provide great opportunities for short duration studies of local variations in microclimate and soil characteristics, and of hydrological and geomorphic processes. In thinking about the types of project that may be feasible in any particular forest area, the variety of forest types and terrain present must be considered.

In most localities contrasts exist among ridge crest, slope, valley floor and floodplain sites. In many localities, patches of disturbed forest are close to natural forest, providing opportunities for comparative studies of the impacts of people on the forest.

Inevitably, many forest study sites are remote and conditions for setting up instruments for field monitoring are difficult. The high humidity often makes electronic instruments inoperable and battery life short. Experience suggests that only instruments that have proven reliability under rain forest conditions are worth using by expeditions, and that rechargeable batteries, replaced every few days, are the only way of guaranteeing continued operation of battery-powered recording instruments.

Even though tropical rain forests are wet environments, long periods without rain can occur, e.g. at the Danum Valley Field Studies Centre in Borneo, annual rainfall averages some 2800 mm, or more than 230 mm per month. However, in April 1991, the rainfall was only 24.2 mm, nearly all the days that month being dry. Projects relying on measuring rainfall and water flows may be frustrated and expedition planners should have alternative projects in case the weather is unusually dry.

Any expedition intending to work in a tropical rain forest should obtain and study Tim Whitmore's book *An Introduction to Tropical Rain Forests* (1989) and, for those going to Asia or the Pacific, *Tropical Rain Forests of the Far East* (1984).

CLIMATIC, MICROCLIMATIC AND METEOROLOGICAL OBSERVATIONS

The difference in microclimate between the interior of the forest and the open provides many opportunities for comparative investigations. Temperature and humidity show diurnal (daily) fluctuations, with open areas warming up far more than the forest interior during the daytime, whereas humidity within the forest is much more constant. Simple thermometers and hand-held whirling hygrometers read at half-hourly or hourly intervals could provide basic information. Clockwork, chart-recording, wet and dry bulb thermometers, or more sophisticated data loggers would be preferable.

The vertical zonation of temperature and humidity in the forest is of great interest. One simple way of obtaining a vertical temperature profile is to use a catapult to send a line across a high bough of a tree, and then to haul up a rope on to which are fixed maximum and minimum thermometers at intervals of, say, 5 or 10 m. The rope can be hauled down as frequently as required for the instruments to be read.

Within the forest the spatial variability of throughfall (the rain penetrating through the vegetation canopy) is even greater and the value of observations of both throughflow and stemflow (the rain running down tree trunks) is high. To be useful, such measurement schemes have to be carefully designed. The average of the catch of all the gauges, randomly relocated, can be compared with the catch in the open, to give an indication of the percentage of the rainfall intercepted, i.e. not arriving at the throughfall collectors.

A small part of the rain falling on the forest runs down the tree trunks as stemflow. This is measured by fitting collars around trees to divert the stemflow into collectors. Such collectors are easily made from epoxy resin, which can be moulded into the required shape before it hardens.

Meteorological observations of the types described above could be used to establish differences in the conditions in primary forest, secondary forest, grassland and open areas or, in an undisturbed forest, among natural gaps, closed forest, slopes and river banks.

Soil temperatures are poorly known, and soil thermometers could be used to assess thermal conditions at sites along a soil catena. Temperature and humidity observations are safe topics, whereas with rainfall there is always a risk that the expedition may coincide with an unusually dry period.

STUDIES OF SOILS AND FOREST FLOOR CONDITIONS

The forest floor is a key part of the forest ecosystem. The plant debris that falls to the floor is attacked by decomposing organisms, releasing nutrients that are added to soil, taken up by plants or lost to drainage waters.

Litterfall rates can be assessed by establishing simple traps under various types of forest canopy. Essentially a litter trap is a fine gauze mesh suspended between four posts about 50 cm above the forest floor. Usually about 1 m^2 in area, the traps should be emptied every day, or at frequent intervals, otherwise insects will have already destroyed part of the material. The litter may be dried, in the sun if an oven is not available, and weighed, to obtain an estimate of the litterfall per m^2. A simple classification into leaf, twig, other plant fragments, insects and other debris may be attempted. As litterfall varies seasonally and depends on wind and rainfall, time is a constraint on the value of results from short-term studies. Comparisons of litterfall in different types of forest, and particularly between natural forest and secondary regrowth, would be most useful

A wide range of assessments of the physical properties of soil may be made in the field, including measurements of infiltration, bulk density and permeability. Infiltration can be measured using a double-ring infiltrometer; this is made of two rings, possibly of strong plastic of the type used for gas mains, or of metal, 20 cm deep and one ring about 30 cm diameter and the other 36 cm. The two rings are driven into the ground. A known quantity of water is applied to the ground surface inside the inner ring, and the time taken for that water to infiltrate is calculated. The result is converted into the depth of water infiltrating in one hour. Infiltration rates vary greatly with soil types, from 10 to 27 cm/h over a range of soils in Puerto Rico.

Bulk density measurements require access to a balance, but are easily accomplished if one is available. Permeability may be measured using a field permeameter, which is easily constructed in a workshop but would have to be thoroughly tested before being taken on an expedition. Details of these techniques are readily available in texts such as those edited by Goudie (1989) and Landon (1984). If a field test requires a ready water supply or needs basic laboratory facilities, its accessibility must be ascertained in advance.

Soil description in the field, by excavating soil pits and examining soil profiles (a major activity on many expeditions), is particularly valuable on a downslope catena if linked to studies of litter and slope hydrology. The percentage silt–clay content of a soil, a good indicator of water-holding capacity, can be obtained by sieving if a 63 mm mesh sieve, lid and collecting pan are taken in the field, and water and a good balance are available.

Investigations of soil erosion and nutrient loss

Concern about the impacts of forest disturbance has led to many proposals to measure erosion and the removal of chemical elements in solution. In planning to undertake such studies, the episodic nature of erosion must be recognised. A few heavy rainstorms may carry away nearly all the soil eroded in a year. An expedition may be lucky to sample such a major event, but probably it will not do so. Comparative studies are therefore more appropriate for short-term projects. The second factor is the level of logistic support for such investigations. Chemical analyses are possible only if portable field analytical kits are available, if local laboratories are able to assist, or if an adequate number of samples can be taken back to the home country base for analysis. Concentrations of most dissolved substances in tropical rain and river waters are extremely dilute, usually less than 10 mg/l for most elements. If field analysis is planned, the field kit must be able to give reproducible results for such low concentrations. As samples deteriorate with storage without refrigeration, transport back to home base for additional analyses may be unwise. Analytical specialists or geochemists should be consulted before finalising the programme.

Notwithstanding the necessary cautionary approach to erosion studies, highly valuable projects can be carried out, particularly when disturbed areas subject to large amounts of erosion are compared with one another. A typical project might be to investigate erosion on abandoned logging tracks, one of which has uninterrupted flow of surface water downslope, the other having barriers to trap sediment and impede water flow. Water can be tested for temperature, pH and conductivity using small field meters, but many such instruments are unreliable under humid tropical conditions. The pH cells must be kept in standard solutions except when in use. Conductivity reflects the chemical composition of water and thus is useful, for reconnaissance studies, to test whether tributary streams have similar solute contents or to identify where major changes in water quality occur. Levels of dissolved oxygen indicate the potential of streams for aquatic life, but dissolved oxygen meters must be calibrated with standard solutions. Before buying or borrowing such field instruments, advice should be sought on their suitability and robustness under tropical conditions.

Stream water quality is an excellent environmental indicator. In the Amazon basin, rivers are described as black water, white water and clear water rivers. The acidic, dilute, black waters are found in many sandy podzolic areas of the tropics, including large areas of coastal peats and freshwater swamp forests. Classifying rain forest aquatic environments in this way, by water testing over a wide area, is an ideal expedition project. A field check using pH and conductivity meters is advisable before selecting sites for water sampling in a study designed to highlight environmental contrasts

If a major storm occurs during an expedition period, the opportunity should be seized to resurvey any stream for which debris data had already been collected.

INVESTIGATIONS OF RIVER MORPHOLOGY AND FLUVIAL PROCESSES

Rivers draining rain forests differ enormously according to the geological history of their catchment areas and the climatic regime of their part of the tropics. Detailed observations of the form and process of river channels in the tropics are needed to improve flood prediction and forecasts of sediment load. Simple surveys of channel cross-sections and mapping of channel bed materials and vegetation provide useful information. Mapping of the distribution of sediments of different sizes in the channel and on gravel bars helps to establish the material available for transport during high flows. Measurements of sizes and the lithology of pebbles, preferably using the 4- to 6-cm long axis class, help to establish the way that rock fragments change shape and lithologies are eliminated with downstream transport. Tropical vegetation grows rapidly, but on river channel margins it shows a zonation related to flood frequency. Mapping this riverine vegetation, with simple descriptions in terms of life-forms, gives a good indicator of annual and extreme flood heights.

Small streams are often encumbered with large amounts of broken tree trunks and branches, some of which form debris dams that trap sediment being transported downstream. Such debris dams are washed out during the biggest storm events or exist until they rot away. Although they have been well studied in the wet temperate forests of the west coast of North America, they have seldom been investigated in the tropics. Simple surveys of the amounts of coarse woody debris and numbers of debris dams in streams of different sizes, and the lithology, gradient and disturbance by people would add to the understanding of how biological and hydraulic factors work together in rain forest streams. Many have hypothesised that tropical streams exhibit minor fluctuations during calm weather as a result of the daily evapotranspiration cycle. If a simple river level gauge, in the form of a graduated board or staff, can be erected on a stream bank, hourly observations, day and night, could enable any such fluctuation to be tested. This work could be coupled to air, soil and water temperature measurements, pH and dissolved oxygen determinations to reveal whether there are significant diurnal variations. Ideally such observations could be repeated on streams of differing characteristics. Many of these projects involving aspects of the hydrological cycle and fluvial processes could be combined together in a team study, with individuals having responsibility for different components, such as rainfall and interception, slope processes, channel form and river water quality. Detailed studies of soil properties and hillslope hydrology have been integrated in this way.

THE IMPACT OF VARYING DEGREES OF FOREST DISTURBANCE

Although much remains to be learnt about the undisturbed natural rain forest, even more information is required about what happens in areas that have been logged, cleared, replanted or abandoned to secondary regrowth. Usually patches of such modified forest are readily accessible and provide good opportunities for short-term comparative investigations. In particular, areas of soil compaction, such as logging roads, places where logs were assembled, abandoned construction sites and tracks along which logs were dragged, may be investigated to determine the density of the vegetation at a certain time after cessation of the disturbance, the proportion of water still running over the surface, the amount of organic matter on the soil surface, the size and length of any rills and gullies, and any other evidence of the rate of recovery from disturbance. All the projects relevant to natural forest are relevant to disturbed or secondary forest.

PROJECTS STUDYING BIODIVERSITY

The tropical forest is one of the richest habitats for plant and animal diversity (see Prance, 1982) and some simple but informative work can be undertaken comparing animal and/or plant diversity across different habitats, altitudes, vegetation formations, regions or countries. This kind of project lends itself to longer-term recording and can be the aim of successive expeditions from the same institute/university. Nadkarni and Longino (1990) compared invertebrates in the canopy and in the forest floor litter in montane forests in Costa Rica. Samples of the litter were sifted for the following groups and the numerical dominance was counted: mites, adult beetles, holometabolous insect larvae, ants, collembola, amphipods and isopods. All were easy to identify by the non-specialist. Similar work was carried out by Collins (1979a) and his team in Gunung Mulu National Park. The foraging activity of insects such as termites can be intensively studied for short periods and can result in data worthy of publication (e.g. Collins, 1979b).

Getting into the canopy of forest trees has long been both a physical challenge and scientifically rewarding. Several accounts and techniques have been documented, which are best summarised by Mitchell (1986), but see also Whitacre (1981). Assessment of arthropod diversity in the canopy has been a subject of much debate over the past 10 years (Stork, 1988), but comparative quantitative studies of tropical insects, especially in relation to plant host specificities, are projects worthy of consideration. The technique using knock-down insecticide fogging, which can be set up in the forest canopy by a competent expedition, can collect large numbers of insects on sheets laid out on the forest floor.

STUDIES ON FOREST ECOLOGY

Biomass variation in different forest formations usually needs longer periods than are available on short-term expeditions but restricted comparisons of interest can be made. Forest inventories measuring all trees with a DBH (diameter at breast height) of 10 cm or more in a 1-hectare linear transect (10 × 1000 m) to show diversity, frequency, density and dominance can be a worthy objective (Boom, 1986). It must be linked with collecting good herbarium specimens of each species (which can be from 90 to more than 200 species per hectare) for later determination. This is a project that would benefit by having a local forester or botanist join the team, or at least have someone in the local forest herbarium identify the species. This also provides a good opportunity to collect information on local names and uses of the plants. Invariably only a number of the trees will be flowering, and very difficult even to see, let alone collect. Good herbarium material will always be welcomed by the local national herbarium, which should always be offered the first set of any collected material. Herbaria here in the UK (Natural History Museum, Royal Botanic Gardens, Kew and Edinburgh) will always be pleased to have duplicates of any named material, but may not be in a position to identify these plants for you.

Studies in leaf morphology in relation to forest type and altitude zonation, especially if a field microscope can be taken in, can show how plants adapt themselves to different environmental extremes. Other projects could compare structure and even physiology (e.g. photorespiration rates) of plants in sun and shade (see also Medina et al., 1977). These physiological investigations can use simple apparatus, which should be tried out before leaving the UK.

PROJECTS INVOLVING STUDIES OF EPIPHYTES

The tropical forest abounds in epiphytes, e.g. orchids, many ferns, the screw-pine family (Pandanaceae), the pineapple family (Bromeliaceae), aroids of all forms (climbers such as *Philodendron* to single but often enormous plants of the genus *Anthurium*) and many others, but the diversity is not that great and, given good herbarium material collected in the field, the species can usually be identified. Lichens and bryophytes are other epiphytes that are more abundant in the canopy and on the more stunted trees of the upper montane forest (elfin forest). There are also specialised lichens and bryophytes, mainly liverworts, which grow specifically on leaves of young trees and larger herbs in the lower montane and lowland TRF. These folicolous species can be difficult to identify but are easy to collect and with specialist help can be tackled on one's return. Specific studies on the distribution of mosses and lichens in relation to host specificity and position on the trunks of those host species could add substantial knowledge to an underworked field.

Such projects must be backed up by well-prepared specimens to identify the components of these relationships, so familiarity with collection and preservation techniques is needed.

Vascular plant epiphytes often have complex structures to catch or retain leaf litter, e.g. special leaves as in some ferns such as the stagshorn (*Platycerium*) and basket ferns (*Drynaria*). Both the roots of the fern and other epiphytes get nutrients from these aerial peat pockets. Ants and termites play a significant role in establishing these aerial gardens (Huxley, 1980), carrying up sand grains and other detritus. In a large mature crown a very wide spectrum of plants from woody rhododendrons to small sedges can live undisturbed for many years. Roots also have complex structures, which help in water absorption/retention, and interesting anatomical studies can also be instigated.

PROJECTS THAT STUDY THE INTERRELATIONSHIPS OF PLANTS AND ANIMALS

The interdependence of plants and animals presents interesting problems, e.g. flower pollination and seed dispersal. Studies on nectar production and its relation to microclimate in bird-pollinated species, e.g. *Heliconia* spp., can elucidate the role that the plant itself plays in the feeding rate of pollinators. The study of ant plants (see Huxley, 1978, 1980) and the role of ants in preventing herbivory opens a wide field for observation when you are camping in the forest.

ECOLOGICAL PROJECTS ON VERTEBRATES

Any programme involving larger vertebrate populations usually requires more time in the field than is available to the average expedition. Exceptions to this rule will be found when naturalists or zoologists of considerable experience are attached to the expedition (compare Medway and Wells, 1971; Medway, 1972). Studies on range, feeding habits and breeding behaviour of birds are frequent objectives for expeditions and ornithological teams should contact specialists at Birdlife International. Baiting, capture and recapture of various animals (fish, amphibians and other trappable vertebrates) can give useful information on population size or location distribution patterns.

Work on plotting and describing amphibian breeding sites, collecting tadpoles and adults, together with sound recordings of their mating calls, opens up a number of avenues for projects, especially where a range of altitude can be covered.

PROJECTS ON THE ECOLOGY AND BEHAVIOUR OF INVERTEBRATES

Many projects on smaller animals, especially invertebrates, e.g. insects, can be carried out. Studies of invertebrates should be linked with the existing programme of a professional entomologist in order to maximise the data obtained, because there will be many species new to science in a project of this nature.

Studies on activity patterns or reproduction rates, e.g. in relation to temperature and other microclimate states, can be carried out in the few months available (e.g. Larsson, 1990), as can studies on the feeding habits of invertebrates (e.g. Monk and Samuels, 1990). Territory ranges of certain flying insects can also be studied. Investigations into the faunal composition of forest water bodies, e.g. tree holes, water caught or secreted in leaf bases, specialised organs such as flower bracts in gingers and related plants, and insectivorous pitchers of the genus *Nepenthes*, stimulate one to think of the use of such water bodies to the plants themselves, and the interrelationships of the animals that live there. Life cycles are often extremely short where the water bodies are ephemeral and animals can often be bred through to the adult stage in the few weeks available on an expedition. Interesting short-term studies were made on *Heliconia*, a banana-like plant of the American tropics that has hard horny persistent floral bracts. These hold liquid, most of which is secreted by the plant (Vandermeer et al., 1972; Bronstein, 1986). Identification to species is not necessary to give the spectrum of life forms found and their adaptation to the microhabitats that these sites provide.

THE IMPORTANCE OF PREPARATION

Work in tropical forests is hot and uncomfortable, despite the splendid ecological variety and complexity of the environment. Everything possible must be done before departure to ensure that projects will work and instruments will function. All equipment and techniques should be thoroughly tested in a forest environment near the home base before departure. Check carefully the assistance available in the field and find out what restrictions or regulations there may be on taking samples out of the field area or the country, or on importing them to the home country.

FURTHER INFORMATION

Useful addresses and websites

Centre for International Forestry Research (CIFOR). Website: www.cifor.cgiar.org/

Edinburgh Centre for Tropical Forests, Pentlands Science Park, Bush Lane, Penicuik, Edinburgh EH26 0PH. Tel: +44 131 440 0400, fax: +44 131 440 0440, website: www.nmw.ac.uk/ectf

European Tropical Forest Research Network, c/o Tropenbos International, PO Box 232, 6720 AE, Wageningen, website: www.etfrn.org

Global Canopy Programme, John Krebs Field Station, University of Oxford, Wytham, Oxford OX2 8QJ. Tel: + 44 1865 724 222, website: www.globalcanopy.org

Global Forest Watch at www.globalforestwatch.org

New York Botanical Garden, Bronx, NY 10458-5126, USA. Tel: +1 718 817 8700, website: www.nybg.org

Organization for Tropical Studies:

North American Office: Box 90630, Durham, North Carolina 27708-0630 USA. Street address: OTS, Duke University, 410 Swift Ave. Tel: +1 919 684 5774, fax: +1 919 684 5661, email: nao@duke.edu

Costa Rican Office: Apartado 676-2050 San Pedro, Costa Rica. Street address: 400 mts Oeste del Colegio Lincoln, diagonal a plaza Los Colegios, Moravia. Tel: +506 240 6696, fax: +506 240 6783, email: cro@ots.ac.cr

La Selva Biological Station: Apartado 676-2050, San Pedro, Costa Rica. Tel: +506 766 6565, fax: +506 766 6535, email: laselva@sloth.ots.ac.cr

Las Cruces Biological Station: Apartado 73-8257, San Vito, Coto Brus, Costa Rica. Tel: +506 773 4004, fax: +506 773 3665, email: lascruces@hortus.ots.ac.cr

Palo Verde Biological Station: Apdo. 49-5750, Bagaces, San Pedro, Costa Rica. Tel: +506 661 4717, fax: +506 661 4712, email: paloverde@ots.ac.cr, website: www.ots.ac.cr

Programme for Belize, 1 Eyre Street, PO Box 749, Belize City, Belize. Tel: +501 275 616, website: www.pfbelize.org

Pro-Natura International, Pro-Natura USA, 8123 Heatherton Lane, 104, Vienna, VA 22180 USA. Tel: +1 703 641 5900, website: www.pronatura.org.br

Rainforest Information Centre, PO Box 368, Lismore, NSW 2480 Australia. Website: www.rainforestinfo.org.au

Rainforest Concern, 27 Lansdowne Crescent, London W11 2NS. Tel: +44 20 7229 2093, fax: +44 207 221 4094, website: www.rainforestconcern.org
Works with the Ecuadorean Maquipucuna Foundation.

Rio Mazan Project, The Greenhouse, 48 Bethnel Street, Norwich, Norfolk NR2 1NR. Tel: +44 1603 611953, fax: +44 1603 666879
A small independent charity working for the conservation of Andean forests.

Royal Botanic Gardens Edinburgh, Inverleith Row, Edinburgh EH3 5LR. Tel: +44 131 552 7171, fax: +44 131 552 0382, website: www.rbge.org.uk

Royal Botanic Gardens, Kew, Richmond, Surrey TW9 3AB. Tel: +44 20 8332 5000, fax: +44 20 8332 5197, website: www.rbgkew.org.uk

Smithsonian Institution, 1000 Jefferson Drive, PO Box 37012, SI Building Room 153, MRC 010, Washington DC, 20013-72 USA. Website: www.si.edu

Tambopata Reserve Society (TreeS – UK), c/o John Forrest, 64 Belsize Park, London NW3 4EH. Tel: +44 20 7722 8095, website: www.geocities.com/treesweb/index

Tropenbos Foundation, Lawickse Allee 11, PO Box 232, 6700 AE Wageningen, The Netherlands. Website: www.tropenbos.nl

Tropical Biology Association, Department of Zoology, University of Cambridge, Downing Street, Cambridge CB2 3EJ. Tel: +44 1223 336619, fax: +44 1223 336619, website: www.zoo.cam.ac.uk/tba

Tropical Conservation and Development Program, Center for Latin American Studies, University of Florida, 319 Grinter Hall, PO Box 115530, Gainsville, FA 32611-5530, USA. Website: www.latam.ufl.edu

UK Tropical Forest Forum, Jane Thornback, c/o Natural Resources Institute, Central Avenue, Chatham, Kent ME4 4TB. Tel: +44 20 8332 5717, fax: +44 20 8332 5278, website: www.forestforum.org.uk

World Rainforest Movement, Unit 1c, Fosseway Business Centre, Stratford Road, Moreton in Marsh, Glos GL56 9NQ. Tel: +44 1608 652893, fax: +44 1608 652878, website: www.wrm.org.uk

World Resources Institute, G Street, NE (Suite 800), Washington DC 20002, USA. Tel: +1 202 729 7600, fax: +1 202 729 7610, website: www.wri.org

Bibliography (field and laboratory techniques)

Barlow, K. (1999) *Bats: Expedition field techniques*. London: RGS–IBG Expedition Advisory Centre.

Barnett, A. (1995) *Primates: Expedition field techniques*. London: RGS–IBG Expedition Advisory Centre.

Barnett, A. and Dutton, J. (1995) *Small Mammals (excluding Bats): Expedition field techniques*. London: RGS–IBG Expedition Advisory Centre.

Bennett, D. (1999) *Reptiles and Amphibians: Expedition field techniques*. London: RGS–IBG Expedition Advisory Centre.

Bibby, C., Jones, M. and Marsden, S. (1998) *Bird Surveys: Expedition field techniques*. London: RGS–IBG Expedition Advisory Centre.

Coad, B. (1995) *Fishes: Expedition field techniques*. London: RGS–IBG Expedition Advisory Centre.

Goudie, A. (ed.) (1989) *Geomorphological Techniques*, 2nd edn. London: Unwin Hyman.

Haynes, R. (ed.) (1982) *Environmental Science Methods*. London: Chapman & Hall.

Hurst, J. (1998) [Conservation] *Education Projects: Expedition field techniques*. London: RGS–IBG Expedition Advisory Centre.

Kapila, S. and Lyon, F. (1994) *People-Oriented Research: Expedition field techniques*. London: RGS–IBG Expedition Advisory Centre.

Landon, J.R. (ed.) (1984) *Booker Tropical Soil Manual*. London: Longman.

McGavin, G. (1998) *Insects: Expedition field techniques*. London: RGS–IBG Expedition Advisory Centre.

Mitchell, A. and Secoy, K. (eds) (2002) *Global Canopy Handbook: Techniques for access and study in the forest roof*. Oxford: Global Canopy Programme.

Royal Botanic Gardens, Kew. *The People and Plants Handbook*. London: Royal Botanic Gardens, Kew.

Bibliography

Anderson, J.A.R. and Chai, P.P.K. (1982) Vegetation. In Jermy, A.C. and Kavanagh, K.P. (eds), Gunung Mulu National Park, Sarawak, an account of its biota, etc. *Sarawak Museum Journal* 30(51): 195–206.

Anderson, J.M. and Spencer, T. (1991) *Carbon, Nutrient and Water Balances of Tropical Rain Forest Ecosystems Subject to Disturbance: Management implications and research proposals*. Paris: Unesco.

Bawa, K. and Hadley, M. (1990) *Reproductive Ecology of Tropical Forest Plants*. Unesco Man and Biosphere Series, Vol. 7. Paris: Unesco.

Boom, B.M. (1986) A forest inventory in Amazonian Bolivia. *Biotropica* 18: 287–94.

Bronstein, J.L. (1986) The origin of bract liquid in a neotropical *Heliconia* species. *Biotropica* 18: 111–14.

Castner, J.L. (1990) *Guide to Research Facilities in Central and South America*. Gainsville, FA: Feline Press.

Collins, N.M. (1979a) A comparison of the soil macrofauna of three lowland forest types in Sarawak. *Sarawak Museum Journal* 27(48): 267–82.

Collins, N.M. (1979b) Observations on the foraging activity of *Hospitalitermes umbrinus* (Haviland), (Isoptera: Termitidae) in the Gunung Mulu National Park, Sarawak. *Ecological Entomology* 4: 231–8.

Collins, N.M. (1990) Some research priorities in the biology of tropical forests. In: Collins, N.M., Sayer, J.A. and Whitmore, T.C. (eds), *The Conservation Atlas of Tropical Forests: Asia and the Pacific*. London: Macmillan.

Fosberg, F.R., Garnier, B.J. and Kuchler, A.W. (1961) Delimitation of the humid tropics. *Geographical Review* **51**: 333–47.

Gentry, A. (ed.) (1990) *Four Neotropical Rain Forests*. New Haven, CT: Yale University Press.

Halle, F., Oldeman, R.A.A. and Tomlinson, P.B. (1978) *Tropical Trees and Forests: An architectural analysis*. Berlin: Springer.

Hawkes, J.G. (1980) *Crop Genetic Resources Field Collection Manual*. Birmingham: IBPGR and EUCARPIA (also available from the Expedition Advisory Centre).

Holdridge, L.R. (1967) *Life Zone Ecology*. San Jose, CR: Tropical Science Center.

Holttum, R.E. (1954) *Plant life in Malaya*. London: Longman-Green.

Huxley, C.R. (1978) The antplants *Myrmecodium* and *Hydrophytum* (Rubiaceae) and the relationship between their morphology, ant occupants, physiology and ecology. *New Phytology* **80**: 231–68.

Huxley, C.R. (1980) Symbiosis between ants and epiphytes. *Biological Review* **55**: 321–40.

Janzen, D.H. (1975) *The Ecology of Plants in the Tropics*. London: Arnold.

Jermy, A.C. (1981) *Youth and Exploration: Tropical rain forests*. London: BP International.

Jordan, C.F. (1989) *An Amazonian Rainforest: The structure and function of a nutrient-stressed ecosystem and the impact of slash-and-burn agriculture*. Unesco Man and Biosphere Series, Vol. 2. Paris: Unesco.

Larsson, F.K. (1990) Thermoregulation and activity of the sand wasp (*Steniolia longirostra* [Gay]) in Costa Rica. *Biotropica* **21**: 65–8.

Lee, R. (1978) *Forest Microclimatology*. New York: Columbia University Press.

Leigh, E.G., Rand, A.S. and Windsor, D.M. (1982) *The Ecology of a Tropical Forest*. Washington DC: Smithsonian Institution Press.

Leith, H. and Werger, M.J.A. (eds) (1989) *Tropical Rain Forest Ecosystems. Ecosystems of the world*, vol. 14B. Amsterdam: Elsevier.

Longman, K.A. and Jenik, J. (1974) *Tropical Forest and its Environment*. London: Longmans.

Mabberley, D.J. (1991) *Tropical Rain Forest Ecology*. Glasgow: Blackie & Son Ltd.

Martin, C. (1991) *The Rainforests of West Africa: Ecology, threats, conservation*. Basel: Birkhauser.

Medina, E., Mooney, H.A. and Vusquez-Yanes, C. (eds) (1977) *Physiological Ecology of Plants of the Wet Tropics*. The Hague: Junk.

Medway, Lord (1972) The Gunung Benom Expedition, 1967. 6: the distribution and altitudinal zonation of birds and mammals on Gunung Benom. *Bull. British Museum Natural History* **23**: 105–54.

Medway, G. and Wells, D.R. (1971) Diversity and density of birds and mammals at Kuala Lumpur, Pahang. *Malaysian Nature Journal*. **24**: 238–47.

Meggers, B.J. (ed.) (1973) *Tropical Forest Ecosystems in Africa and South America: A comparative review*. Washington DC: Smithsonian Institution.

Mitchell, A.W. (1986) *The Enchanted Canopy: Secrets from the rain forest roof*. London: Collins.

Monk, K.A. and Samuels, G.J. (1990) Mycophagy in grasshoppers (*Orthoptera: acrididae*) in Indo-Malayan rain forests. *Biotropica* **22**: 16–21.

Nadkarni N.M. and Longino, J.T. (1990) Invertebrates in canopy and ground organic matter in a neotropical montane forest in Costa Rica. *Biotropica* **22**: 286–9.

Odum, H.T. and Pigeon, R.F. (eds) (1970) *A Tropical Rain Forest*. Tennessee: US Atomic Energy Commission.

Ooi Jin Bee (1990) The tropical rain forest: pattern of exploitation and trade. *Singapore Journal of Tropical Geography* **2**: 117–42.

Prance, G.T. (1982) *Biological Diversification in the Tropics*. New York: Columbia University Press.

Rabinowitz, A. (1993) *Wildlife Field Research and Conservation Training Manual*. NYZS The Wildlife Conservation Society, 185th St & Southern Blvd. Bronx, NY 10460-1099, USA.

Raich, J.W. (1989) Seasonal and spatial variation in the light environment in a tropical dipterocarp forest and gaps. *Biotropica* **21**: 299–302.

Reich, P.B. and Berchert, R. (1988) Studies with leaf-age in stomatal functions and water status of several tropical tree species. *Biotropica* **20**: 60–9.

Richards, P.W. (1952) *The Tropical Rain Forest*. Cambridge: Cambridge University Press.

Schimper, A.F.W. (1903) *Plant Geography upon a Physiological Basis*. Oxford: Oxford University Press.

Southwood, T.R.E. (1978) *Ecological Methods, with Particular Reference to the Study of Insect Populations*, 2nd edn. London: Chapman & Hall.

Stork, N.E. (1988) Insect diversity: facts, fiction and speculation. *Biological Journal of the Linnaeus Society* **35**: 321–37.

Sutton, S.L., Whitmore, T.C. and Chadwick, A.C. (eds) (1983) *Tropical Rain Forest Ecology and Management*. Oxford: Blackwells.

Vandermeer, J.H., Addicott, J., Anderson, A. et al. (1972) Observations on *Paramecium* occupying arboreal standing water in Costa Rica. *Ecology* **53**: 291–3.

Whitacre, D.F. (1981) Additional techniques and safety hints for climbing tall trees, and some equipment and information sources. *Biotropica* **13**: 286–91.

Whitmore, T.C. (1984) *Tropical Rain Forests of the Far East*. Oxford: Oxford University Press.

Whitmore, T.C. (1989) *An Introduction to Tropical Rain Forests*. Oxford: Clarendon Press.

24 EXPEDITION RESEARCH PROJECTS IN WETLANDS

Edward Maltby

Wetland is a collective term for a very wide range of ecosystems, the formation of which has been dominated by water and the processes and characteristics of which are largely controlled by water. A wetland is essentially a place that is wet enough for a long enough period to develop specially adapted vegetation and other organisms, and generally comprises mineral substrates or soils with particular morphological and physicochemical characteristics. Wetlands occur in a wide range of geographical locations and cover an estimated 6 per cent of the world's land surface (Maltby and Turner, 1983).

Investigation of the characteristics and functioning of wetlands is a high priority

Figure 24.1 *The mangrove environment is physically testing but offers exciting opportunities for research (© M. Huxham)*

not only because of the increasing evidence of their value in environmental support, but also because of the accelerating loss of the resource base. The physical, chemical, biological, hydrological and ecological processes that occur within these ecosystems are complex and sometimes difficult to measure and evaluate. However, many relatively simple and portable techniques are available for scientific study (e.g. Faulkner et al., 1989) and there is a particular need in less developed countries to evaluate more directly the relationships between wetlands and human use.

WETLAND DIVERSITY

Wetlands vary according to their genesis, size, geographical location, hydrological regime, chemistry, vegetation, and soil or sediment characteristics. Such diversity has complicated investigation of fundamental processes that may be common to many different systems, and has contributed to the lack of development of any unified discipline of wetland science (see Maltby, 1991a, 1991b). They include some of the most, as well as some of the least, productive ecosystems in the world; they occur from mountains to coasts, and range from freshwater to hypersaline systems, inorganic to organic, oligotrophic to eutrophic, acidic to alkaline, and from forested systems to those lacking any higher plants.

The main types of wetland are described in Maltby (1986). A comprehensive classification can be found in Cowardin et al. (1979) and an up-to-date review of wetlands can be found in Finlayson and Moser (1991). Conservation and management issues are covered in Dugan (1990) and more recently in Keddy (2000).

WETLAND FUNCTIONS, PRODUCTS, ATTRIBUTES AND VALUES

Wetlands perform functions as a result of the interactions among soil, water, plant and animal species. Products such as fisheries, wildlife and forest resources may be generated and attributes conferred such as biodiversity and cultural uniqueness. Functions, products and attributes are all valuable to society but the extent of this varies from wetland to wetland. A useful overview of this topic is given in Dugan (1990). Much more detailed discussion of functions can be found in Adamus and Stockwell (1983), Maltby (1986) and Mitsch and Gosselink (1986). An alternative view of functioning is found in Maltby (1991a) in which wetland roles are characterised as producer, store/sink, pathways and buffers.

WETLAND RESEARCH

Given the diversity of wetlands, there is a large scope for study which can either concentrate specifically upon functional values described above, or examine more specific processes and interrelationships of the biological, physical and chemical components of these ecosystems. A comprehensive account of the appropriate techniques cannot be given here; instead some suggestions are given for suitable areas of study with more detailed consideration of certain aspects of wetland research. It is hoped that this will serve as an introduction to the literature and the work that is possible.

A range of useful work might include:

- relationships between hydrological regime, vegetation and soil characteristics (e.g. water table–redox profile–vegetation distribution)
- investigation of zonation of use of wetlands by human communities and wildlife (patterns of use in space/time), e.g. Marchand and Udo De Haes (1989)
- assessment of the ecological/environmental functioning and values of products obtained from wetlands, e.g. wood products, animals, crops (Barbier, 1989; Nather Khan, 1990; Othman and Shalwahid, 1990; Maltby, 1991b, 1992). A good example of such on-going work can be found in the Royal Holloway Institute for Environmental Research's Darwin Initiative-funded project on wetland restoration in the Mekong Delta of Vietnam. For further details visit the Institute's website and follow links to the Darwin Southeast Asian Wetland Restoration Initiative
- effects of river regulation, irrigation and other human intervention on wetland characteristics (e.g. flooding extent, vegetation change, water table) (Drijver and Marchand, 1985)
- historical changes in wetlands (e.g. interviews, aerial photographs, maps, flood records) (Hollis, 1986).

WETLAND HYDROLOGY

Wetlands are by definition dependent on the presence of water for all or part of a year either just below the soil/sediment surface or above it. It is vital in the transport of materials to, from and within the wetland while providing the habitat for often rich, diverse plant and animal communities; as a result most wetland research and study require at least a fundamental grasp of the site's hydrology. Therefore the objectives of hydrological work can cross a spectrum from basic budget studies complementing other projects to self-contained research topics such as flood mitigation and water quality regulation. An excellent recent review is found in Bullock and Acreman (2003).

The overall water budget is a very useful approach to studying wetland hydrology and is described in Mitsch and Gosselink (2000). It provides an overall view of the transfer of water in a system over a year; subsequent division of the budget into smaller time periods will describe the hydrological regime, especially when coupled with knowledge of a site's water storage capacity (see later). Knowledge of a wetland's

Figure 24.2
Collecting sediment samples as part of a mangrove impact assessment study on Rodrigues Island (© RGS–IBG/Tom Hooper)

hydrological regime is important in the understanding of functions such as flood alteration, groundwater recharge/discharge, nutrient dynamics, food chain support and habitat provision.

The following is an equation describing a wetland's overall water budget (Mitsch and Gosselink, 2000):

$$dV = Pn + Si + Gi - ET - So - Go \pm T$$

where V is the volume of water storage in the wetlands, dV is the change in volume of water storage in the wetlands, Pn is the net precipitation, Si is the surface inflow, including flooding streams, Gi is the groundwater inflow, ET is the evapotranspiration, So is the surface outflow, Go is the groundwater outflow, T is the tidal inflow ($+$) or outflow ($-$).

Obviously completion of the formula requires a considerable input of data. These can be collected in the field using standard hydrological techniques (Wilson, 1983). However, it is rare to find a study that does monitor the entire hydrological suite. Often the "missing" data can be obtained from government sources, previous studies at the site or recording stations in a similar climatic region. LaBaugh (1986) gives an overview of these problems and Hollis (1986) provides a good example of their application.

Some wetland functions are purely hydrological in their nature, such as flood storage. Wetlands often serve as natural storage or collecting points for run-off or river discharge; therefore the potential exists to prevent or reduce flood peaks moving downstream, by temporarily detaining the water, by retaining the water from surface run-off or by the reduction of floodwater velocity. This is potentially of great socioeconomic value (see US Army Corps of Engineers, 1972; Novitzki, 1979). Larson et al. (1989) provide a useful overview of this function and list a number of predictors of opportunity and effectiveness, based on map-and-field observation techniques. Quantitative techniques range from direct measurement of flood events to more morphological assessments of a wetland's capacity to mitigate floods. Outlet condition is a key indicator. A comparison of the wetland inlet cross-sectional area with that of its outlet gives an indication of the wetland's capacity to attenuate upstream discharges. Standard surveying techniques and cross-section measurement can be used; if discharge measurements can be made at the inlets and outlets then, using the Manning "n" equation, a rating curve can be constructed enabling regular run-off measurements to be taken using a stage reading (Wilson, 1983 gives a clear description of this technique). Using maps and field surveying, the volume of the wetland basin or its storage capacity can be estimated and compared with expected volumes of floodwater. Alternatively trash lines of previous flood events can be mapped at the edge of the wetland and used as reference points for previous floodwater levels.

The water quality of resident and discharging water is intimately linked to a wetland's hydrology. Considerable influence on local and even regional water supply quality may be exerted by a wetland receiving nutrients from run-off and groundwater sources. Interactions of wetland ecosystems with groundwater are often very important to their chemical budget. Good overviews of wetland hydrogeological relationships are provided by Carter and Novitzki (1988) and Brown et al. (1985). Larson et al. (1989) again give a useful rapid functional assessment of the hydrogeological wetland function, which helps to provide an initial overview for a study. It is important to establish the quantity, quality, direction and regime of groundwater flow. Already published hydrogeological and geological data can (where available) provide a history of change and display annual trends. More specific variation can be investigated by sampling water in the field using existing wells and field installation of piezometers (see Siegel, 1988a, 1988b; Faulkner et al., 1989; Roulet, 1990) and comparing these with other hydrological data collected.

Water quality is also dictated by surface interactions between water and sediment/soil. The examination of these interactions has been covered by workers in wetland wastewater treatment as well as natural systems. If only an establishment of the net change in the water quality is required, a "black-box system" of monitoring inlet and outlet water quality variation coupled with discharge measurement can be considered. If more detailed process studies of the spatial and temporal relationships is required, sampling along transects within the wetland and the measurement of direction and discharge of surface water is necessary. Hill and Warwick (1987, 1988) and Kadlec (1988, 1990) describe these techniques.

WETLAND SOILS AND VEGETATION

Wetland soils and vegetation interact to produce the biological uniqueness of a site as well as performing socioeconomically valuable functions such as the removal of nitrogen compounds from run-off, the trapping of sediment from waterways, and the provision of local medicines, foodstuffs and building materials. Mitsch and Gosselink (2000) provide useful introductions to the physical functions whereas Marchand and Udo De Haes (1989) provide good insight into the social values of wetlands. A few of the projects that might be attempted are outlined briefly below.

One of the most interesting features of wetlands is the remarkable zonation that occurs in plant communities. This is particularly well illustrated in the case of coastal and river-marginal wetlands, especially in the tropics, e.g. zonation occurs in mangrove species and these species often give way in succession to nipa forest and peat swamp forest communities. We still know relatively little about the exact environmental relationships in these successions and of the dynamics of plant communities and change. Despite the relative inaccessibility of densely vegetated areas such as these, some locations exhibit change over short distances and therefore lend them-

Figure 24.3 *Using ground-penetrating radar to measure soil density in peat. Remember, if you are using sophisticated equipment in wetland environments it is absolutely essential that this is tested thoroughly beforehand and if necessary protected from the wet (© Jo Holden)*

selves to more complex investigation. Studies can range from simple mapping exercises to more complex investigation of the successions and their relationships to soil parameters (e.g. pH, salinity) or hydrological regime (e.g. degree of inundation, periodicity of flooding).

Soils often provide important clues to wetland processes, character and history. Waterlogged soils have classic morphological features, such as the presence of a surface peaty layer and distinct mottling in the mineral horizons. The latter is caused by variations in the oxidation/reduction state of the soil, which affects the chemical status of particular elements resulting in distinct colour changes. The most notable is that which occurs between ferric and ferrous; in reduced pockets the Fe^{2+} is often bluish/green in colour which contrasts sharply with the rusty red/brown of the Fe^{3+} oxidised state. The depth and degree of development of such "mottling" patterns give a good indication of water table regime in the soil and particularly the depth to permanent water table (the mottling regime generally disappears in permanently saturated horizons). Investigation is relatively simple using an auger and soil colour chart. The extent and depth of such horizons can be mapped and related to hydrology, vegetation and the impact of anthropogenic alterations to a system.

WETLAND FIELD LOGISTICS

All environments pose logistical problems to expeditions and wetlands are no exception. At all stages of planning it is wise to keep these in mind. In wetlands they are generally caused by one or a combination of the wet and boggy conditions, and the temporal variation of these, the often dense and sometimes impenetrable vegetation, the lack of good communications to and within a site, and often the remoteness of these ecosystems from civilisation. The more common problems being:

- site accessibility
- mobility of expedition equipment
- the availability of portable and sufficiently rugged (and waterproof) equipment
- the availability of basic data representative of the site, e.g. rainfall, river discharge, maps (soil, vegetation and geology) and even indigenous population statistics
- the presence of local accommodation or even land to camp on; risk of diseases such as schistosomiasis
- local facilities for the storage of degradable samples (e.g. cold storage for soils being analysed at a later date)
- local facilities for sample analysis.

Despite these problems, much help can be gained from non-governmental organisations and local agencies involved or associated with wetland research, conservation and management. Examples of the former are the Royal Society for the Protection of Birds, Wetlands International and the WWF International Mire Conservation Group. The latter are often government agencies involved in agriculture, forestry, fisheries, hydrology/water resources and conservation, or very often universities. Contact with these, preferably before the arrival of the main expedition (either indirectly or directly during a reconnaissance trip) will often yield much help in background data collection, field logistics, local information and sometimes even liaison in the field.

FURTHER INFORMATION

Useful websites
EVALUNET project web page: http://www1.rhbnc.ac.uk/rhier/evaluweb/index.shtml
DARWIN SEAWRI web page: http://www.rhul.ac.uk/Environmental-Research/Research/Darwin/Darwin.html
International Mire Conservation Group: www.imcg.net
Ramsar Convention on Wetlands: www.ramsar.org
River Basin Initiative: www.riverbasin.org
Royal Holloway Institute for Environmental Research: http://www.rhul.ac.uk/Environmental-Research/index.html
Society of Wetland Scientists: www.sws.org
US Fish and Wildlife Service National Wetlands Inventory: www.nwi.fws.gov
US Wetlands Regulation Center: www.wetlands.com/regs/tlpgeooa.htm
Wetlands International: www.wetlands.org
Wetlands of the Central and Southern California – a methodology for classification: http://lily.mip.berkeley.edu/wetlands
Wetland definitions from American and Canadian Agencies: www.ecn.purdue.edu/agen521/epadir/wetlands/definitions.html
Wetland links – sorted by category: www.mindspring.com/~rbwinston/wetland.htm
Wildfowl and Wetlands Trust: www.wwt.org.uk

Bibliography
Adamus, P.R. and Stockwell, L.T. (1983) *A Method for Functional Assessment:* Vol 1. *Critical Review and Evaluation Concepts.* FHWA-IP 82 83. Washington DC: US Department of Transport Federal Highway Administration.
Barbier, E.B. (1989) *The Economic Value of Ecosystems:* I. *Tropical Wetlands.* Gatekeeper Series LEEC 89-02. London: IIED,
Brown, R.G., Stark, J.R. and Patterson, G.L. (1985) Groundwater and surface water interaction in Minnesota and Wisconsin wetlands. In: Hook, D.D. (ed.), *The Ecology and Management of Wetlands,* Vol. 1, *The Ecology of Wetlands.* London: Croom-Helm, pp. 176–80.
Bullock, A. and Acreman, M. (2003) The role of wetlands in the hydrological cycle. *Hydrological and Earth System Sciences* 7: 358–89.

Carter, V. and Novitzki, R.P. (1988) Some comments on the relation between groundwater and wetlands, Ch. 7. In: Hook, D.D. (ed.), *The Ecology and Management of Wetlands*, Vol. 1, *The Ecology of Wetlands*. London: Croom-Helm.

Cowardin, L.M., Carter, V., Golet, F.C. and La Roe, E.T. (1979) *Classification of wetlands and deepwater habitats of the United States*. US Fish & Wildlife Service Publication, FWS/OBS 79/31, Washington DC.

Drijver, C.A. and Marchand, M. (1985) *Taming the Floods: Environmental aspects of floodplain development in Africa*. Centre for Environmental Studies, State University of Leiden.

Dugan, P. (ed.) (1990) *Wetland Conservation*. Gland, Switzerland: IUCN.

Faulkner, S.P., Patrick, W.H. Jr and Gambrell, R.P. (1989) Field techniques for measuring wetland soil parameters. *Soil Science Society of America, Journal*. **53**: 883–90.

Finlayson, M. and Moser, M. (eds) (1991) *Wetlands*. London: Facts on File.

Hamilton, L.S. and Snedaker, S.C. (eds) (1991) *Mangrove Area Management Handbook*. IUCN, Unesco, East-West Centre. Available from: Environment and Policy Institute, East-West Centre, 1777 East-West Road, Honolulu, Hawaii 96848.

Hill, A.R. and Warwick, J. (1987) Ammonium transformations in springwater within the riparian zone of a small woodland stream. *Canadian Journal of Aquatic Science* **44**: 1948–56.

Hill, A.R. and Warwick, J. (1988) Nitrate depletion in the riparian zone of a small woodland stream. *Hydrobiologia* **157**: 231–40.

Hollis, G.E. (ed.) (1986) *The Modelling and Management of the Internationally Important Wetland at Garaet El Ichkeul, Tunisia*. IWRB Special Publication No. 4. Slimbridge: IWRB.

Kadlec, R.H. (1988) Monitoring wetland responses. In: Zelazny, J. and Reierabend, J.S. (eds), *Increasing Our Wetland Resources*. Proceedings of a Conference, The National Wildlife Federation-Corporate Conservation Council, pp. 114–20.

Kadlec, R.H. (1990) Overland flow in wetlands: vegetation resistance. *Journal of Hydraulic Engineering*, **116** (5): 691–706.

Keddy, P.A. (2000) *Wetland Ecology, Principles and Conservation*. Cambridge: Cambridge University Press.

Kent, D.M. (ed.) (2000) *Applied Wetlands Science and Technology*. Boca Raton, Fl: Lewis Publishers.

LaBaugh, J.W. (1986) Wetland ecosystem studies from a hydrologic perspective. *Water Research Bulletin* **22**(1): 1–10.

Larson, J.S., Adamus, P.R. and Clairain, E.J. Jr (1989) *Functional Assessment of Freshwater Wetlands: A manual and training outline*. Glaud, Switzerland: WWF Publication (89-6): 62.

Maltby, E. (1986) *Waterlogged Wealth – Why waste the world's wet places?* London: Earthscan.

Maltby, E. (1991a) Wetlands – their status and role in the biosphere. In: Jackson, M.B., Davies, D.D. and Lambers, H. (eds), *Plant Life under Oxygen Deprivation*. The Hague: SPB Academic Publishers, pp. 3–21.

Maltby, E. (1991b) The world's wetlands under threat – developing wise use and international stewardship. In: Hansen, J.A. (ed.), *Environmental Concerns*. London: Elsevier, pp. 109–36.

Maltby, E. (1992) Wetlands and their values. In: Finlayson, M. and Moser M. (eds), *Wetlands*. London: Facts on File, pp. 8–26.

Maltby, E. and Turner, R.E. (1983) Wetlands are not Wastelands. *Geographical Magazine* LV: 92–7.

Marchand, M. and Udo De Haes, H.A. (eds) (1989) *Traditional Uses: Risks and potentials for wise use*. Proceedings of the International Conference, The People's Role In Wetland Management. Leiden, The Netherlands.

Mitsch, W.J. and Gosselink, J.G. (2000) *Wetlands*. 3rd edn. New York: Van Nostrand Reinhold Co.

Nather Khan, I.S.A. (1990) Socio-economic values of aquatic plants (freshwater macrophytes) of Peninsular Malaysia. *Asian Wetland Bureau Report*. No 67c, University of Malaya, Kuala Lumpur.

Novitzki, R.P. (1979) Hydrologic characteristics of Wisconsin's wetlands and their influence upon floods, streamflow and sediment. In: Greeson, P.E., Clark, J.R. and Clark, J.E. (eds), *Wetland Functions and*

Values: The state of our understanding. Minneapolis, MI American Water Resources Association,: pp. 377–88.

Othman, M. and Shalwahid, H. (1990) *A Preliminary Economic Valuation of Wetland Plant Species in Peninsular Malaysia.* Kuala Lumpur, Malaysia: Asian Wetland Bureau.

Roulet, N.T. (1990) Hydrology of a headwater basin wetland: groundwater discharge and wetland maintenance. *Hydrological Process* **4**: 387–400.

Shaw, E.M. (1983) *Hydrology in Practice.* New York: Van Nostrand Reinhold Co.

Siegel, D.T. (1988a) The recharge-discharge function of wetlands near Juneau, Alaska: part 1: hydrogeological investigations. *Groundwater* **26**: 427–34.

Siegel, D.T. (1988b) The recharge-discharge function of wetlands near Juneau, Alaska: part 2: geochemical investigations. *Groundwater* **26**: 580–6.

US Army Corps of Engineers (1972) *Charles River Watershed, Massachusetts.* Waltham, MA: New England Division.

Wilson, E.M. (1983) *Engineering Hydrology,* 3rd edn. London: Macmillan Education Ltd.

25 EXPEDITION RESEARCH PROJECTS IN SAVANNAH REGIONS

Malcolm Coe and Andrew Goudie

ECOLOGICAL STUDIES

If you are planning an expedition to a semi-arid tropical savannah you will have plenty of places to choose from, whether it is your intention to study animals, plants or even the ecology of the local people. As a rough guide, these exciting but harsh environments cover up to 65 per cent of the surface of Africa, 60 per cent of Australia and 45 per cent of South America (Huntley and Walker, 1982). They are characterised by an intensely seasonal climate, in which the wet seasons are often unpredictable. When the rains do arrive they will frequently do so as gentle "grass rains" to start with, followed by irregular, intense storms, during which up to 16 per cent of the annual total may fall in a single day (Coe, 1990). Between the rainy seasons, savannah environments are usually intensely dry, during which the daily temperature range may approach 40°C. During the dry season virtually no plant growth will take place and large quantities of dead organic matter will accumulate on the ground surface until the next rains, when decomposition will take place very rapidly, releasing nutrients for future photosynthetic activity.

Top tip

It is worth remembering that most of the material collected on your expedition will have to be deposited in the local museum or university, but you may obtain permission to return some samples to your home base for identification. Voucher specimens must always be returned to your host country after the study is completed unless an arrangement has been made to retain duplicates here.

Having selected an area that you wish to visit, the first thing that you need to do is to identify the organisms that you wish to study and to be quite certain that they will be active when you arrive at your study site. There is little point in visiting a savannah to study frogs or flowering plant phenology in the middle of the dry season, when the organisms that you hope to study will be inactive. This does not mean, however, that destinations in East and Central Africa, South America or Australia are not worth visiting in the long vacation, which frequently coincides with their dry season, because there are still plenty of things that you can do there. By contrast the semi-arid savannahs of West Africa and India receive much of their annual rainfall between June and September, which makes them very suitable for the study of animals and plants during their period of maximal activity. Even though you may have the whole animal and plant kingdoms potentially available for study, it is wise to choose your organisms with care, ensuring that your generally inexperienced team can identify them and obtain a reasonably complete data-set in the time available. Such projects may range from general topics, to more detailed and specialised ones, depending on your interests, but the paramount rule must always be "keep it simple".

Top tip

In conducting a general survey it is vital that you locate literature sources that will enable you to identify your material accurately. This can usually be done through a university library, either at home or in your host country. Cooperation with your overseas counterparts is very valuable, because they may often be able to provide you with taxonomic assistance.

General surveys usually require large numbers of people in the field, together with the attendant problems of transport and logistic support. It is often therefore simpler to have a specific objective of studying a single group of organisms, or even a single species, providing you are sure that it will be abundant. There is little point in setting out to study a single, rare, endangered species if you are going to spend most of your time simply trying to find it. It is not possible to outline every organism that you could study, but the following headings will outline the general principles of deciding "what to study" and "how to study it" (Magurran, 1988; Wilson et al., 1996).

Vegetation studies

As the activity and distribution of most animals are, to a large degree, dependent on their habitats, the study of vegetation is often a vital preliminary component of most ecological studies. Simple vegetation maps are of great value to conservation authorities, and can be carried out by small teams. Before you leave, you may be able to obtain or arrange to view aerial photographs or even satellite images, which will act as a good baseline from which to carry out your survey.

Top tip

In conjunction with a vegetation study it is of great interest to select a genus or species of plant in your research area and to investigate the animals associated with it. Comparative investigations of a number of related plant species often

---

yield information that is of great ecological, evolutionary and conservation value.

Within such a programme of fieldwork, you may wish to use simple techniques to study the structure of the local vegetation. This can be accomplished by using standard techniques, but it is worth remembering that studies of woody vegetation can still be carried out in a dry season, which is impossible with herbaceous components of the vegetation.

Quadrats are quite valuable in studying the vegetation of small areas, but stratified transects are often a more efficient method of investigating habitat components on larger study sites. Tree and shrub density can be studied using the "point-centred quarter" method of Curtis (1959). Additional information on tree bole cover can be estimated using the Bitterlich stick (Cooper, 1963; Agnew, 1968). Having carried out these measurements on your transect(s), it is quite easy to convert this to a strip of predetermined width to obtain greater detail on structure and composition. The number of transects that you need will depend on the size of the area under study and its habitat heterogeneity.

Under suitable climatic conditions, studies of the phenology (leaf production, flowering and fruiting) of local trees and shrubs can be of immense interest and value to local and international agro-forestry bodies.

Methods of vegetation study and analysis may be consulted in Greig-Smith (1983), Gauch (1982) and Ludwig and Reynolds (1988).

Animal studies

There are a vast number of different methods available for studying and sampling animals, which are well summarised in Southwood (2002). A good point to remember is that small animals provide much larger data-sets than large animals. With only 6 or 8 weeks actually working in the field, it will not be possible to obtain complete data on, for example, the African elephant, whereas the same period spent studying rodents would yield really valuable information, even if it reflects the situation in only part of a single season. The general picture for the major animal groups is summarised below.

Invertebrates

Arthropods are ideal objects for study by small expeditions, because they are generally abundant, and many are active even in the dry season as adults or their immature stages. They may be sampled using sweep nets or beating trays, mark–release recapture, pit-fall traps or by "fogging" with pyrethroid insecticides. Interpretation of the data collected by these methods may be consulted in Southwood (2002). Social insects such as termites and ants are particularly valuable as objects of study, because their discrete nests enable the investigator to concentrate their studies in a small area,

289

whether this concerns mound ventilation, foraging behaviour, or even the other organisms that use or share the mounds or nests.

Expeditions working on sand or fine alluvial soils will find the tracks of both invertebrates and vertebrates on the soil surface, whereas slightly raised ridges will indicate the presence of sub-surface predatory adult arthropods or their larvae. These "signs" provide an excellent opportunity to study the activity and foraging behaviour of these creatures.

Vertebrates

Fish surveys are of great value to the museums and fisheries authorities in many less developed countries. They may be sampled directly using nets or lines, when the necessary permits have been obtained, or you may simply be able to cooperate with local fishermen to study their catches. Even many of the most arid regions have permanent or seasonal rivers, lakes and seasonal water bodies that are worthy of investigation (Coad, 1998).

Top tip

Studies of dangerous reptiles should be avoided, unless you have experience in handling them, or are accompanied by a local expert.

Reptiles and amphibians are interesting creatures to study, although the latter are much more seasonal than the former. Simple species lists, and their local abundance and distribution in relation to local habitat structure can provide valuable and often unique data. It may be necessary to kill some animals if you wish to study their feeding behaviour, but this will usually require special permission and should always be kept to a minimum (for all animals).

Behavioural studies of activity rhythms, in relation to sex, size and diurnal climatic variables, are easy to accomplish and provide interesting and valuable data. If it is your intention to study local movements of your study animals it will be necessary to mark them. Good guidelines for humane methods of marking animals may be found in Stonehouse (1978).

Birds are popular study creatures because they are generally easy to observe and identify. Behavioural studies are often limited by the seasons, but liaison with local ornithologists should help you to decide which species can be studied during the period of your fieldwork. Species lists and their relative abundance are of considerable interest, especially if it is possible to relate this information to local habitat and climatic variables.

Mammals provide good opportunities for the field biologist, but you should remember that large mammals will be less common than small ones. If you have transport and the local authorities are interested in obtaining information on the abundance and distribution of large mammals, this can be accomplished by carrying out regular road transects in a vehicle or even by walked transects. This information may be related to factors such as habitat type, habitat structure and time of day. Clearly, in a few weeks of fieldwork you will obtain only a picture that has relevance to the season in which you carry it out. Riney (1982) and Bothma (1989) provide valuable information on the study of large mammals and their habitats. GPS/GIS (global positioning system/geographic information system) techniques are increasingly being employed in these studies, providing information on position and even abundance (Packer et al., 1998).

Small mammals provide an opportunity to observe a fairly diverse fauna in most savannah environments. Although it may be necessary to kill some animals if you wish to study feeding or reproduction, a great deal of valuable work can be carried out on distribution and abundance, using simple mark–release recapture techniques. Special precautions should always be taken in handling small mammals in respect of rabies and other potentially hazardous parasites and diseases.

TABLE 25.1 **TOP REFERENCES FOR THOSE WANTING TO STUDY THE EFFECT OF ORGANISMS ON SAVANNAHS**

Organism	Reference
Termites	Goudie (1988)
Ants	Humphreys (1981)
Worms	Humphreys (1981)
Birds	Mitchell (1988)
Anteaters	Mitchell (1988)
Elephants	Laws (1970)

SOURCES OF ADDITIONAL INFORMATION

Geomorphological research
In this section, attention is drawn to certain geomorphological phenomena that are of special interest, and references will be given to work that has previously been undertaken on such phenomena in savannah environments. A good general introduction to savannah landforms is given by Thomas (1994).

Past flood estimation
From time to time savannah areas, in spite of the fact that they are not normally as wet as the humid tropics, are subjected to very powerful storms, such as tropical cyclones or other major atmospheric disturbances. Such storms can cause extensive

Figure 25.1
Sediment coring is physically hard work but can reveal much about the palaeoecology of a region (© Andrew Plater, Liverpool University)

flooding and, for engineering purposes (e.g. bridge or dam construction), it is valuable to have an estimate of the sort of flood discharges that can come down a particular river under such circumstances. Unfortunately, in many areas long-term gauging records are not available, and in some cases the record may have been disrupted by past flood events themselves! For this reason geomorphologists and hydrologists have developed techniques for estimating past flood discharges that do not depend on gauging records.

A good example of this technique, together with details of how calculations are made, is provided by Gillieson et al. (1991). Another related method of estimating the discharges of past flood events is to look at evidence for bent or damaged trees along a gorge and to try to estimate the date of the damage by dendrochronology (tree-ring analysis) (see Hupp, 1988).

Dune system descriptions

Savannah areas, being at the transition between dry (desert) and moist (rain forest) environments, have been subjected to major climatic changes during the course of the Quaternary era. During dry phases, when desert margins extended towards the equator, desert dunes were more extensive and they now underlie large areas of savannahs, as in the Kalahari, West Africa, North-west India and much of Australia.

From a geomorphological and palaeoclimatological viewpoint, it is of value to describe the form, sedimentology and age of such ancient dune systems. Examples of the type of work that can be undertaken are provided in Thomas and Shaw (1991).

Extinct lake surveys

The imprint of past moister conditions is equally evident in lake basin areas. When conditions were wet the lakes reached higher levels (leading to the creation of shore-lines) and had different chemical and biological conditions. Thus, it is important to seek evidence for high lake shorelines or to obtain cores from sediments laid down during different stages of their history. Such cores can be extracted from lake beds by a range of coring devices and using rafts. The analysis of the cores is a complex matter, requiring input from pollen analysts, chemists, diatomists, etc. Dating is also an expert matter, but there is no doubt that, if we are to understand the past history of savannah areas, these are among the most productive methods. A good example of recent work in this area includes that of Hooghiemstra (1989).

Studies of erosion

Savannah areas have been the subject of considerable erosional activity, because of either the intrinsic nature of tropical rainfall or the effects of vegetation degradation promoted by human activities. Such erosion may be evident as a general reduction in the level of the land surface or through the development of erosional scars (gullies). It is important to know the age, rates and causes of such erosion. A range of useful work can be done in the field:

- Surveying of gully systems to compare their extent with those shown on old maps and air photographs.
- Archaeological examination of gully systems to determine their age and history.
- Measurement and dating of degree of root exposure of trees by dendrochronology.
- Estimation of rates of sediment accumulation in reservoirs behind dams of known age.
- Instrumentation (e.g. with erosion pins) and detailed survey of gully systems so that sequential measurements can be undertaken by future teams.
- Experimental run-off and sediment generation using rainfall simulation techniques on different land surfaces.

The following publications give good examples of the type of work that can be undertaken: Price-Williams et al. (1982), Biot (1990) and Dunne et al. (1979).

Indeed, soil erosion is but one manifestation of the possible role of humans in modifying and degrading savannah environments. Savannah areas are one of the prime environments that are subject to desertification – the spread of desert conditions into areas where under normal climatic circumstances they would not exist. Deforestation, over-grazing and related processes expose savannah surfaces to wind and water erosion, landslip formation, accelerated sedimentation and various other deleterious geomorphological processes. There is a great need for "ground truth" on the status of areas subject to desertification and for field surveys connected with current and past remote sensing imagery (including air photographs, some of which may go back four or more decades). An excellent general discussion of desertification is provided by Grainger (1990), whereas Mortimore (1989) demonstrates what can be done by painstaking research in a specific savannah area of West Africa. This includes the use of air photography and ground survey to monitor accelerated deflation and dune reactivation. Social aspects are expanded in Chapter 4.

Organisms other than humans play a major role in the moulding of savannah environments, be they small (e.g. termites) or large (e.g. elephants), yet their contribution to landform development has not received the attention that it deserves. The whole area of what is called "biogeomorphology" has been reviewed by Viles (1988).

In some of the world's savannah areas there is extensive development of various phenomena associated with the solution of limestone bedrock. Savannah karst phenomena have probably not received as much attention as those in more humid areas, but major cave systems (possibly dating to earlier more humid phases) do exist. The karstic phenomena of the Napier range of north-west Australia were the subject of the classic investigation of Jennings and Sweeting (1963) and comparable work needs to be done in other areas. Moreover, because of the high rates of evaporation in such areas, a whole suite of limestone precipitation forms develop, called tufas or travertines (Viles and Goudie, 1990) and these deserve further study, particularly with regard to the role of organisms (such as mosses) in their formation.

Landform surveys

In the eyes of many visitors to savannah environments the most typical landforms are miscellaneous types of isolated hill (inselbergs, bornhardts, koppies, tors, etc.), which have developed in a range of rock types, including granites, migmatites and sandstones. The development of such features is closely related to the type, structure and mineralogy of the rocks in which they are developed and there is considerable scope for trying to establish the precise relationship between rock type and inselberg form and distribution. An example of this type of work as part of an undergraduate project is described in Gibbons (1981), whereas some of the methods of determining rock properties are described in Pye et al. (1986).

Another characteristic landform type of savannah areas is the Dambo (see Thomas and Goudie, 1985). These are small channels, seasonally waterlogged, grassy valleys, often with rather rectangular patterns. They are especially widespread in Central Africa (e.g. in Zimbabwe, Malawi and Zambia) but are also known from West

Africa (the *fadama* of Nigeria and the *bolis* of Sierra Leone) where they are as much a part of the landscape as the magnificent inselbergs and endless plains. Little is known about the distribution and form of these features in some parts of Africa, and our knowledge of such features in northern Australia, India and South America is slim indeed. They present considerable research opportunities and offer scope for collaboration with plant scientists.

To conclude, savannah areas, in spite of their very considerable extent and importance, have not always received the same level of attention from geomorphologists that they deserve. They do not have some of the specifically climatic-related landforms of some other major biomes (e.g. the active dunes of hyperarid regions or the glaciers of cold areas). Nevertheless, they present many challenging geomorphological problems and phenomena and, especially in the dry season, can offer a congenial and productive environment in which to work.

FURTHER ADDITIONAL INFORMATION

Bibliography
Agnew, A.D.Q.A. (1968) Observations on the changing vegetation of Tsavo National Park (East). *East African Wildlife Journal* 6: 75–80.
Biot, Y. (1990) The use of tree mounds as benchmarks of previous land surfaces in a semi-arid tree Savanna, Botswana. In: Thornes, J.B. (ed.), *Vegetation and Erosion*. Chichester: Wiley, Chapter 26
Bothma, J. du P. (ed.) (1989) *Game Ranch Management*. Pretoria: J.L. van Schaik.
Coad, B.W. (1998) *Fishes: Expedition field techniques*. London: RGS–IBG Expedition Advisory Centre.
Coe, M. (1990) The conservation and management of semi-arid rangelands and their animal resources. In: Goudie, A.S. (ed.), *Desert Reclamation*. Chichester: John Wiley, pp. 219–49.
Coe, M. and Collins, N.M. (eds) (1986) *Kora: An ecological inventory of the Kora National Reserve, Kenya*. London: Royal Geographical Society.
Coe, M., McWilliam, N., Stone, G. and Packer, M. (1999) *Mkomazi: The ecology, biodiversity and conservation of a Tanzanian savanna*. London: Royal Geographical Society.
Cole, M. (1986) *The Savannas: Biogeography and geobotany*. London: Academic Press.
Cooper, C.F. (1963) An evaluation of variable plot sampling in shrub and herbaceous vegetation. *Ecology* 44: 565–8.
Curtis, J.T. (1959) *The Vegetation of Wisconsin, Madison*. Madison: University of Wisconsin Press.
Dunne, T., Dietrich, W.E and Brunenoo, M.J. (1979) Recent and past erosion rates in semi-arid Kenya. *Zeitschrift für Geomorphologie Supplementband* 29: 130–40.
Gauch, H.G. (1982) *Multivariate Analysis in Community Ecology*. Cambridge: Cambridge University Press.
Gibbons, C.L.M.H. (1981) Tors in Swaziland. *Geographical Journal* 147: 72–8.
Gillieson, D., Ingle Smith, D. Greenaway, M. and Ellaway, M. (1991) Flood history of the limestone ranges in the Kimberley Region, Western Australia. *Applied Geography* 11: 105–23.
Goudie, A.S. (1988) The geomorphological role of termites and earthworms in the tropics. In: Viles, H.A. (ed.), *Biogeomorphology* Oxford: Basil Blackwell, pp. 166–92.
Grainger, A. (1990) *The Threatening Desert: Controlling desertification*. London: Earthscan.
Greig-Smith, P. (1983) *Quantitative Plant Ecology*, 3rd edn. Berkeley, CA: University of California Press.

Harris, D.R. (1980) Tropical savanna environments: definition, distribution, diversity and development. In: Harris, D.R. (ed.), *Human Ecology in Savanna Environments*. London: Academic Press, pp. 3–27.

Hooghiemstra, H. (1989) Quaternary and Upper Pliocene glaciation and forest development in the tropical Andes: evidence from a long high resolution pollen record from the sedimentary basin of Bogota, Colombia. *Palaeogeography, Palaeoclimatology, Palaeoecology* **72**: 11–26.

Humphreys, G.S. (1981) The rate of ant mounding and earthworm casting near Sydney, New South Wales. *Search* **12**: 129–31.

Huntley B.J. and Walker, B.H. (eds) (1982) *Ecology of Tropical Savannas*. Berlin: Springer-Verlag.

Hupp, C.R. (1988) Plant ecological aspects of flood geomorphology and paleoflood history. In: Baker, V.R., Kochel, R.C. and Patton, P.C. (eds), *Flood Geomorphology*. New York: Wiley, Chapter 20.

Jennings, J. and Sweeting, M.M. (1963) The limestone ranges of the Fitzroy Basin, Western Australia. *Bonner Geographislher Abhandlungen* **32**: 60pp.

Laws, R.M. (1970) Elephants as agents of habitat and landscape change in East Africa. *Oikos* **21**: 1–15.

Ludwig, J.A. and Reynolds, J.F. (1988) *Statistical Ecology: A primer on methods and computing*. New York: John Wiley.

Magurran, A.E. (1988) *Ecological Diversity and Its Measurement*. Princeton, New Jersey: Princeton University Press.

Mitchell, P.B. (1988) The influences of vegetation, animals and micro-organisms on soil processes. In: Viles, H.A. (ed.), *Biogeomorphology*. Oxford: Basil Blackwell, pp. 43–82.

Mortimore M. (1989) *Adapting to Drought. Farmers, famines and desertification in West Africa*. Cambridge: Cambridge University Press.

Packer, M.J., Canney, S., McWilliam, N.C. and Abdallah, R. (1998) Ecological mapping of a semi-arid savannah. In: *Mkomazi: The ecology, biodiversity and conservation of a Tanzanian savannah*. London: Royal Geographical Society (with IBG), pp. 43–68.

Price-Williams, D. Watson, A. and Goudie, A.S. (1982) Quaternary colluvial stratigraphy, archaeological sequences and palaeoenvironment in Swaziland, southern Africa. *Geographical Journal* **148**: 50–67.

Pye, K., Goudie, A.S. and Watson, A. (1986) Petrological influence on differential weathering and inselberg development in the Kora area of Central Kenya. *Earth Surface Processes and Landforms* **11**: 41–52.

Riney, T. (1982) *Study and Management of Large Mammals*. Chichester: John Wiley.

Southwood, T.R.E. (2002) *Ecological Methods*. London: Chapman & Hall.

Stonehouse, B. (ed.) (1978) *Animal Marking: Recognition marking of animals in research*. London: Macmillan.

Thomas, D.S.G. and Shaw, P. (1991) *The Kalahari Environment*. Cambridge: Cambridge University Press.

Thomas, M.F. (1994) *Geomorphology in the Tropics*. Chichester: John Wiley.

Thomas, M.F. and Goudie, A.S. (eds) (1985) Dambos: small channelless valleys in the tropics. *Zeitschrift für Geomorphologie Supplementband* **52**: 1–222.

Viles, H.A. (ed.) (1988) *Biogeomorphology*. Oxford: Basil Blackwell.

Viles, H.A. and Goudie, A.S. (1990) Reconnaissance studies of the Tufa deposits of the Napier Range, N.W. Australia. *Earth Surface Processes and Landforms* **15**: 425–43.

Wilson, D.E., Cole, F.R., Nichols, J.D. et al. (1996) *Measuring and Monitoring Biological Diversity: Standard methods for mammals*. Washington DC: Smithsonian Institution Press.

Websites

Cooperative Research Centre for Tropical Savannas Management, Australia: http://savanna.ntu.edu.au/
International Association of Geomorphologists: www.geomorph.org
Virtual Geomorphology: http://main.amu.edu.pl/~sgp/gw/gw.htm
Mpala Research Centre, Kenya: www.nasm.si.edu/ceps/mpala/main.html

26 EXPEDITION RESEARCH PROJECTS IN ARID LANDS

Paul Munton and Andrew Warren

In the arid zone, loosely defined as having rainfall below 300 mm/year, life survives at the limits of its capacity for adaptation. Survival of both plants and animals requires them to be specialists and species must have the capacity to cope with, or avoid, stress resulting from lack of moisture in their environment. Behavioural adaptations of animals and physiological adaptations of plants and animals able to survive in the arid zone are therefore of special interest. The study of their physiology raises general issues about the limits of adaptation of which species are capable. Although areas subject to drought, the arid zone and semi-arid zones, comprise between 30 and 45 per cent of the Earth's landmass, they have been relatively little studied. Nevertheless there are several web links to organisations working in the arid zone, and at the end of this chapter web addresses are given for the Royal Botanic Gardens, Kew, the Convention on Biological Diversity and the Arid Lands Information Centre, Arizona. For an introduction to the arid zone, see Heathcote (1983) and the Action Plan resulting from the UN Conference on Desertification (UN, 1977); a number of research papers have also been produced by Unesco on specific topics. A recent arid zone resource study in Jordan sets out problems typical of the field (Dutton, Clark and Battikhi, 1988).

Top tip
You must be thoroughly conversant with the use of all equipment before the expedition. Heat, dust, sand and being bumped around in vehicles are not conducive to the efficient working of delicate instruments. You must know what is likely to go wrong with your equipment, have spare parts and back-up repair facilities that will get the equipment back into the field in time for you to gather sufficient data.

First be warned. The arid zone is characterised not only by low rainfall, but also by rainfall that occurs unpredictably in space and time. As moisture is the main limit to biological productivity, there are long periods in arid areas when very little is happening that can easily be studied. The average 12-week expedition, unless timed on the basis of very good intelligence, will probably fall into such a time of low productivity so there will be little life to study. You will find in these long dry periods that annual plants are absent or vestigial, most perennial plants bear no flowers or seed, and are not easily identified, reptiles will be aestivating and difficult to find, and densities of small mammals and non-migratory insects will be much reduced and therefore difficult to catch.

Do not be too put off, however; instead be like the plants, animals and people of the desert – opportunist. Make the most of what is available. If a high level of biological productivity would help your project, try to time the expedition for the period of year when rain is most likely to have occurred in the previous 2 months. Look for boundaries with more productive environments, consider studies on migrants who use, but are not totally dependent on, the local resources, don't be afraid to intervene with some water of your own, combine with specialists in other subject areas to broaden research opportunities. Look at aspects of human survival in the arid zone, such as building and architecture, patterns of water use and agriculture. Seek ideas from other expeditions into arid zones. The Oman Wahiba Sands project (1988) and Jordan Badia project of the Royal Geographical Society (with the Institute of British Geographers) (RGS–IBG) carried out substantial work on a number of different projects using several teams with different skills.

PLANT PHYSIOLOGY

An understanding of drought tolerance, or its development, in economically important plant species is recognised as a key research area, but it remains relatively unstudied. This may be because countries of the arid zone are generally too poor to finance such research, and larger countries with arid zones concentrate their research in more productive cooler and wetter areas.

Perennial plants make especially good subjects for physiological research because they are adapted to survive dry periods, not as seeds but as reactive organisms. When present, they are easy to find. At first glance many perennial trees and grasses may appear to be dead, but close examination will show one or two grey-green stems at the centre of a grass clump or a few green leaves distributed at low density over a tree. These perennials will have a low, but measurable, level of metabolism (Laurie, 1988). Such plants are stressed by low moisture levels and perhaps, in addition, by the saline soils on which they are dependent. Strong sunlight is also stressful and, in many tropical arid areas, intensity of insolation is so great that it is not easily replicable in the laboratory without very costly equipment, so fieldwork on the effects of sunlight is especially important. The stresses may be lessened by fog or dew during rainless periods.

Figure 26.1 *Field research into the sustainability of wild harvesting of thyme, High Atlas Mountains, Morocco (© Rachel Kaleta)*

It is possible to design research work using any combination of these parameters. Furthermore, you can subject plants to different artificial watering regimes and measure their metabolic response. Be sure that you can secure a source of water in the field and that you have enough rain gauges and control plots to cope with changes caused by natural rainfall, which will inevitably occur when you have your experimental system, based on artificial irrigation, neatly set up!

The main problem with such work is that it requires sophisticated equipment. If you wish to have any results at the end, it is unwise to use an expedition to test such equipment and you must be satisfied that any equipment is hardy before thinking about planning a project on based it.

MAPPING AND DESCRIPTIVE WORK

Descriptive work is often appropriate for short expeditions. It is the basis upon which more complex scientific research can be built and is always useful to host governments and other agencies who often lack basic knowledge of natural resources available to people and the way in which people use those resources. Mapping can also contribute to knowledge of the distribution of wild plant and animal species and how they are used (Cope, 1980); of soils, their distribution and condition; of water, its availability and use; of livestock varieties, their distribution and how they use an area;

and of the presence of local people and their flocks and herds at various times. Expeditions can also be used for ground proofing interpretations from satellite images, although vegetation mapping from satellite images in the arid zone continues to be difficult because of the sparsity of vegetation and the overwhelmingly high reflectivity of the background soils. The Royal Botanic Gardens at Kew, England, and other major collectors will give advice and perhaps lend equipment for plant collection. On insect collection and identification, see Buttiker and Buttiker (1988).

WORKING WITH LOCAL PEOPLE

Working with local people is essential to the success of most projects, but especially when they involve the study of the use of natural resources by wildlife or by people and their flocks and herds. Local people have the most intimate knowledge of the areas in which they live and they can help fill in the gaps, giving information on what happens at times when your project members are not present to observe directly. Gaining access to such information may not be straightforward. There are a number of reasons for this; language is one obvious problem, but differences between your way of conceiving of the world and that of the local people may be even more important. People may not want to reveal the nature and extent of their resources, such as the number of animals they possess, just as you may not want to reveal the balance of your bank account. Out of politeness people will often tell you what they think you want to hear, or, if they think you are involved with government, may give you answers which have more to do with their point of view on local political issues than with the reality on the ground. Barley (1989) gives an amusing account of the problems and misconceptions of the would-be anthropologist grappling with another culture.

Work with local people should always be matched with direct observation of what actually appears to be happening on the ground. This stimulates dialogue and may reveal important information which local people regard as too "obvious" to be worth mentioning. It will also reveal if there is a difference between what people say or believe they are doing and what they actually do – a common human trait in any society.

ANTHROPOCENTRIC STUDIES

The dry period is the period that determines the minimum productivity of an arid area and this in turn determines human use of an area. Perennial plants, but especially trees, are often the key resource in determining how animals, plants, people and their livestock use areas. If there is any productivity in an area, it will be productive throughout dry periods. This is especially so for trees and shrubs, which are often large enough to buffer the effects of long dry periods by tapping into a deep water table, storing water or absorbing dew through their leaves. Groups of trees or even a single tree will attract its own fauna of insects, mammals and birds worthy of study.

The relationship of pasturalists to their environment has been a major item of study since the droughts in the semi-arid Sahelian zone and the southward progress of the Sahara aroused the interest of the world community, manifested, for example, in the World Conference on Desertification in 1977 (UN, 1977). Since then, and in spite of substantial research, the problems of arid and semi-arid zone pasturalism have remained largely unsolved (e.g. Timberlake, 1985).

Local people may move around to use different parts of an area's resource at different times of the year. These sorts of interaction are complex and normally take place over the cycle of the year, or longer irregular periods, depending on the pattern of rainfall. There are often rapid changes after rain, the speed of change being related to the speed of change in the productivity of plants. Nevertheless, the task of mapping woodlands or perennial grasses is often a very useful exercise for the host country and becomes even more useful if the use of these resources by humans, their livestock and wild animals can be related to their distribution in time and space. Even if your project is of short duration, use of resources over a short space of time is worthy of study, especially when supplemented with information from local people, about how the area is used during the part of the year when you cannot be there, and how their use changes after rainfall. Information obtained in this way can be substantial and lead to a real understanding of how people use an area and contribute to a knowledge of survival strategies of peoples in the arid zone (Pratt and Gwynne, 1977; Munton, 1988; Webster, 1988). Resource use can also be related to the social structure of human groups, kinship, access to resources, and self-regulation of resource use through time, space and mode of use.

Again artificial watering can be useful to stimulate the growth of ephemeral plants, identify them and measure the rate at which forage becomes available for livestock or wild animals under different watering regimes. Be aware that the bare area under trees and shrubs in a dry period may produce a dense cover of ephemeral grasses after rain.

VEGETATION AND LAND FORMS

Vegetation is often important in determining erosive processes and may contribute to the form of the land. It is especially important in sand binding and catching blowing sand, especially in dry periods when sand and dust are more mobile. The combined work of a botanist and a geomorphologist has potential for useful studies in dry periods (Buckley, 1987).

Edge habitats

Pay special attention to the borders between different sorts of habitats, e.g. arid lands are often situated adjacent to very productive cool seas (there is often a causal link). Warm-blooded animals, especially scavenging desert birds and desert foxes and cats, which do not have an option to aestivate and are unable to migrate long distances, often occur in higher densities at such border areas, making them more amenable to study. The densities and behaviour of animals on these edges can usefully be compared with the ecology and behaviour of the same species in true arid habitats (Skinner et al., 1984). Radio tagging is often feasible in the open arid habitats (Amlaner and Macdonald, 1980). You should be aware, however, that your own presence and the rubbish that you produce represent a major rise in local biological productivity and the opportunist birds, mammals and reptiles of the desert will change their behaviour to make the maximum use of what you bring them. You may not therefore be studying their "normal" behaviour but the behaviour determined by your presence. Other edge areas are human settlements, especially camps of nomads. Even a group of leafy trees will provide an edge where it may be possible to observe how behaviour and ecology of the same species differ on either side of the boundary or how species of antelope (such as gazelle) make use of patchy habitat (Brown, 1988).

Overall the arid zone is a fascinating area to study and the difficult physical conditions will always make work challenging and demanding. Many people are dependent on the arid zone for their survival so any information or research work is always worthwhile from the human point of view, as well as for the survival of animal and plants that have evolved to live in its difficult environment.

DESERT GEOMORPHOLOGY

The attraction of doing geomorphological research in arid regions is that geomorphology is laid bare for all to see, particularly when it comes to the work of the wind. If you are lucky and choose a windy season, you will see dunes moving at up to a metre a day, and ripples forming and reforming in front of your eyes. Even if you are not looking at active landforms, stream channels, dunes, inselbergs, soil profiles and so on are all extremely accessible.

Useful texts on desert landforms, which will provide many ideas, are Thomas (1989) and Cooke et al. (1992).

Top tip

It is always as well to be prepared for the worst (or the best, depending on how you look at it). If you are lucky enough to see a flash flood, or an extreme dust or sand storm, you might as well measure it, because you, as a geomorphologist, will have been lucky. Ian Reid and Lynn Frostick were lucky enough and have provided a model of what might be done in relation to a flood (Reid and Frostick, 1987). If it is very windy, one can look at the effects on the form and movement of dunes as I did in the Wahiba Sands (Warren, 1988).

Figure 26.2 *Aerial reconnaissance is an excellent way of getting an overall view of the landscape, in this case flying by light aircraft over the Nubian Desert (© John Radford)*

Examples of projects include the relation of the size and shape of ripples to where they occur on a sand dune. The theory of ripples has recently been thrown wide open by the research of Bob Anderson in California (1987b, 1990). He has also thrown open the discussion of slip-faces, which are active and surprisingly complicated features (Anderson, 1987a). If you can get hold of a few simple sand traps or even anemometers, you could add to the understanding of dune shapes. A day or two's measurements (if conducted properly) can add a lot to what is known in this area (Weng et al., 1991). If you have the time and facilities (and the weight allowance) to bring home sand samples for analysis, then a large area of investigation opens up (Warren, 1971; Sarre and Chancey, 1990).

Anderson (1986) has also opened up the research of wind erosion, suggesting projects on the form and distribution of pebbles and rocks eroded by the wind. There are two schools of wind erosion now – one believing that sand is the main agent, the other that dust may erode (Breed et al., 1989). Ron Cooke's (1970) work on desert pavements (the stony layer at the surface of most deserts) has shown that they too are remarkably complicated and fascinating phenomena. Although a short expedition does not have the time to do their dynamics justice, there is still a lot of interest in their morphology, distribution and associated soils, which can be studied over a short period (McFadden et al., 1987).

Stream channels in deserts and semi-arid areas are another area that is wide open for new ideas and data collection, as the work of Graf has shown in New Mexico (various references in the "Bibliography"). Surveys of channels can often reveal why entrenchment has occurred, be it because of an infrequent flood (Reid and Frostick, 1987) or some form of interference by people (Cooke and Reeves, 1976). Surveys of channels, if accompanied by observations of the alluvium into which they are cut, can also establish a history of cut and fill (Vita-Finzi, 1969).

Research on desert inselbergs has shown that measurements of rock strength can help to explain some of their characteristics (Selby, 1980, 1982). Other projects could be undertaken on the relationship between the size of debris (often huge boulders) and the form of mountain slopes (Cooke and Reeves, 1972).

There are many ways in which desert landforms impinge on the lives of people. Dunes move over fields, desert floods wash away roads, etc. Studies of this kind of problem can draw on Ron Cooke's work on these effects for examples (Cooke, 1974a, 1974b, 1982; Cooke et al., 1978, 1982). If you are lucky enough, you may find an old map or air photograph of the position of a sand dune, and see how far it has gone, rather as one of Brigadier Bagnold's sand dunes was rediscovered 57 years later and several kilometres away (Haynes, 1989).

FURTHER INFORMATION

Useful websites

Arid Lands Information Centre, Arizona: http//ag.arizona.edu/OALS/aols/alic/alic.html
Convention on Biological Diversity: http//www.biodiv.org/links/default.aspx?sbj=dls
Global Drylands Partnership: www.undp.org/seed/unso/globalpartnership/gdp.htm
RIOD International NGO Network on Desertification: www.riodccd.org
Royal Botanic Gardens, Kew: http://www/rbgkew.org.uk/scihort/eblinks/arid/html
United Nations Convention to Combat Desertification: www.unccd.int
UNDP Office to Combat Desertification and Drought (UNSO): www.undp.org/seed/unso/

Bibliography

Allen, T. and Warren, A. (eds) (1993) *Deserts: The encroaching wilderness.* A Mitchell Beazley World Conservation Atlas. London: Mitchell Beazley with IUCN.

Amlaner, C.J. and MacDonald, D.W. (eds) (1980) *A Handbook on Biotelemetry and Radio Tracking.* Oxford: Pergamon Press.

Anderson, R.S. (1986) Erosion profiles due to particles entrained by wind: application of an aeolian sediment transport model. *Geological Society of America, Bulletin* **97**: 1270–8.

Anderson, R.S. (1987a) The pattern of rainfall deposition in the lee of aeolian dunes. *Sedimentology* **34**: 175–88.

Anderson, R.S. (1987b) A theoretical model for aeolian impact ripples, *Sedimentology* **34**: 943–56.

Anderson, R.S. (1990) Aeolian ripples as examples of self-organization in geomorphological systems. *Earth Science Reviews* **29**: 77–96.

Anon (1988) The Scientific Results of the Royal Geographical Society's Oman Wahiba Sands Project 1985–1987. *Journal of Oman Studies* Special Report No. 3.

Barley, N. (1986) *The Innocent Anthropologist.* London: Penguin.

Breed, C.S., McCauley, J.F. and Whitney, M.I. (1989) Wind erosion forms. In: Thomas, D.S.G. (ed.), *Arid Zone Geomorphology.* London: Belhaven, pp. 284–307.

Brown, K. (1988) Ecophysiology of *Prosopis cineraria* in the Wahiba sands, with reference to its reafforestation potential in Oman. *Journal of Oman Studies* Special Report No. 3: 257–70.

Buckley, R. (1987) The effect of sparse vegetation on the transport of dune sand by wind. *Nature* 325: 426–8.

Buttiker, W. and Buttiker, S. (1988) The invertebrate collection of the Oman Wahiba Sands Project. *Journal of Oman Studies* Special Report No. 3: 313–16.

Cooke, R.U. (1970) Stone pavements in deserts. *Annals of the Association of American Geographers* 60: 560–77.

Cooke, R.U. (1974a) Applied geomorphological studies in deserts; a review of examples. In: Hails, J.R. (ed.), *Applied Geomorphology.* Amsterdam: Elsevier, pp. 183–225.

Cooke, R.U. (1974b) The rainfall context of arroyo initiation in southern Arizona. *Zeitschrift für Geomorphologie, Supplementband* 21: 63–75.

Cooke, R.U. (1982) The assessment of geomorphological problems in dryland urban areas. *Zeitschrift für Geomorphologie, Supplementband* 44: 119–28.

Cooke, R.U. and Reeves, R.W. (1972) Relations between debris size and slope of mountain fronts and pediments in the Mojave Desert, California. *Zeitschrift für Geomorphologie* 16: 76–82.

Cooke, R.U. and Reeves, R.W. (1976), *Arroyos and Environmental Change in the American South-West.* Oxford: Oxford University Press.

Cooke, R.U. and Warren, A. (1973) *Geomorphology in Deserts.* London: Batsford.

Cooke, R.U., Goudie, A.S. and Doornkamp, J.C. (1978) Middle East – review and bibliography of geomorphological contributions. *Quarterly Journal of Engineering Geology* 11: 9–18.

Cooke, R.U., Brunsden, D., Doornkamp, J.C. and Jones, D.K.C. (1982) *Urban Geomorphology in Drylands United Nations University.* London: Oxford University Press.

Cooke, R.U., Brunsden, D., Doornkamp, J.C. and Jones, D.K.C. (1985) Geomorphological dimensions of land development in deserts – with special reference to Saudi Arabia. *Nottingham Monographs in Applied Geography,* Vol. 4. University of Nottingham.

Cooke, R.U., Warren, A. and Goudie, A.S. (1993) *Desert Geomorphology.* London: University College Press.

Cope, T.A. (1980) The chronology of Old World species of Graminae. *Kew Bulletin* 35: 135–71.

Dutton, R.W., Clark, J.I. and Battikhi, A.M. (1988) *Arid Land Resources and their Management: Jordan's desert margin.* London: Kegan Paul International.

Frostick, L.E. and Reid, I. (1982) Alluvial processes, mass wasting and slope evolution in arid environments. *Zeitschrift für Geomorphologie, Supplementband* 44: 53–68.

Graf, W.L. (1977) The rate law in fluvial geomorphology. *American Journal of Science* 277: 178–91.

Graf, W.L. (1978) Fluvial adjustments to the spread of tamarisk in the Colorado Plateau region. *Geological Society of America Bulletin* 89: 1491–501.

Graf, W.L. (1979) The development of montane arroyos and gullies. *Earth Surface Processes* 4: 1–14.

Graf, W.L. (1981) Channel instability in a sand river bed. *Water Resources Research* 17: 1087–1094.

Graf, W.L. (1982) Distance decay and arroyo development in the Henry Mountains region, Utah. *American Journal of Science* 282: 1541–1554.

Graf, W.L. (1983a) Variability of sediment removal in a semi-arid watershed. *Water Resources Research* 19: 643–652.

Graf, W.L. (1983b) Downstream changes in stream power in the Henry Mountains, Utah. *Annals of the Association of American Geographers* 73: 373–387.

Graf, W.L. (1988a) *Fluvial Processes in Dryland Rivers.* Berlin: Springer-Verlag.

Graf, W.L. (1988b) Definition of flood plains along arid-region rivers. In: Baker, V.R., Kochel, R.C. and Patton, P.C. (eds), *Flood Geomorphology.* Chichester: Wiley, pp. 231–242.

Harrison, D.L. (1968) *The Mammals of Arabia,* 3 vols. London: Ernest Benn.

Haynes, C.V. Jr (1989) Bagnold's barchan: a 57 year record of dune movement in the eastern Sahara and implications for dune origin and palaeoclimate since Neolithic times. *Quaternary Research* **32**: 153–67.

Heathcote, R.L. (1983) *The Arid Lands: Their use and abuse.* London: Longman.

Laurie, S. (1988) Water relations and solute content of some perennial plants in Wahiba Sands, Oman. *Journal of Oman Studies* Special Report No. 3: 271–6.

Lindsey, J.F. (1973) Ventifact evolution in Wright Valley, Antarctica. *Geological Society of America Bulletin* **84**: 1791–8.

McFadden, L.D., Wells, S.G. and Jercinovich, M.J. (1987) Influences of aeolian and pedogenic processes on the origin and evolution of desert pavements. *Geology* **15**: 504–8.

Munton, P.N. (1988) Vegetation and forage availability in the sands. *Journal of Oman Studies* Special Report No. 3: 313–16.

Pratt, D.J. and Gwynne, M.D. (1977) *Rangeland Management and Ecology in East Africa.* London: Hodder and Stoughton.

Reid, I. and Frostick, L.E (1986) Slope processes, sediment derivation and landform evolution in a rift valley basin, northern Kenya. In: Frostick, L.E. et al. (ed.), *Sedimentation in the African Rifts.* Geological Society Special Publication, 25. Oxford: Basil Blackwell, pp. 99–111.

Reid, I. and Frostick, L.E. (1987) Flow dynamics and suspended sediment properties in arid zone flash floods. *Hydrological Processes* **1**: 239–53.

Sarre, R.D. and Chancey, C.C. (1990) Size segregation during aeolian saltation on sand dunes, *Sedimentology* **37**: 357–65.

Scientific Results of the Royal Geographical Society's Oman Wahiba Sands project 1985–87. *Journal of Oman Studies,* Special Report No. 3.

Selby, M.J. (1980) A rock mass strength classification for geomorphic processes, with tests from Antarctica and New Zealand. *Zeitschrift für Geomorphologie* **24**: 31–51.

Selby, M.J. (1982) Form and origin of some bornhardts of the Nambi Desert. *Zeitschrift für Geomorphologie* **26**: 1–15.

Skinner, J.D., Van Aarde, R.J., and Van Jaarsveld, A.S. (1984) Adaptations in three species of large mammals (*Antidorcas, marsupialis, Hystrix africaeaustralis, Hyaena brunnea*) to arid environments. *South African Journal of Zoology* **19**: 82–6.

Thomas, D.S.G. (ed.) (1989) *Arid Zone Geomorphology.* London: Belhaven.

Thomas, D.S.G. and Middleton, N.J. (1994) *Desertification: Exploding the myth.* Chichester: John Wiley.

Timberlake, L. (1985) *Africa in Crisis – the causes, the cures of environmental bankruptcy.* IIED London: Earthscan.

UN (1977) *United Nations Conference on Desertification: Round-up, plan of action and resolutions.* New York: United Nations.

Vita-Finzi, C. (1969) *The Mediterranean Valleys: Geological changes in historical time.* Cambridge: Cambridge University Press.

Warren, A. (1971) The dunes of the Tenere Desert. *Geographical Journal* **137**: 458–61.

Warren, A. (1988) The dynamics of network dunes in the Wahiba Sands: a progress report. The Scientific Results of the Royal Geographical Society's Oman Wahiba Sands Project 1985–7. *Journal of Oman Studies,* Special Report, No. 3: 169–81.

Webster, R. (1988) The Bedouin of the Wahiba Sands: pastoral ecology and management. *Journal of Oman Studies* Special Report No. 3: 443–51.

Weng, W.S., Carruther, D.J., Hunt, J.C.R., Warren, A., Wiggs, G.F.S. and Livingstone, I. (1991) Air flow and sand transport over sand dunes. *Acta Mechanica,* Supplementum 2, 1–22.

Global Drylands Partnership (GDP) challenge papers:
  Poverty and the Drylands
  Strategies for the Sustainable Development of Dryland Areas
  Biodiversity in the Drylands
  Vulnerability and Adaptation to Climate Change in the Drylands

Unesco Arid Zone Research Series:
  VII *Human and Animal Ecology. Reviews of Research*, 1957, 244pp.
  XIV *Salinity Problems in the Arid Zones. Proceedings of the Tehran Symposium*, 1961, 395pp.
  XV Plant – *Water Relationships in Arid and Semi-arid Conditions. Proceedings of the Madrid Symposium*, 1962, 352pp.
  XXV *Methodology of Plant Ecophysiology. Proceedings of the Montpelier Symposium*, 1965, 531pp.
  XXVII *Evaporation Reduction, Physical and Chemical Principles and Review of Experiments*, 1965, 79pp.
  XXVIII *Geography of Coastal Deserts*, 1966, 140pp.
  XXIX *Physical Principles of Water Percolation and Seepage*, 1968, 465pp.

27 EXPEDITION RESEARCH PROJECTS IN TUNDRA AND PERIGLACIAL REGIONS

John Matthews

Tundra refers to the treeless landscape beyond the tree-line in polar regions characterised by mosses, lichens and dwarf shrub vegetation. The term "tundra" is derived from a Finnish word, and was originally used to describe areas of the Arctic north of the boreal forest or "taiga", but it is now increasingly used also to describe similar areas in the Antarctic and sub-Antarctic. In addition, the alpine zone above the tree-line in mountain areas is often termed "alpine tundra" and the transition zone or ecotone with forested regions, where there are scattered trees, is called "forest tundra" (Ives and Barry, 1974).

Tundra is characterised by a periglacial environment with a non-glacial cold climate, where freezing and thawing of the ground are a dominant influence on landscape and life. There is, however, no one-to-one agreement between periglacial envi-

Figure 27.1 *Collecting meltwater from Antarctica as part of a long-term project to model climate change (© Alun Hubbard)*

ronmental conditions and vegetation. Most tundra is underlain by permafrost (perennially frozen ground), only the surface metre or so of which (the active layer) melts during a relatively short summer season (Harris, 1986). Extensive areas of the boreal forest in Siberia and North America are also underlain by permafrost, which may be continuous, discontinuous or sporadic, depending on the severity of the periglacial climate. Permafrost develops where the loss of heat from the ground caused by winter cooling exceeds the heat absorbed during the summer months. Where conditions are not so cold, non-permafrost periglacial environments are characterised by seasonally frozen ground. This occurs, for example, south of the permafrost limit in the Arctic. The most severe periglacial climates give rise to polar deserts, which are characterised by a much sparser and simpler vegetation than the tundra (Alexandrova, 1988).

Today, some 20 per cent of the land area of the Earth is periglacial (i.e. has a cold climate but is not necessarily adjacent to an ice sheet or glacier) and about twice this area was affected beyond the limits of the glaciers and ice sheets at the maximum of the last glaciation (French, 1996). Such regions include much of Canada, Alaska, the fringes of the Antarctic (Hansom and Gordon, 1998), and numerous islands at high latitudes in both the Northern and Southern Hemispheres. Research projects might involve aspects of the present natural landscape (possibly involving landforms and geomorphic processes, geology, microclimate, soils, plant communities, animal ecology, contemporary environmental change and human impact), or palaeoenvironmental reconstruction (the reconstruction of past conditions from sedimentary sequences or other "natural archives").

GEOMORPHOLOGICAL ASPECTS

A wide range of distinctive periglacial landforms could be investigated. These are described by French and Slaymaker (1993) and Ballantyne and Harris (1994). Some forms are rather spectacular, such as rock glaciers and pingoes, the latter being conical-shaped hills up to 50 m high which grow as a result of high water pressures in partly frozen ground. These should be distinguished from other types of frost mound, such as palsas: peat-covered mounds up to about 5 m high in areas of discontinuous permafrost, which form as a result of the growth of ice lenses as water freezes. A wide variety of sorted and non-sorted patterned ground phenomena (which result from frost processes in the active layer) may provide a larger sample size for investigation. Expedition research could focus on specific types, such as earth hummocks (thufur), sorted circles, solifluction lobes or stone stripes. Alternatively, the whole assemblage of forms could be investigated in an area, paying particular attention to their position in the landscape and/or altitudinal zones. Such forms could be mapped, measured by cross-profiles, excavated to study their internal structure, sampled with particular reference to sediment characteristics and related to site conditions. If there is the possibility of a return visit, there would also be considerable potential for monitoring movement and associated environmental conditions, which would shed further light on their origin and development, e.g. the re-survey of peg lines or the position of surface stones and the re-excavation of buried markers or flexible tubes are common approaches to monitoring mass movement on periglacial slopes.

Surprisingly little is known about most periglacial processes and their effects. Several areas of controversy are discussed in the reviews in Clark (1988). The effectiveness of both physical weathering (frost shattering) and chemical weathering under periglacial conditions provides a good example. Whether or not such processes are enhanced beneath snow patches has not been resolved and observations on where frost shattering is important or on the nature and rate of chemical weathering could make a major contribution to knowledge. Field observations and measurements should, where possible, be followed up by relevant laboratory analyses of samples collected in the field (e.g. Goudie, 1990; Gale and Hoare, 1991).

Fluvial processes are another neglected area of periglacial geomorphology. Although expeditions will rarely be in the field for the whole melt season, even a short period of monitoring can detect interesting patterns in stream discharge and sediment yield. Fluvio-periglacial landforms may possess distinctive characteristics (see, for example, McEwan and Matthews, 1998). The effects of lake and sea ice, wind action and thermokarst development (from the melting of ground ice) are other possible topics for investigation.

PALAEOENVIRONMENTAL RECONSTRUCTION

The reconstruction of past environments can add an interesting dimension to scientific research. Lowe and Walker (1997) and Roberts (1998) should be consulted for good introductions to the major geological and climatic changes of, respectively, the late Quaternary (about the last 100,000 years) and the Holocene (the last 11,500 years). Cold, frozen or waterlogged conditions, all common in tundra and periglacial regions, are particularly conducive to the preservation of evidence for palaeoenvironmental reconstruction. There are, however, possible disadvantages in the slow rates of organic production and the high potential for the disturbances of sedimentary sequences. Nevertheless, plant remains in the form of macrofossils (tree trunks, wood fragments, leaves and seeds) and microfossils (pollen) may be found in boggy areas, lakes and soils. Animal fossils, such as the bones of vertebrates, mollusc shells and insect remains, may similarly provide vital evidence. If good sections cannot be found, sedimentary sequences may be revealed by excavation or coring. Whereas excavation with spades may be possible for some drier terrestrial sites, the use of specialist coring devices is usually necessary for bogs (mires) and lake sediments, the latter also requiring rubber boats or rafts. Tree boring for dendrochronological investigations is also possible in the forest tundra zone. Many techniques are described by Berglund (1986).

BIOLOGICAL ASPECTS

One advantage to carrying out research in tundra and periglacial environments is the relative simplicity of the ecosystems. There are fewer species of plants and animals than in temperate and tropical regions. Identification of species, many of which are circum-polar in distribution, is relatively straightforward. However, it is worth considering a project on one of the lower plant groups, such as mosses and lichens, which are often neglected but which comprise the most important component of many tundra plant communities. The types of vegetation and their ecology in different parts of the world are described in some detail in the work of Walter and Breckle (1986) and Wielgolaski (1997). The range of ecological research themes that might be attempted is exemplified by the individual chapters in Sonesson (1987) and Woodin and Marquiss (1997).

There are strong environmental controls on plant distribution in both polar and alpine tundra (Körner, 1999), and the effects of environmental gradients are often clearly visible in the vegetation landscape at both large and small scales (Dahl, 1986), e.g. a distinct zonation of plant species and communities occurs around late-lying snow patches, which reflects the length of the snow-free season and other factors. There is plenty of scope in the tundra for mapping plant communities, relating their distribution to site conditions, or carrying out detailed measurements of produc-tivity and environmental controls, such as heat, moisture, nutrients, light and wind.

Studies on the population ecology of individual species can yield important new data. Diverse adaptations to periglacial environments are reflected in plant morphology, dynamics and physiology. Different adaptations and/or slightly different environmental requirements may permit the coexistence of species within the same community. Various modes of vegetative reproduction (e.g. vivipary, bulbils and layering) are well developed in tundra species, although sexual reproduc-tion by seed is more important in the High Arctic, on glacier forelands and in early successional stages at more favourable sites.

There is also plenty of scope for studies of the ecology of small mammals, birds and invertebrates. Some larger mammals are, of course, dangerous; others, however, such as beaver and reindeer, can be safely investigated. Remmert (1980) and Stonehouse (1989) give introductory accounts of Arctic animal ecology, and several chapters in Bliss et al. (1981) give useful insights into more specialised studies of particular groups.

SOILS

Soils of the tundra and cold regions generally have not been as intensively studied as those of temperate and tropical regions because of their limited agricultural poten-tial. Low temperatures, deep freezing and the existence of permafrost produce unique soil properties and distinctive soil profiles. An introduction to soils in the Arctic is provided by Fitzpatrick (1997) and an introduction to the soils in a Norwe-gian mountain tundra area by Ellis (1980).

Figure 27.2 *A periglacial slope in the Norwegian Alpine zone (© John Matthews)*

INTERACTIONS IN THE LANDSCAPE

The integrated study of whole ecosystems (e.g. Bliss et al., 1981) or whole landscapes (e.g. Oechel, 1989) is probably beyond the scope of most expeditions. However, the idea of investigating interactions between, for example, plants and animals, plants and soils, or plants and geomorphic processes, has much to recommend it. Investigations of the last mentioned type could be described as studies in landscape ecology, geoecology or biogeomorphology. An example is provided by the interaction of vegetation and frost disturbances of various sorts (Komárková and Weilgolaski, 1999). Many more interactions and disturbances are described in the context of recently deglaciated terrain by Matthews (1992, 1999).

HUMAN ACTIVITY AND IMPACTS OF GLOBAL WARMING

Last, but not least, human activity in tundra and periglacial regions should not be ignored. Small populations have left a legacy of archaeological sites, many of which have not yet been excavated (see, for example, Jacobs and Sabo, 1979). All of the indigenous peoples of the Arctic have been greatly affected by the exploration and exploitation of Arctic resources (Sugden, 1982; Harris, 1986). Numerous sociological and economic problems have arisen for the Inuit of North America and similar groups as a result of this intrusive human impact. The unique engineering difficulties and conservation issues arising from the mining, oil and gas industries, road construction, water supply and waste disposal could, in turn, be investigated (e.g. Williams, 1979). Even the effects of expeditions have been the subject of serious study (Gellatly et al., 1986).

The effects of global warming could also be investigated because most climatic models predict that high latitudes will be most affected by the continuing increase in the concentration of greenhouse gases in the atmosphere. Continuing global warming is likely to affect both the geomorphology (e.g. Koster, 1994) and the vegetation (e.g. Chapin et al, 1992).

CONCLUSION

In conclusion, there are varied and important opportunities for scientific research in tundra and periglacial regions. Included are some of the most remote and inhospitable places on Earth. Although expedition research in these environments is often uncomfortable and sometimes dangerous, there are relatively few health hazards and there will be no shortage of excitement.

FURTHER INFORMATION

Key references

Alexandrova, V.D. (1988) *Vegetation of the Soviet Polar Deserts.* Cambridge: Cambridge University Press [Translated from Russian].

Ballantyne, C.K. and Harris, C. (1994) *The Periglaciation of Great Britain.* Cambridge: Cambridge University Press.

Bliss, L.C., Heal, O.W. and Moore, J.J. (1981) *Tundra Ecosystems: A comparative analysis.* Cambridge: Cambridge University Press.

Chapin III, F.S., Jefferies, R.L., Reynolds, J.F., Shaver, G.R. and Svoboda, J. (eds) (1992) *Arctic Ecosystems in a Changing Climate: An ecophysiological perspective.* San Diego: Academic Press.

Clark, M.J. (ed.) (1988) *Advances in Periglacial Geomorphology.* Chichester: Wiley.

Dahl, E. (1986) Zonation of Arctic and alpine tundra and fellfield ecobiomes. In: Polunin, N. (ed.), *Ecosystem Theory and Application.* Chichester: Wiley, pp. 35–62.

Ellis, S. (1980) Soil-environmental relationships in the Okstindan Mountains, north Norway. *Norsk Geografisk Tidsskrift* 34: 167–76.

Fitzpatrick, E.A. (1997) Arctic soils and permafrost. In: Woodin, S.J. and Marquiss, M. (eds), *Ecology of Arctic Environments.* Oxford: Blackwell Science, pp. 1–39.

French, H.M (1996) *The Periglacial Environment*, 2nd edn. London: Longman.

French, H.M. and Slaymaker, O. (eds) (1993) *Canada's Cold Environments.* Montreal: McGill Queen's University Press.

Gellatly, A.F., Whalley, W.B., Gordon, J.E. and Ferguson, R.I. (1986) An observation on trampling effects in North Norway: thresholds for damage. *Norsk Geografisk Tidsskrift* 40: 163–8.

Hansom, J.D. and Gordon, J.E. (1998) *Antarctic Environmental Resources: A geographical perspective.* Harlow: Longman.

Harris, S.A. (1986) *The Permafrost Environment.* London: Croom Helm.

Ives, J.D. and Barry, R.G (eds) (1974) *Arctic and Alpine Environments.* London: Methuen.

Jacobs, J.D. and Sabo III, G. (1978) Environments and adaptations of the Thule culture on the Davis Strait coast of Baffin Island. *Arctic and Alpine Research* **10**: 595–615.

Komárková, V. and Wielgolaski, F.E. (1999) Stress and disturbance in cold region ecosystems. In: Walker, L.R. (ed.), *Ecosystems of Disturbed Ground*. Amsterdam: Elsevier, pp. 39–122.

Körner, C. (1999) *Alpine Plant Life*. Berlin: Springer.

Koster, E.A. (1994) Global Warming and Periglacial Landscapes. In: Roberts, N. (ed.), *The Changing Global Environment*. Oxford: Blackwell Science, pp. 127–47.

McEwen, L.J. and Matthews, J.A. (1998) Channel form, bed material and sediment sources in the Sprangdøla, southern Norway: evidence for a distinct periglacio-fluvial system. *Geografiska Annaler* **80**(A): 17–36.

Matthews, J.A. (1992) *The Ecology of Recently-deglaciated Terrain: A geoecological approach to glacier forelands and primary succession*. Cambridge: Cambridge University Press.

Matthews, J.A. (1999) Disturbance regimes and ecosystem response on recently-deglaciated terrain. In: Walker, L.R. (ed.), *Ecosystems of Disturbed Ground*. Amsterdam: Elsevier, pp. 17–37.

Oechel, W.C. (ed.) (1989) Ecology of an arctic watershed: landscape processes and linkages. *Holarctic Ecology* **12**(3): 227–334.

Remmert, H. (1980) *Arctic Animal Ecology*. Berlin: Springer-Verlag.

Roberts, N. (1998) *The Holocene: An environmental history*, 2nd edn. Oxford: Basil Blackwell.

Stonehouse, B. (1989) *Polar Ecology*. Glasgow: Blackie.

Sugden, D. (1982) *Arctic and Antarctic: A modern geographical synthesis*. Oxford: Basil Blackwell.

Walter, H. and Breckle, S.W. (1986) *Ecological Systems of the Geobiosphere 3. Temperate and polar zonobiomes of northern Eurasia*. Berlin: Springer-Verlag.

Wielgolaski, F.E. (ed.) (1997) *Polar and Alpine Tundra*. Amsterdam: Elsevier.

Williams, P.J. (1979) *Pipelines and Permafrost: Physical geography and development in the circumpolar north*. London: Longman.

Woodin, S.J. and Marquiss, M. (eds) (1997) *Ecology of Arctic Environments*. Oxford: Blackwell Science.

References for key techniques and methodology

Berglund, B.E. (1986) *Handbook of Holocene Palaeoecology and Palaeohydrology*. Chichester: Wiley.

Gale, S.J. and Hoare, P.G. (1991) *Quaternary Sediments: Petrographic methods for the study of unlithified rocks*. New York: Halsted Press.

Gardiner, V. and Dackombe, R. (1983) *Geomorphological Field Manual* London: George Allen & Unwin.

Goudie, A. (1990) *Geomorphological Techniques*, 2nd edn. London: Unwin Hyman.

Kent, M. and Coker, P.D. (1992) *Vegetation Description and Analysis: A practical approach*. New York: Wiley.

Lowe, J.J. and Walker, M.J.C. (1997) *Reconstructing Quaternary Environments*, 2nd edn. Harlow: Longman.

Moore, P.D. and Chapman, S.B. (eds) (1986) *Methods in Plant Ecology*, 2nd edition. Oxford: Blackwell Science.

Sonesson, M. (ed.)(1987) *Research in Arctic Life and Earth Sciences: Present knowledge and future perspectives*. Copenhagen: Munksgaard (Ecological Bulletins, No. 38).

Key scientific journals

Arctic
Arctic, Antarctic and Alpine Research
Boreas
Ecography (formerly *Holarctic Ecology*)
The Holocene
Permafrost and Periglacial Processes

Polar and Glaciological Abstracts
Polar Record
Polar Research

Key organisation

Scott Polar Research Institute. Website: www.spri.cam.ac.uk
This organisation has several research groups investigating a range of issues in both the environmental sciences and social sciences in the Arctic and Antarctica:

- Glaciology and Climate Change Group
- Glacimarine Environments Group
- Polar Landscape and Remote Sensing Group
- Polar Social Science and Humanities Group
- NERC Centre for Polar Observation and Modelling Group

SECTION 6

EXPEDITION TRANSPORT

28 VEHICLE-DEPENDENT EXPEDITIONS

Tom Sheppard

There are two key rules to be obeyed when fitting out a vehicle for an expedition: never exceed the vehicle's design limitations in terms of

1. payload, or
2. terrain.

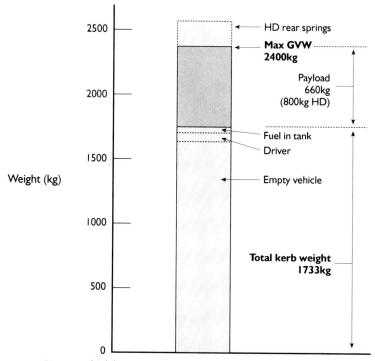

Figure 28.1 *How a vehicle's GVW (gross vehicle weight) is made up. Never exceed it by adding too much payload. Figures shown are for Defender 90*

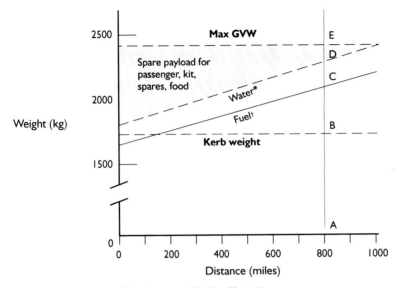

*Water: for two crew, 7.5 ltr/day, 150 miles/day.
†Fuel: assumes 15 mpg V8 to highlight load. Diesel would yield 25–30 mpg.

Figure 28.2 *Payload versus range. This graph, for a V8 Defender 90, shows vividly how operating range (distance between replenishment points) erodes spare payload. In a fairly extreme case, on an 800-mile leg, A–B is kerb weight, B–C is fuel load, C–D is water required, leaving only 150 kg (D–E) for a 75-kg passenger and remaining kit. In real life you would go diesel, shorten legs and/or get a bigger vehicle such as the Land Rover Defender 110 or a robust 4 × 4 pickup such as a Toyota Hilux. Fuel calculations assume reserve, i.e. distance + 100 miles + 25 per cent. Bigger vehicle equals more payload but less power:weight ratio (see Figure 28.6).*

Payload and range

You must stick to load limits. Gross vehicle weight, or GVW, is the "never exceed" or maximum permitted weight of a vehicle. It is made up of the empty (or "kerb") weight and the load, i.e. fuel, driver, passengers and cargo. In general a big vehicle can carry more cargo for a much longer distance than a small one.

For a given vehicle and crew, distance (and days) between replenishment points dictates the load of essentials such as fuel and water that you need to carry. The available spare payload for food, spares, camping gear and other cargo follows from there and will diminish as the length of the journey – and thus the fuel load – goes up.

Terrain

The kind of tracks or terrain that you have to traverse, although less easy to quantify,

319

Figure 28.3
*The third generation
Range Rover (top), like
the up-spec Land
Cruiser and Jeep
Cherokee (bottom), are
rare examples of high-
comfort 4 × 4s with
exceptional off-road
ability. Payload,
however, is limited. (©
Tom Sheppard)*

is another dominating factor in defining the task that your expedition vehicle needs to do. There could well be routes straightforward enough to cover in a normal two-wheel drive (4 × 2) van or pickup. If the terrain is more demanding or is uncertain you will want more ground clearance or off-road capability to use or have in reserve.

No-frills functionality

There is no law against having fun in a 4 × 4 or against being luxuriously comfortable when logistics permit but, although the divisions are not hard and fast, it is worth having in the back of your mind the following broad categories of vehicle:

- lightweight "fun" vehicles (RAV4, Jimny, Honda HRV)
- luxury vehicles with not much payload (Range Rover, Cadillac Escalade)

- "working" vehicles for expeditions: no frills, plenty of payload (Defender, Pinzgauer, Toyota Type 75 and 78, simple-spec Land Cruiser, 4 × 4 pickups).

HOW MANY VEHICLES, LOAD, TYPE?

Spreading the load – and the risk

One big vehicle or two small ones? Some journeys will naturally be multi-vehicle, for others there may be a choice. Influences will be:

- degree of mobility required by any subgroups
- cost of multi-vehicle ferry fares, etc.
- difficulty of terrain: large vehicles tend to be less athletic than small ones – but two small ones can tow one large one when it sticks
- safety in case of breakdown of a single vehicle.

Do not let this last consideration stem from a feeling that breakdown and damage are inevitable. Rather take the opposite view that the implications of breakdown or

Figure 28.4 *Outer limits. Long-range load carrier pushing its luck – two-wheel drive and no support vehicles. Ample manpower, however, is beneficial in boggings (© Tom Sheppard)*

damage on an expedition can be so dire or so expensive that they must not be allowed to occur. Driving, general care and maintenance standards must be that good.

Nevertheless, random mechanical failures do occur and a back-up vehicle and one to help in towing out a stuck vehicle will be invaluable. The payoff in peace of mind is high. If you can, never take fewer than two vehicles – three is best because all the load from one incapacitated vehicle can in many cases be transferred and spread without overloading the other two.

Overloading – excuses

Overloading your vehicle must not be considered an option. "Ah, but there are margins …", say some. And margins are exactly what you want on an expedition over difficult terrain in foreign parts. "Ah, but I've seen vehicles with roof racks up to here …", say some, and they will also probably have seen the same vehicles rolled on to their sides as a result of the high centre of gravity or with cracked pillars because of the fatigue loads on elements not designed for the stress. "Ah, but you can fit stronger springs …", say some, and in doing so merely transfer more road shocks into an already overladen chassis. Don't be misled by the appearance of rally vehicles operating on a wing and a prayer and having huge back-up safety infrastructures. Ordinary expeditions are not like this, and should operate with maximum safety margins. So, when considering your vehicle requirement, do not let the idea of upping the load even enter your head.

Figure 28.5 *Classic example of a fuel-carrier trailer off-loading the main vehicle. Note also how the under-tyred tug has bogged whereas the lightly laden trailer on larger tyres at low pressures rides over the soft sand (© Tom Sheppard)*

Trailers – 50 per cent more axles

You may well encounter the problem of having a greater load and bulk than the vehicle's size and payload maximum. A vehicle with a trailer is less agile than one without but a given load may be spread over six instead of four wheels. So long as two or three people are available to manhandle it, a trailer can be a solution – provided that it is really robust (e.g. an ex-military three-quarter-ton trailer such as those used behind army Land Rovers). Towing arrangements will have to be similarly upgraded – usually a NATO towing pintle. Be sure that the trailer's shock dampers are in first-rate condition when you buy it; they will keep lateral roll in check.

Keep the trailer load light – not more than about 60 per cent of the rated load. This will not only put the trailer under less mechanical stress but also enable lower tyre pressures to be used, thus reducing sinkage, drag and load on the tug if soft going. Importantly, running light will also reduce the ratio of gross weights between tug (the towing vehicle) and trailer, which has considerable effect on the stability and agility of the ensemble. Keep the centre of gravity of the load in the trailer low down, and keep the high-mass items close to the trailer axle and central to reduce the moment of inertia. Ensure that there is a down-load at the trailer tow bar of about 50–75 kg. Remember that this "nose weight" is bearing down on the towing vehicle's rear end and will result in a reduction in the tug's payload. To accommodate this and the effect of overhang – as a rule of thumb – remove twice this figure from the towing vehicle's payload. Thus, if the download is 50 kg, take 100 kg off the listed maximum payload of the tug.

THE EXPEDITION VEHICLE: INGREDIENTS

What to consider

Be aware of the ingredients of a competent off-road vehicle – what ingredient yields what reward and at what cost – and so bring together a ghost specification that you can template on to what the market is offering at any given time. See Table 28.1.

The features mix

Bearing in mind that you could start off looking at 4 × 2 vans and pickups, scan the attributes in Table 28.1 and get a feel for what features are to your advantage and why. Some customers want a 4 × 4 only for the safety that it gives on snow and slippery surfaces, without having the need for extra low gears – some even find these "confusing". Hence special specifications for special markets evolve without a two-speed transfer box – a whole raft of "soft roaders" such as the Land Rover Freelander, Nissan X-Trail and Honda CRV being cases in point. It is unlikely that a vehicle without a low-range transfer gearbox will be satisfactory for an expedition.

TABLE 28.1 A COMPARISON OF VEHICLE ATTRIBUTES IN ORDER OF PROGRESSION FROM SIMPLE 4 × 2 PICKUP

Feature	Benefits	Disadvantages
Leaf-springs	Low cost, simplicity, easy replacement. Springs act as means of locating axles	Inter-leaf friction gives stiff ride, poor traction; limited wheel movement. If springs very long and one- or two-leaf, less of a problem
Large diameter wheels	Improved under-axle ground clearance. Goes *over* pot holes rather than into them	No functional disadvantage
Torsion-bar front springs	Smoother ride than leaf-springs. Better traction and braking. More wheel movement?	Usually associated with independent front suspension so less ground clearance
Beam axles	Good under-axle clearance, wheels always perpendicular to ground	Clearance above axle needed for wheel movement makes vehicle tall. High unsprung weight difficult to damp
Coil springs all round	Smoother ride than leaf-springs. Better traction and braking. Usually a lot more wheel movement so best off-road capability; best traction on uneven ground	More expensive than leaf-springs as a result of need for alternative axle location links. If too short and stiff, ride is still poor (e.g. Lada)
No anti-roll bars	Permits full axle articulation – twist relative to body – off-road wheel movement enhanced	Body roll. Designer's nightmare to balance on- and off-road performance
Short wheelbase	Improved off-road capability but only noticeable in extreme conditions	Usually associated with lower maximum payload than long wheelbase versions
Large approach, departure, ramp angles, "high stance"	Off-road agility without danger of grounding body parts. Short tail overhang specially valuable exiting ditches	High centre of gravity can cause body roll
High payload	Obvious advantage when there are long distances between provisioning points	Stiffer springs give less pliant ride. Classic division between luxury/working
Automatic transmission	Helps driver. Smoothest gear changes safeguard driveshafts, precludes lost traction through jerkiness. Very good	Cost mainly, some weight. Perceived loss of manhood by some. Prop shaft may need disconnecting for towing
"Part-time" (selectable) four-wheel drive (4 × 4)	Huge improvement over two-wheel drive in soft sand, mud, snow, etc	Compared with 4 × 2, cost. Must be selected when needed and de-selected on hard surfaces. Full-time 4 × 4 better
Part-time pseudo (or "automatic") 4 × 4 (common in "soft roaders")	As above but speed differences between front and rear axles have to be sensed before 4 × 4 is engaged with viscous coupling	As above but things have to get bad before they get better, i.e. some wheelspin. Not totally positive drive
Two-speed transfer box	In effect a second set of extra-low gears for off-road use. Highly desirable for expeditions	Cost and complexity but a must-have for any serious expedition
On-the-move range change (Lo to Hi range)	Invaluable when you have to start in Lo and need to change to Hi without stopping. Can't be done with "electronic" range changes	No real disadvantage except some skill/technique required to do it on most vehicles. G-Wagen has synchro
"Full-time" 4 × 4 (permanent) with centre differential	Much better than part-time or "automatic" because it is there all the time, ready for anything. Best kind has manually lockable centre differential	Compared with part-time 4 × 4, more cost as centre differential needed. Must remember to unlock diff if on hard ground unless VC controlled
Locking axle differentials	Overcomes those "one spinning wheel" situations superbly to preclude getting stuck	Cost. Risk to half-shafts if not properly engineered. Must remember to de-select
Traction control	Foolproof way round wheelspin. Automatic	Electronic-dependent, brake heat, wear. Not as good as manual-select diff-locks
Portal axles (Pinzgauer)	Dramatic increase in under-axle clearance for rough ground and deep ruts	Very expensive to produce, high centre of gravity, higher unsprung weight
Roll control	Limits on-road body roll while preserving off-road articulation	Electro-hydraulic systems complex, expensive. "Mechanical" ones are simpler but still electrics-dependent for selection

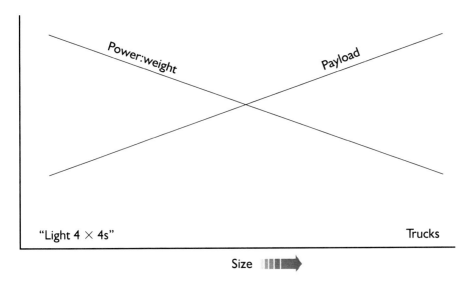

Figure 28.6 *The size trade-offs: power:weight ratio and payload*

Size, power:weight ratio, payload

In general a medium truck (say 4 tonnes) can support itself and a team over a greater distance and/or longer time than a "light 4 × 4", such as a Cherokee or a Defender. It is a direct function of payload (the fuel or supplies it can carry) and there is no substitute for detailed and accurate calculations of requirements in this area. These calculations would be based on the number of people to keep in the field for how long at given consumption rates, the distance between resupply points and whether there are basics such as water close to the worksite or along the route. The general picture looks like that in Figure 28.6.

Power:weight – where it matters

If payload is what gets you long distances then, all other things being equal, power:weight ratio is what gets you up a sand dune, a steep loose slope or through a boggy patch for which you might have to use speed. The power:weight ratio is the number of bhp per tonne of GVW; as Figure 28.6 shows, small vehicles with big engines have lots of it (4.0-litre Jeep Wrangler, RAV4, Range Rover); trucks don't. So long as there is grip – which there isn't on a sand dune, a loose slope or a sticky mud patch – a vehicle with a low power:weight ratio will, by sheer low gearing, crawl over rocks or slowly up a steep climb. Power:weight ratio equals dynamic off-road capability and is usually, as Figure 28.6 implies, a direct trade-off for size, carrying capacity and – importantly – fuel consumption. Over-powered vehicles are very uneconomical.

325

Tyres

Steep, loose, short inclines are still going to need power to achieve the momentum that they need to "ballistic" to the brow; on the level or with moderate slopes, however, wide, sometimes deflated, tyres can help a low-powered vehicle "soft shoe" through without sinkage or excessive demands on torque. Enormous all-wheel drive trucks on big soft tyres can be seen performing astonishingly well in the Sahara despite low power:weight ratios. But remember, low tyre pressures demand low speed; reinflation is essential if rock or hard going is encountered.

Tubeless tyres will be virtually impossible to refit in the field – massive amounts of compressed air are required. Consult with tyre specialists (the manufacturers if need be) and fit tubes before departure.

TRANSMISSION SYSTEMS

Two-wheel drive or four-wheel drive?

Will a 4 × 2 do? A robust 4 × 2 with big wheels and a couple of willing crew to push is surprisingly capable if firm roads – surfaced or unsurfaced – are available. "Dual cab" or "crew cab" pickups (with two rows of seats and four doors ahead of a smaller

Figure 28.7 *Classic application of the crew-cab 4 × 4 Toyota pickup in short-range expedition role, very popular in South Africa. Note that any heavy cargo will all be over the rear wheels (© Tom Sheppard)*

load bed) are a popular solution where small groups are concerned. If considering this approach in a 4 × 2 (or even in the 4 × 4 versions), check carefully the payloads, ground clearance, wheel size and wheelbase before making your choice. A 15-inch rim size should be regarded as the minimum to provide adequate ground clearance and tyre footprint.

Driving standards
Remember, equally, that the wrong tyres, tyre pressure and driving technique can see a bogged 4 × 4 passed by a well-operated 4 × 2. Having said that, a novice in a 4 × 4 will do better than a novice in a 4 × 2. So will an expert. It is clear, therefore, that existing and potential driving skills come into the list of parameters to assess when considering vehicle choice.

So, 4 × 2 or 4 × 4?
The choice has to be to go for 4 × 4 if you can. Cost could well be the deciding factor but even this can be accommodated to some extent in the various types of 4 × 4 available today.

Types of four-wheel drive
Some argue – shakily – that four-wheel drive uses more fuel so the facility should be used only when needed. The result is a bunch of different driveline design philosophies – and terminologies – when looking at 4 × 4 vehicles:

- "Part-time" 4 × 4 (selectable) – many pickups, simple, straightforward
- "Auto" 4 × 4 (so-called "when needed") – "soft-roaders", Freelander, Honda CRV
- "Full-time" 4 × 4 (permanent) – all current Land Rover and Toyota models
- A blizzard of trade names such as Selectrac, Super Select, Quadra-trac, Control Trac to cause you further confusion, but they all fit into one of the above three categories
- Not all of the above systems (RAV4, Honda CR-V, Freelander, Nissan XTrail, BMW X5, Volvo XC90, etc.) are combined with a two-speed, selectable transfer gearbox that gears the final drive down by a factor of two or more, giving you a "second set" of extra low gears. For expeditions, use a 4 × 4 with a low-range transfer box.

Part-time 4 × 4 OK – but not on tarmac
A selectable 4 × 4 system, the simple type found on most pickups, is not as desirable as a permanent 4 × 4 for everyday driving but, for most expedition applications, on/off tracks with 4 × 4 selected, performance will be identical. As it lacks a centre differential, however, for accommodating different axle speeds front and rear in

Figure 28.8 *A 4 × 4 is best. Here, as with rock crawling or steep washouts, low-range gears are virtually essential (© Tom Sheppard)*

turns, it should never be used in four-wheel drive on tarmac or other hard grippy surfaces.

AUTO OR MANUAL

Elegant and gentle

Automatic transmissions are neither a sissy option nor the passport to high-fuel consumption that they once were considered to be. Looking at the revolutions per minute (rpm) an automatic transmission asks of the engine on a given off-road section, in comparison to a white-knuckle manual driver, it is easy to see that the opposite is sometimes the case. On a track where a soft patch takes you by surprise, the down change can be lightning quick and in rough-track "forest floor" situations the auto really shines, reliably executing countless gentle gear changes in long-day conditions where, with a manual, driver fatigue may rear its head. An automatic will make immaculate UP changes on steep loose inclines when you are in danger of provoking wheel spin. An automatic will also reduce shock loading on drive shafts and differentials. Many vehicles used as standard in the armed forces worldwide are equipped with automatic transmission – the operational advantages in terms of driver workload and vehicle durability are seen to win over the slightly increased costs.

Do use low range

Don't let an auto lull you into laziness, though. Be sure to get into low ratio when required or you will be using the torque converter like a slipping clutch and will induce overheating of the transmission fluid. Remember also to select "1" for steep descents.

Reliability, service

The reliability of automatics is, if anything, better than that of a manual box and clutch – partly because they are difficult to mishandle. One expert summed it up: "If an auto is OK for the first year, it'll live for ever." As vehicle dealers encounter so few faults with autos and may lack experience, go to a specialist automatic gearbox engineer for a pre-expedition service or if you want a second-hand vehicle checked; they are working on them every day and will know what to look for. Be sure that your vehicle has a transmission oil cooler (mounted up front where the engine fan is). Some automatics have clunky changes; clunks are not why you opt for an auto. Try before you buy.

PETROL OR DIESEL

Characteristics

Generally, petrol engines are lighter, more powerful, more thirsty and cheaper (to buy) than diesels. Diesels, however, have advanced dramatically in recent years in terms of power, responsiveness and efficiency. Above all they are a lot more economical. They are also "greener". Turbo-charged and intercooled diesels cost more than a simple diesel but offer improved power and lower fuel consumption, the very essence of the expedition requirement. Not all turbo-charged diesels are intercooled – a cost compromise again.

High power conversions

Beware of "performance conversions". On expeditions, durability and reliability are all and high-power conversions will usually compromise the structural margins of an engine.

Fuel: availability, load, cost

Know your route. If use of a petrol engine is a real possibility, know its generic type (leaded, unleaded, etc.) and find out the grade of fuels available en route. Petrol engines have to be designed (or tuned) to the fuels available and most modern engines cannot be tuned to use very low-grade gasolines; they will destroy themselves if you try. But the bottom line is that wherever you are you'll always find usable diesel; finding the right type of petrol may be a lot harder. And diesel is always cheaper.

Figure 28.9 *Classic robust simplicity – the Australian spec Toyota Type 78 even comes with long-range fuel tanks as standard. This one has coil springs at the front. Never imported into UK, earlier all-leaf-sprung Type 75 in diesel form is highly valued in continental Europe for expedition work (© Tom Sheppard)*

Reputations: assessing reliability

Solid repairable simplicity has an attractive ring. But there are parts of some modern engines that cannot be repaired in the field and electronic control units (ECUs), engine management chips, fuel injection pumps and injectors are among them. Repair by replacement is often the order of the day even in maintenance centres and then sometimes accompanied by electronic analysis or facilitated by the use of special tools. If your engine does have an ECU check out the failure case. Is there a "limp home" mode? That said, the reliability of such components is generally very high.

Question specialists about known faults. Early Land Rover 300 Tdi engines could suffer random cambelt failure with catastrophic consequences, but a modification to preclude this is available. Be sure that yours has it. Traditionally diesels "have a long life" because we all think back to trucks that go on for ever. Modern but not too modern (i.e. beyond their teething troubles) is probably the best phrase to have in mind.

The choice

All things being equal, a modern, established, turbo diesel is probably the best engine for an expedition vehicle – especially where distances are large. Diesels tolerate poor

Figure 28.10 *Give careful thought to recovery equipment: "sand channels" of some kind to put under the wheels for flotation in slippery mud or soft sand; shovels and long tow ropes so the tug does not get stuck in the same hole as the stricken vehicle. Winches are heavy and expensive; rarely useful except with multi-vehicles in forests or jungle (© Tom Sheppard)*

fuel better than petrol engines. The days of diesel vehicles being underpowered are gone.

SIMPLICITY, SERVICE, SPARES

Keep it simple
Keeping the specification simple will certainly be a good start. Deleting air conditioning is a case in point where hot climates are concerned – a lot of cost, climatic shock every time you get in and out of the vehicle, a lot of weight, a lot of complexity.

Service, spares
You will still have to make your own assessment of the reliability record of your chosen vehicle. Is the design "bedded in" or a brand-new model? What if there really is a problem? Are there dealers in the area that you are visiting – or the country? Is there a course you can go on before departure? Are there good service manuals? What is a sensible spares pack? What about the need for special tools?

Figure 28.11 *Be ready to improvise. Electronic control unit (ECU) defaulting to "limp home" and suspected fuel contamination demanded drain and filter for all fuel on board – an all-day job. Rocks provide handy ramp for jerry can access; failing that a dig-down hole would have sufficed (© Tom Sheppard)*

Operation: mechanical sympathy

Ponder the fact that very rarely do things break of their own accord. More usually they are broken by people – usually through insensitive driving, overloading or inattention. The acquisition of mechanical sympathy, that sensitivity to the operation of equipment, especially in conditions of stress, is an attribute almost beyond price when it comes to keeping going on an expedition. Of course there will come times when you must operate the vehicle to its full potential but even here it can be done with sensitivity – feel for it, care for it, don't break it. Prepare the vehicle impeccably, secure the load; drive impeccably too.

INSPECTION, PREPARATION

New buy? Independent inspection

Have an independent inspection carried out before purchase. Be there, if possible, when it is done. Depending on your vehicle and its servicing intervals, have a major service done (again, be there if possible) before leaving. As well as all-round oil changes (with synthetic for the engine), consider renewing hoses, accessory drive belts, camshaft drive belt (if applicable), brake shoes or pads, battery(s) and tyres. Even if new tyres are not needed, have the old ones removed and replaced using proper rubber lubricant; it will pay at your first repair. Learn how to change a tube; it is easier than it looks.

ONCE YOU HAVE IT, TAKE CARE OF YOUR VEHICLE. IT IS THE LIFE BLOOD OF YOUR EXPEDITION

This chapter is based on the author's books, Vehicle-dependent Expedition Guide *(second edition now available) and* Off-roader Driving.

FURTHER INFORMATION

Further reading

Jackson, J. (2003) *The Off-Road 4-Wheel Drive Book: Choosing using and maintaining go-anywhere vehicles.* Sparkford, Yeovil, Somerset: Haynes.

Scott, C. (2000) *Sahara Overland: A route and planning guide.* Hindhead: Trailblazer Publications.

Shackell, C. and Bracht, I. (1993) *Africa by Road: 4WD – Motorbike – Bicycle – Truck: The Bradt travel guide.* Chalfont St Peter, Bucks: Bradt.

Sheppard, T. (2003) *Vehicle-dependent Expedition Guide.* 2nd edn (field manual). Hitchin: Desert Winds.

Sheppard, T. (1999) *Off-roader Driving.* Hitchin: Desert Winds.

Land Rover Driver Training

Land Rover have been supplying expedition vehicles to the Royal Geographical Society (with the Institute of British Geographers or IBG) since 1977. Land Rover's world-renowned reputation as manufacturers of permanent four-wheel drive vehicles make them an ideal choice for use by scientific and adventurous expeditions. However, motor incidents are a major cause of injury abroad. To help raise the safety and effectiveness of small teams operating in remote areas, the RGS–IBG's Expedition Advisory Centre and Land Rover have teamed up to provide a practical course covering key driving and safety techniques.

For further information, see www.rgs.org/eacseminars

29 CANOEING AND RIVER-RAFTING EXPEDITIONS

Peter Knowles

M ost people have dreamt at some time of descending a great river from the mountains to the sea. This chapter is for those interested in making this dream a reality – or for those expeditioners who want to use river transport to achieve more worthy scientific aims.

We have all been brought up on tales of the fur traders and explorers in their Indian canoes, and Eskimo hunters in their kayaks. Less well publicised are the more modern journeys by kayaks and canoes across the Atlantic, a solo circumnavigation

Figure 29.1 *River travel gives opportunities to meet local people and should have minimal impact on the environment (© Peter Knowles)*

of Australia, expeditions down some of the world's largest rapids, descending 20-metre waterfalls, and many thousands of other journeys done for pure enjoyment.

Although it is probably true to say that most major river systems of the world have now been explored, new areas and countries are opening each year and presenting fresh opportunities waiting to be grasped, e.g. Paul Grogan and Richard Boddington did a first descent of the Amur river, from its source in Mongolia 4500 km to the Pacific Ocean in the year 2000 – this expedition would not have been possible before political tensions eased between China and Russia.

In North America, the sport of river running using large inflatable rafts is now very popular and a multimillion dollar business. Commercial rafting has spread worldwide and there are now operations in most countries – from Bali to Siberia; these companies are usually good sources for advice on local rivers.

ADVANTAGES OF RIVER EXPEDITIONS

In many areas, rivers and waterways have been the traditional and national means of travel, and local people normally "relate" to river travellers. Water transport enables you to cover a lot of ground relatively fast and easily and with more equipment in comparison to travel on foot. River expeditions "leave no footprints" and have minimal effect on local ecology, while canoes and rafts are quiet and a good means of observing wildlife. Rivers, and especially white water sections, are normally highly photogenic, although photos are best taken from the bank because of camera shake. Modern canoes, kayaks and rafts are very strong and light, and can be used on a surprisingly wide range of water, from rocky trickles to the open sea. Rivers and waterways are normally excellent ways of getting into the countryside – away from towns and roads. Probably the best reason for a canoeing and rafting expedition is that it is normally highly enjoyable.

LIMITATIONS OF RIVER EXPEDITIONS

Obviously you are limited to where the water is, and to what is navigable, and it should be noted that river trips are normally one way – downhill. A river expedition is usually more committing and may need more planning and logistics than the corresponding foot expedition. Some experience and/or training, especially for harder rivers, is recommended. The logistics of obtaining your boats and equipment and transporting these to your chosen river may be costly.

DIFFICULTY OF WATER

Rivers are graded for difficulty on an international scale of river difficulty from I to VI.

- Grade I is just moving water – easy and safe, something like the lower river Wye.
- Grade VI is almost impossible and carries extreme danger to life.

We use the term "white water" normally to mean grades II–VI inclusive. High water will make a river considerably more difficult – usually one grade higher and probably two grades higher in flood conditions.

Experience needed

Most people should be proficient at paddling on grade I and II water after a few days of a course in kayaks or canoes. For grades III and upwards, you are looking at considerably more training and experience. It normally takes a few seasons before a kayaker is competent and at home on grade IV. The river leader should be able to paddle one grade higher than the rest of the group. Loaded boats make the subjective difficulty at least half a grade harder, as does very cold water.

Inflatable rafts require a slower reaction time and less skill to paddle than kayaks or canoes; however, because they are less manoeuvrable, they perhaps require more ability at reading the rapids, so the helmsman in charge needs to be experienced.

Size of river

Small white water rivers are often described as "technical". They require quick decisions and faster reactions than a larger river. More damage to equipment can be expected on a small rocky river than a larger one of the same grade.

Big white water rivers often have horrendous-looking rapids that require more confidence than skill. These big rapids are often graded on the main hazards – normally obvious: huge unremitting stoppers, whirlpools, dangerous undertows, etc. These can often be skirted, but if you get it wrong the consequences can be very serious – involving the total loss of, rather than damage to, equipment and personnel.

LAKES, ESTUARIES AND SEA

Kayaks, and to a lesser extent canoes, have been used for some impressive ocean journeys. However, this is too specialised a subject to cover here (there are some excellent books available on sea canoeing), except for some simple brief notes in case your planned journey involves some stretches of open water.

Kayaks are very seaworthy and ideal for hopping from inlet to inlet, but, on any large body of water more than a mile or so across, *wind* becomes the crucial factor, and may make a lake or sea crossing dangerous for an inexperienced party. You should build a safety factor of a few days into your plans to allow for adverse weather conditions. Particularly dangerous is an offshore wind, which makes conditions appear deceptively calm inshore but blows you out to sea if anything goes wrong.

CHOICE OF BOAT

In North America a canoe is something that you kneel or sit in and paddle with a single blade (e.g. an "Indian" canoe); in the UK we use the term "open canoe" to define these types. A kayak is a closed boat that you sit in and paddle with a double blade. In Britain and many other countries, the term "canoe" is used as a generic term to cover both kayaks and canoes.

- Rigid kayaks are fast and manoeuvrable but cannot carry much gear.
- Open canoes are excellent and versatile and are often a better choice for most expeditions.
- Inflatable rubber rafts, designed for river running, are very robust and are good craft for white water. However, they are slow on flat water and difficult if not impossible to paddle against a head wind. Popular sizes are 4 m to 5 m long, and they cost in the region of £3000.
- Folding kayaks, e.g. Klepper, have always been popular for river trips because of their ease of transport. They are more robust than you might suppose, but they are expensive.
- Inflatable canoes and kayaks (called "duckies" in the USA) can be used for river trips, and are great fun, but they are poor load carriers and are slow and susceptible to wind. The cheaper ones are made of thin plastic and are very prone to damage.

Local craft

A classic mistake is to go to great trouble to select, buy and transport your boats from Britain, only to find that the local people have evolved a much better river craft which you end up hiring or buying. If you read expedition accounts you will be amused to see that some well-known personalities seem well practised at this. Your first choice should always be to look at the costs and practicalities of using local craft. In 1987, a Cambridge student followed this advice and descended the Sepik river of New Guinea in a dug-out canoe that he purchased at the top and sold when he reached the sea – at a profit. If you do decide to save money by using a local craft, then don't stint on the equipment – really good waterproof bags and efficient paddles will make all the difference to the success of your trip.

As mentioned earlier, there are commercial raft and canoe operators now in most countries who may be willing to hire you equipment and local experienced guides. This is an option that you should definitely consider – particularly from the safety viewpoint.

EQUIPMENT

Water-borne expeditions always have a high loss and damage element which tests

Figure 29.2 *A grade IV rapid on the Tamur river, Nepal (© Peter Knowles)*

equipment to the limits and beyond. Take good, strong, tough equipment. Expect and make allowances for breakage and loss, and don't forget a large repair kit. Take a visit to your nearest specialist retailer to get expert advice.

The design of buoyancy aids (known as PFDs in the USA) has improved a great deal over the last decade. You should choose one with plenty of flotation rather than the ones with less buoyancy that are designed for competition use.

For white water, personal protection and padding are often needed – crash helmets, strong padded footwear and buoyancy aids, which provide good protection for the chest. Wetsuits are often chosen in preference to drysuits for the extra protection that they give against knocks and abrasion.

Excellent waterproof bags with roll-over tops of different sizes are available from specialist retailers. Peli camera boxes and similar specialist boxes can be recommended (these use "O" rings). Big blue barrels are good for large rafting expeditions. Various waterproof specialist cameras and housings are available – the majority are designed for sub-aqua use and cumbersome for river trips. Better suited are the "water-resistant" types of automatic 35 mm compact cameras.

If you are doing a white water expedition you should carry rescue throw bags and practise with these beforehand. These also make good washing lines when you make evening camp. It is a feature of all expeditions that equipment can have several functions. When you camp, the waterproof bag that kept your gear dry can become a water container, your paddles become tent poles; your wetsuit becomes a mattress, and your wetsuit boots a particularly nasty bear deterrent!

For kayaks, think about split paddles that can be stowed in the boat. These are much more convenient and are also essential if a set of paddles is lost while bank support is far away.

Packing of boats is a science that you should get expert advice on. All equipment should be tied or otherwise secured in the boat. Weight should be evenly distributed so that the boat is evenly balanced and any heavy items should be low down in the centre of the boat. Last, but perhaps most important, you should not stow the gear such that it impedes you getting out of your canoe or kayak.

RESEARCH AND PLANNING

Where to go
Unfortunately, rivers throughout the world are progressively being dammed and tamed, so, to find a reasonably "wild" river or water journey (other than the open sea), you will have to travel. Obtain as much background information as possible – use your local library, the internet, travel books, guide books, maps, magazine articles, rainfall and temperature tables.

Don't be too rigid in your early days of research into your choice of river – it may be too difficult, uninteresting or too expensive; your research may lead you to much more interesting alternatives. At an early stage find out the rough cost of travel to and within your proposed country so that you know what is within your budget, e.g. travel within Alaska is very expensive so the cost for a trip there could typically be £2000 per person. I recently did a 3-week trip to Ecuador – wild rivers deep in the jungle – for a total budget of £800, including to and from the UK.

It is not generally realised how much detailed information is available on rivers in most countries – river guide books are now available for the whole of Europe and North America and even for countries such as Pakistan, Peru, Russia and Nepal. The internet is, of course, another great source.

When planning, remember that Britain is small. In Canada, for example, the rivers are ten times bigger, the rapids ten times rougher, the mosquitoes ten times bigger, the walk out ten times further and, if things go wrong, your problems are ten times worse. This usually means ten times the cost.

Map study
Study topographic maps carefully. Trace the river and mark off where the contours cross. Draw a longitudinal profile. What is the average gradient? Are there some steep places? Does the map give any clues in these places? Is the valley nice and open (not too worrying) or do the contours or other clues indicate a gorge? (You need more information.) What are the access points? (New roads may have been built recently.) Note that topographic maps may not be that accurate and may not indicate when or if a river flows underground. Do not trust gradients

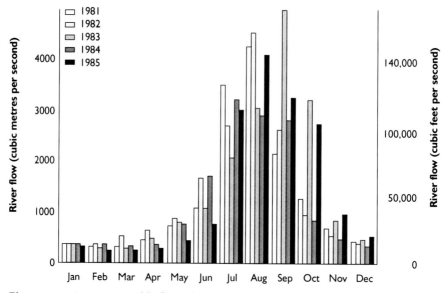

Figure 29.3 *Average monthly flow of the Karnali River at Chisapani for the years 1981–5 (© White Water Nepal, Peter Knowles, 1998)*

absolutely. Local information and reconnaissance are vital in these cases.

The difficulty of a river depends on the gradient, the water volume and the topography of the valley (i.e. is the flow constrained?). As a very rough guide, rivers of less than 4 m/km are probably grade I–II, and those of more than this are of grade II–VI and navigable by experts. I would want to check out in detail any gradient of more than 20 m/km.

Water volume

This is critical. Some questions to ask are:

- Is the river level dependent on rainfall?
- When is the rainy season?
- Is snow melt important?
- When is the river in spate?
- Are sudden fluctuations in the water volume likely (flash floods)?
- Is the river dam-controlled?

Normally most rivers, especially canyons and gorges, are easier in low water conditions.

Geology

Major waterfalls are sometimes marked on maps, but a knowledge of the geology of the area (ask a geologist to help you out) will give you some more clues as to the likely nature of the river and the probability of coming round a corner to find a 20-metre waterfall. Note that a river of long flat stretches interspersed with waterfalls is hard work and not so enjoyable as a river descending steadily with consistent runable rapids.

Information from others

Talk to as many people as possible with a knowledge of the area and the rivers. For a white water river, be a little sceptical of what any non-river runner tells you about the nature of it. For the expert, an enjoyable river is one that the local people often tell you is impossible. Most obvious major rivers have now – I'm sad to say – been run by someone. Your challenge is to find out by whom.

Try to get in touch with local canoeists, river runners or outdoor clubs via the internet and phone calls. After specialist river guidebooks, these are usually the best source of information. Especially in areas like Alaska, local experts can get inundated with general waffly email requests for information from "Dudes and Tourarses"! This is where the telephone is better at establishing contact and credibility. The international telephone system is much more friendly and effective. I also believe that it's more time- and cost-effective. Do identify yourself, tell people about your experience and be specific about the information that you want from them.

Access and support

Ease of access to a river is a critical factor in how serious the expedition is. A river with good road or other access along most of it (especially the difficult sections) means that much of the gear can be carried on the shore if necessary; difficult sections can easily be scouted and shore support, back-up and rescue are much easier. Many expeditions have a "warm-up" before their main trip on a river with such easy access. A warm-up and training on a river before leaving England just before the expedition are useful to resolve initial problems.

The most satisfying river expeditions are probably the ones that are self-contained and do not rely on other outside support. However, the more serious your trip, the more likely it is that you will need some form of support. At the lowest level, this may mean arranging for a vehicle or plane to meet you at the end of the river. It is quite common for a kayak group to have a support vehicle meeting them each night.

I would like to stress how easy it is for the meeting place to be misheard, misinterpreted, misidentified or mistimed. Even on the best-organised expedition, things will go wrong; you may end up with a cold wet group huddling around a fire wondering where the support is. Precautions to take are a super-dependable support driver/leader; both river and support teams carry duplicate maps and mark off meeting points on these; the support team "flag" the river bank to ensure that the river team doesn't miss the "take out", carry emergency supplies on the river, and have previously agreed contingency plans. Walkie-talkie radios can be very useful.

Reconnaissance
Consider how much local reconnaissance will be required before the trip – perhaps from the air or from the ground – and how many days will be needed for this. If you are planning a major expedition, a reconnaissance the year before may be a good cost-effective idea.

Portages
What are the likely and possible portages? How long will these take? What is the terrain and vegetation like? Plan on the worst case. Should you carry special equipment for portaging or a "walk out" if necessary? You may need climbing ropes to get out of gorges.

Contingencies
Do you need to inform anyone of your journey? What do they do if you don't arrive on time? What are the nearest access points if you have an accident? What do you do if you lose a boat? Don't forget adequate insurance.

Size of party
The rule that "less than three should never be" is a good one. It is strongly recommended that raft expeditions should normally have a minimum of two rafts. Groups of more than five or six boats get unwieldy and require more organisation on the river.

How long will it take?
Obviously, you should allow for delays in getting to and from the river, and allow for river portages. The main factors affecting how fast you travel on a river are the speed of the current and the difficulty of the water. Other factors are: how many hours you paddle (length of daylight); how fast you paddle (team morale and training); how often you have to stop to inspect rapids (difficulty of water, but also experience); head winds; bad weather; and the size of party (small groups travel faster).

As a rough guide for canoes and kayaks, it would be reasonable to plan on 20–40 km a day on rivers grade I–II. On grades III–IV, perhaps 10–30 km would be a more typical day. On more difficult water, grades IV–V, 4–10 km might be expected.

On more difficult water, rafts will probably keep up with kayaks. On flatter water grade I, for example, they will prove slow, and 15 miles would be a more typical day's distance. This depends greatly on the speed of the current and winds.

Do not make the mistake of trying to paddle too great a distance in too short a time – one of the main delights of any river trip is the things to see and meeting people on the shore. The Grand Canyon, 220 miles in length, has been paddled in 2 days, but a more typical expedition would plan on 2 weeks, perhaps more, for a 200-mile journey, to allow for rest days, hikes ashore, scouting rapids, portages, etc.

ADMINISTRATION

Grant aid
For canoeing expeditions, the British Canoe Union operate a simple approval scheme and I would advise seeking their support at an early stage of planning. Specific grants are available for worthwhile canoeing expeditions. The earlier you apply for these while having adequate plans for an expedition, the better chance you have of success.

Sponsorship
Very few small river expeditions have been successful at getting major financial sponsorship from commercial companies. Many expeditions have found that without high level contacts in the boardroom this is a waste of time, being very time-consuming, and expensive in terms of postage and phone calls.

More worthwhile in terms of effort are grant aid, equipment sponsorship and personal sponsorship.

Boats make very good billboards for publicity, and a number of expeditions have had individual members personally sponsored by a company (perhaps the individual's employer) with a canoe in their house colours and a logo (they love this), and there are lots of good publicity opportunities in this for the sponsor and you. (In addition, as a publicity expense this is, of course, tax deductible, so you will want to make sure that if possible they are invoiced the full retail cost of the canoe and equipment.)

Permits
River running is recognised by most countries as a legitimate tourist activity, like cycling or trekking. If there are no restrictions on trekking or tourism, then usually there are no restrictions or special permission needed for river running. Note, however, that, if your river is very popular, the tourist authorities may have had to institute a river permit system to restrict the tourist river-running numbers – for the Grand Canyon, the waiting list is now over 10 years.

TRANSPORT
Folding and inflatable boats are ideal for air transport and normally cause few problems. Modern short white water kayaks can usually be carried in the holds of

Figure 29.4 *Rigid canoes and kayaks will fit inside many small planes and can be strapped to the struts of float planes such as Beavers, or suspended underneath a helicopter (© Peter Knowles)*

most large airliners – if the airline carry surf boards and windsurfers, you should have no problems with kayaks. Baggage handlers actually prefer them to boards because they are much more robust.

Manufacturers' recommendations on roof limits are of necessity cautious. Do not be tempted to overload your roof racks because this will seriously de-stabilise your vehicle and affect its road handling. Buy the best possible quality roof rack – ideally, four bars. Clamp these together lengthwise with wooden bars and U bolts. Finally, brace the bars diagonally to the back bumper so that they don't shift forward on emergency stops. Trailers geometrically compound the things that can go wrong – they are fine for smooth highway travel but otherwise are probably best avoided.

With the notable exception of British Rail, canoes, kayaks, rafts, etc. can be carried as accompanied baggage on ferries and most trains, often at no extra charge. Less developed countries often accept them on the roofs of local buses.

SAFETY

Water is a "soft landing", surprisingly forgiving, and river expeditions are much less dangerous than most people imagine. Accidents involving personal injury are more likely to occur off the water, e.g. road accidents, camp-fire burns, sprained ankles, etc.

Where serious accidents have happened on the water, most have been the result of lack of experience and, of course, of planning.

Equipment is usually only a minor factor. The most frequent killers on rivers are high water, cold, entrapment (especially trees) and weirs. On lakes, the biggest hazards are wind, and then cold.

Safety guidelines

- Be conservative in your choice of rivers – particularly the first one.
- Seek advice from local experts on river levels and difficulty.
- Allow yourself plenty of time to paddle the river, so that you are not in a hurry and do not have to paddle when tired.
- Don't paddle what you can't see – never hesitate to scout or portage.
- Do not underestimate how wet and cold you often get on a water expedition (this can be a killer), or the food necessary to "stoke up" energy and warmth.
- On a serious expedition each person when on the water should wear a buoyancy aid, knife and personal survival aid (in case of separation from the rest of the party or loss of boat).

Probably your most important possession is your passport; if this is carried on the river, it should be sealed in a clear waterproof pouch and carried on the person (around the neck or in a secure pocket) – not in a boat where I can assure you bad things can and will happen to it.

We recommend bright colours for boats, paddles and helmets so that you can more readily see a swimmer. Consider marking helmets and important gear such as paddles and cameras with orange day-glo tape: the adhesives company 3M makes a stick-on film called Cal film used for the sides of police cars.

FURTHER INFORMATION

Sources
British Canoe Union
John Dudderidge House, Albolton Lane, West Bridgford, Nottingham NG2 5AS. Website: www.bcu.org.uk
The national body. Provides advice on access and touring. Publishes *Canoe Focus* magazine bi-monthly.

Canoe Association of Northern Ireland
2a Upper Malone Rd, Belfast BT9 5LA.

Canoe Camping Club of Great Britain and Ireland
25 Waverley Road, South Norwood, London SE25 4HT
Some overseas tours.

Chris Film and Video Ltd
Chris Hawksworth, The Mill, Pateley Bridge, Harrogate, North Yorkshire HG3 5QH
Film and video hire of canoeing and rafting expeditions.

International Sea Kayaking Association
John Ramwell, 5 Osprey Avenue, The Hoskers, Westhoughton, Bolton, Lancs BL5 2SL. Tel: +44 1942 842204
Publishes newsletter and a number of information sheets.

Open Canoe Association of Great Britain
First floor flat, 12 Orwell Road, Dovercourt CO12 3LD
For "open canoes" rather than kayaks.

Scottish Canoe Association
18 Ainslie Place, Edinburgh EH3 6AU. Website: www.scot-canoe.org

Welsh Canoeing Association
Frongoch, Bala, Gwynedd, LL32 7NU. Website: www.welsh-canoeing.org.uk

Training courses
Those run by national centres:

Canolfan Tryweryn
Frongoch, Bala LL32 7NU. Tel: +44 1678 521083, website: www.welsh-canoeing.org.uk

Glenmore Lodge
Aviemore PH22 1QU. Tel: +44 1479 861256

Plas Menai, National Outdoor Pursuits Centre for Wales
Llanfairisgaer, Caernarfon, Gwynedd LL55 1VE. Tel: +44 1248 67094, website; www.sports-council-wales.co.uk/plasmenai/welcomeplas.htm

Plas Y Brenin, National Centre for Mountain Activities
Capel Curig, Betws-y-Coed, Gwynedd LL24 0ET. Tel: +44 1690 720214, website: www.pyb.co.uk

Books
There are a huge number of books on canoeing and kayaking. The best British reference was re-published in 2002 – *Canoe and Kayak Handbook* – the official handbook of the British Canoe Union, published by Pesda Press. This has a good bibliography.
    Other recommended up-to-date text books are:
Addison, G. (2000) *Whitewater Rafting: The essential guide to equipment and techniques.* Amsterdam: New Holland Publishers.
Ferrero, F. (1998) *White Water Safety and Rescue.* Pesda Press
Hutchinson, D. (2000) *Expedition Kayaking.* Globe Pequot Press.
    Mostly about sea kayaking.
Watt, A. (2002) Canoe, kayak and raft expeditions. In: Warrell, D. and Anderson, S. (eds), *Expedition Medicine.* London: Profile Books (for RGS–IBG Expedition Advisory Centre), pp. 325–39. Available from: www.rgs.org/eacpubs

Recommended guidebooks and narratives

Addison, G. (2001) *White Water – The world's wildest rivers.*
Inspiring coffee table book.
Cassady J. and Dunlop, D. (1999) *World Whitewater.*
A global guide for river runners.
Deshner, W. (1998) *Travels with a Kayak.*
Prize-winning accounts of kayaking all over the world.
Harrison, J. (2001) *Off the Map.* Summersdale Publishers.
Jordan, T. and Jordan, M. (1987) *South American River Trips.* Bradt Publications.
Kane, J. (1989) *Running the Amazon.*
Knowles, P. (1998) *White Water Nepal: A rivers guidebook for rafting and kayaking.*
Good intro chapters on planning kayak and raft trips.
Manby, D. (1999) *Many Rivers to Run.*
Accounts of river expeditions in exotic countries.

These books are available from Amazon and most are distributed in the UK by www.cordee.co.uk

Magazines

Canoe Focus
Bi-monthly magazine of the British Canoe Union, John Dudderidge House, Albolton Lane, West Bridgford, Nottingham NG2 5AS

Canoeist
Stuart Fisher, 4 Sinodun Row, Appleford, Oxfordshire OX14 4PE. Website: www.canoeist.co.uk
The longest running independent British canoeing magazine – monthly. Sells foreign guide books.

Canoe/Kayak UK
Gunn Publishing, 179 Bath Road, Cheltenham GL53 7LY. Website: www.canoekayak.co.uk
Monthly, started in 2001.

Websites

www.ukriversguidebook.co.uk
A well-managed website that has articles and guides on overseas rivers.

For sea-kayaking see www.nigeldenniskayaks.com

30 CYCLING EXPEDITIONS

Hallam Murray

Weight for weight the cyclist uses less energy to cover a given distance than even the superbly constructed salmon or dolphin, not to mention birds, the great cats, the motor car, or any form of jet or rocket engine. The bicycle is simply the most versatile vehicle known to humans and, with bicycle technology having improved in leaps and bounds, it's hardly surprising that this form of transport has become so popular – for both long distance and locally based expeditions. It can be ridden, carried by almost every other form of transport from an aeroplane to a canoe, and can even be lifted across one's shoulders over short distances. And remember, you have virtually no transport costs, because a bike does not require feeding like a horse or a camel, or petrol to keep its engine running.

With the advent of Kevlar tyres and puncture-resistant inner tubes, it's now theoretically possible to cycle from Alaska to Tierra del Fuego without so much as a single puncture. For a traveller with a zest for adventure, ample time and reasonable energy, there is unlikely to be a finer way to explore.

One of the greatest advantages of cycling is fitness. On a long expedition, a cyclist's immune system soon gets beefed up and, with good hygiene and a little luck, it's possible to cycle for months without having any serious health problems. And with good cycle gearing, climbing hills is no longer a problem, because fitness keeps the cardiac and respiratory changes to a bare minimum and the onset of muscle fatigue is greatly reduced. It is the wind that is the enemy of the cyclist, not mountains. On long expeditions, through such areas as Patagonia or Tibet where strong winds can be a major factor, it is helpful to cycle during the least windy times of day – and this is often predictable. Strangely enough, after a tough day's cycling, it's possible to walk for miles feeling quite refreshed; the muscles we use cycling are quite different from those we use when out of the saddle.

I travelled over 17,000 miles – from California to Tierra del Fuego – largely on unpaved roads. The combined period of these two journeys was 3 years. Much of my time was spent in the High Andes visiting potters and weavers, sometimes in the

Figure 30.1 *Bärli von Toggenburg pushing his bike on bad roads near Mianeh, Iran, during a journey from Switzerland to India to raise money for leprosy sufferers (© Bärli)*

remotest villages with no public transport or electricity, and I was never seriously ill. This was achieved by my fitness, by boiling or pilling all drinking water (*never* getting dehydrated), by cooking most of my own meals (on an Optimus, low-grade petrol cooker), and by eating only meals – when in markets or cafés – that I could see boiling before my eyes. I carried a small but comprehensive first-aid kit on the basis of "a stitch in time", and for me it worked.

If you do choose a bicycle for your expedition, you will often find yourself envied by travellers using more orthodox forms of transport. On a bicycle you can travel at your own pace. Your senses are more in tune with your environment. You can stop at will to admire a view, to talk or to camp. You can explore remote regions and meet people who are not normally in contact with tourists.

CHOOSING A BICYCLE

The choice of bicycle depends on the type and length of the expedition and on the terrain and road surfaces likely to be encountered. Unless you are planning a journey almost exclusively on paved roads – when a high-quality touring bike such as a Dawes Super Galaxy would probably suffice – I would strongly recommend a mountain bike or possibly a hybrid. The good quality ones (and the cast-iron rule is *never* to skimp on quality) are incredibly tough and rugged, with low gear ratios for difficult terrain, wide tyres with plenty of tread for good road-holding, cantilever brakes and a low centre of gravity for improved stability. Heavily laden bicycles take a great pounding when cycling over badly surfaced or unpaved and corrugated roads. Problems with breaking spokes can turn an expedition into a nightmare. So be sure that your wheels are well built with heavy duty spokes – and consider having these double-crossed on the back wheel. You can expect to pay upwards of £500 for a truly robust machine that can cope with desert and mountain and all the other difficult terrains that may get thrown at it.

Although touring and mountain bikes and some spares are available in the larger cities, remember that in the less developed world most indigenous manufactured goods are shoddy and rarely last. So be sure to kit out your bike with accessories of the highest quality *before* you leave, e.g. block and chain, chain rings, pannier racks and panniers. Imported components can be found in the cities of some less developed countries, but they tend to be extremely expensive. In North America, Europe, Japan, Australasia, etc. we are spoilt by quality and often don't appreciate this until it is too late.

Remember too that any broken aluminium components cannot be welded by conventional welding equipment. Some of the aluminium back and front cycle racks are extremely strong and light and well worth considering (Topeak and Blackburn are much to be recommended), but if they do break you may have to get yourself to the nearest airport, where specialist aluminium welding equipment is usually available.

Bikes need to be broken in and well shaken down *before* you leave. Consider making a short trial expedition. If travelling with a group, this also gives you a chance to get to know each other's strengths and weaknesses, and to check for compatibility.

When Michael Edwards and Philip Etherington cycled from Lhasa to Kathmandu, they chose identical bikes to cut down on tools and spares. It's probably a good idea to follow their example.

The best advice for someone without mechanical know-how is to find a top-rate cycle shop – of which there are many in the UK – and to ask for advice from their most experienced cyclist (who is usually to be found in the repair shop). My Ridgeback bike had its wheels re-built by an enthusiast who worked for Evans Cycles – one of the finest cycle shops in Britain. Without his expertise, my wheels would never have got me across Patagonia.

TABLE 30.1 BICYCLE EQUIPMENT

Comfortable seat
Mudguards
Good front and back reflectors
Fluorescent strip and eye-catching gloves (for busy roads)
Cycle lights
Robust cycle helmet with good ventilation (optional)
Water bottle brackets
Secure lock and chain (two keys)
Pump secured by a pump lock (two keys)
A small but comprehensive tool kit to include:
    adjustable spanners, pliers, screwdriver, tyre levers, etc.
    chain rivet and crank removers
    spoke key and possibly a block remover
Spare tyre and two inner tubes
Puncture repair kit with plenty of extra patches and glue
Set of brake blocks
Set of brake and gear cables
Spares of all nuts and bolts
12 heavy-duty spokes (best taped to the chain stay)
Light oil for the chain
Tube of waterproof grease
Stopblock (my choice for the most invaluable accessory and it is easy to make and
    virtually weightless): a stopblock consists of a small rubber wedge to force the
    rear brake lever firmly on, thus preventing a bike from moving on sloping
    ground when propped against a wall, tree or lamppost
Cycle computer: to show speeds, distances and times
Loud bell or horn
Altimeter
Compass
Good maps

LUGGAGE AND EQUIPMENT

Strong waterproof front and back panniers are a must. I chose the Karrimor Icelandic and found these to be excellent. Carradice also sell excellent panniers, which performed impeccably for a recent expedition in Asia.

A top bag-cum-rucksack makes a good addition for use on and off the bike. I used a Cannondale front bag for my maps, camera, compass, altimeter, notebook and small tape-recorder. At the end of 15 months of travelling through tough terrain, all this luggage remained in good condition, which says a lot for the quality of materials and workmanship. My total luggage weighed 27 kg – on the high side, but I never felt seriously over-weight.

"Gaffer" tape is excellent for protecting vulnerable parts of panniers (preferably applied *before* they get worn) and for carrying out all manner of repairs. My most vital equipment is shown in Table 30.2.

TABLE 30.2 **ESSENTIAL LUGGAGE AND EQUIPMENT**

Vango Zephyr tent
Three-season sleeping bag
Optimus petrol stove with small back-up bottle (the best I have ever used, because
    it is light and efficient and petrol can be found almost everywhere); the MSR
    XKG stove is also excellent – it roars at altitude but works cleanly and well with
    petrol
Plastic survival bag for protecting luggage at night when camping
A light, self-inflating sleeping mat can be a godsend for high-quality sleep
Four elastic bungees
Four 1-litre water bottles
Swiss army knife
Torch
Small but comprehensive medical kit
Money belt
Robust camera and slide film
Tape-recorder
Small "World Service" radio
Hat to protect against hours of ferocious tropical sun
Sunglasses
Small presents such as postcards of home, balloons and badges; a rubber mouse
    can do wonders for making contact with children in isolated villages

---

All equipment and clothes should be packed in plastic bags to give extra protection against dust and rain or when crossing rivers. Keep clothing to the minimum, but be prepared to buy extra items en route. A T-shirt or jersey bought from a village market can become a prized possession when back home. Naturally the choice will depend on whether you are planning a journey through tropical lowlands, deserts, high mountains or a combination, and whether rain is to be expected.

Generally, it is best to carry several layers of thin light clothes rather than fewer heavy, bulky ones. Always keep one set of dry clothes, including long trousers, to put on at the end of the day. I would not have parted with my incredibly light, strong, waterproof and wind-resistant GoreTex jacket and over-trousers for neither love nor money. I could have sold them 100 times over and in Bolivia was even offered a young mule in exchange. I took two pairs of Reebok training shoes and found these to be ideal for both cycling and walking. Some cyclists prefer stiff-soled shoes, but this is a case of personal preference.

USEFUL TIPS

Get your bike frame etched with a security number – your local police may be able to do this for you – and take a photo of the bike to keep among your valuables.

Give your bicycle a thorough daily check for loose nuts or bolts or bearings. See that all parts run smoothly. A good chain should last 2000 miles or more but be sure to keep it as clean as possible – an old toothbrush is good for this – and to oil it lightly from time to time. Only the rivets need to be oiled. Try to keep oil off the exposed parts of the chain because this simply attracts dust and dirt.

Try to make the best use of the times of day when there is little wind; mornings tend to be best but there is no steadfast rule. In parts of Patagonia there can be gusting winds of 80 km/h at some times of year. Do your research before you leave.

Your cycle is most vulnerable when it's being carried on other transport. Unless in a canoe, always try to keep the cycle upright, with the weight on its wheels. Bicycle bungees come in really handy for doing this. And travel *with* your bike – and not, say, in the cab of a lorry when your bicycle is being thumped to pieces all alone behind.

Take care to avoid dehydration by drinking regularly. In hot, dry areas with limited supplies of water, be sure to carry an ample supply; it is remarkably easy to run dry. For food, I carried the staples (sugar, salt, dried milk, tea, coffee, porridge oats, raisins, dried soups, etc.) and supplemented these with whatever local foods I could find in the markets. Eating regularly is a good rule. Keep fruit or vegetables or chocolate or biscuits in your front bag so that you can 'graze' as you go along.

Always camp out of sight of a road. You are most vulnerable at night. It is always best *not* to cycle after dark unless you absolutely have to, and remember that night falls very quickly close to the equator so it's best to start looking out for a suitable camp well before dusk.

Remember that thieves are attracted to towns and cities, so, when sightseeing, try to leave your bicycle with someone such as a café owner or a priest. Country people tend to be more honest. Most are friendly and very inquisitive. However, don't take unnecessary risks; always see that your bicycle is secure.

In more remote regions dogs can be vicious; carry a stick or some stones to frighten them off. If you stop and face them with a stick or stone in your hand, invariably they will stop running and barking and hold their distance. You can then move off slowly after a couple of minutes.

Traffic on main roads can be a nightmare; it is usually far more rewarding and safer to keep to the smaller roads or to paths if they exist. (Always make use of capital cities to get hold of the best maps available.)

Most towns have a bicycle shop of some description, but it is best to do your own repairs and adjustments whenever possible. In an emergency it is amazing how one can improvise with wire, string, dental floss, nuts and bolts, odd pieces of tin or "gaffer" tape.

INSURANCE

This is an important area, more especially so when cycling overseas. Remember that, as well as covering against theft, you should also get cover against accidents. Specialist expedition insurances should cover for this. The Cyclists' Touring Club offers a specialised travel insurance for riders, with all the usual travel benefits such as medical expenses, plus a guarantee to transport rider and bike back to the UK in an emergency (see Chapter 14).

FURTHER INFORMATION

A selection of bicycle maintenance books

Bicycling Magazine's *Complete Guide to Bicycle Maintenance and Repair for Road and Mountain Bikes*
Haynes Bike Book
Richard's Bicycle Repair Manual. London: Dorling Kindersley

Useful addresses and websites

Adventure Cycling Association and Magazine. Website: www.adv-cycling.org

Bicycle Association, Starley House, Eaton Road, Coventry CV1 2FM. Tel: +44 2476 553 838, website: www. bikehub.co.uk

This is the national trade body for UK-based manufacturers and importers of bicycles, components and accessories. Its members supply over 80 per cent of all the cycling products available on the UK market. It works by providing a forum for the industry, lobbying government, developing technical standards, assisting exporters and monitoring the worldwide market.

Bicycle Business

Benton Bridge Cottage, Jesmond, Dene, Newcastle on Tyne NE7 7DA. Email: B3@bikebiz.co.uk, website: www.bikebiz.co.uk

British Cycling Federation. Tel: +44 161 230 2301, website: www.bcf.uk.com

CORAX, around the world by bicycle. Website: www3.utsidan.se/Corax-e/

Cyber Cyclery
Website: www.cycling.org
Thousands of bicycle enthusiasts around the world use Cyber Cyclery every day to find a wide variety of biking-related information, resources and services.

Cyclists' Touring Club
Cotterell House, 69 Meadrow, Godalming, Surrey GU7 3HS. Tel: +44 870 873 0060, fax: +44 1483 426994, email: cycling@ctc.org.uk, website: www.ctc.org.uk
Services include country information sheets (covering Europe and much of Africa, the Americas, Asia and Australasia), travel and cycle insurance, and a comprehensive cycling bookshop.

Evans Cycles
Website: www.evanscycles.com
A first-rate cycle shop with eight branches in and around London. They can work on any make of bicycle and will rebuild wheels with heavy duty spokes if required.

L'ordre des Cols Dur
37 Acacia Avenue, Hale, Altrincham, Cheshire WA15 8QY
Club for those interested in cycling in European mountains.

International Bicycle Fund
Sponsors environmentally friendly, rurally based cultural tours in all regions of Africa. Details are available at:
Bicycle Africa Tours. Website: www.ibike.org/bikeafrica
Ibike Cultural Tours. Website: www.ibike.org/ibike

Round-the-World Cyclists Registry
PO Box 1065, Station A, Toronto, Ontario, Canada M5W 1G6

SJScycles.com for good quality bikes and racks

Sustrans
35 King Street, Bristol BS1 4DZ. Tel: +44 117 929 0888, website: www.sustrans.org.uk

Travel with your Bicycle. Website: www.bikeaccess.net/

31 CAMEL EXPEDITIONS

Michael Asher

WHY TRAVEL BY CAMEL?

Having ridden more than 10,000 miles on camels in the deserts of Arabia, Wilfred Thesiger predicted that, though those who came after him would move about in cars and keep in touch with the outside world by radio, "They will bring back results more interesting than my own," he wrote, "but they will never know the spirit of the land nor the greatness of the Arabs."

Thesiger knew that travel by camel offered an intimate relationship with the landscape and its people that could not be experienced through the windscreen of a motor vehicle. To travel by camel is to travel on the desert's own terms, experiencing its realities at first hand. Yet, conversely, as speed inevitably changes one's perception of the environment, the camel's slow pace is actually ideal for those, such as archaeologists, botanists, geologists or other specialists, who need to observe the desert close up. Camels may still be used for major unsupported overland journeys in the desert; more frequently today, however, they are used in combination with motor vehicles for exploring a limited and perhaps less accessible area, or for longer treks with motorised back-up.

The following notes are derived mainly from my experience of almost 20,000 miles of unsupported camel expeditions, alone, with companions, and more recently leading adventure-tour groups, but may be modified according to circumstance.

CHOOSING CAMELS: BUY OR HIRE?

The great advantage for the novice of hiring camels is that it allows him or her to avoid the pitfalls of the market place. The disadvantage is that he or she is obliged to take along the camel's owner who may not be the ideal guide and who may try to influence the route and rate of travel. Buying your own camels makes you totally independent and, assuming that you look after them, you should be able to get some of your investment back at the end of the journey; I reckon generally on retrieving two-thirds of the buying price.

If you buy, choose geldings where possible. They are hardier than bull-camels and do not bite. She-camels can be equally enduring and are equally gentle, but make sure that they are trained for riding before you buy. In many parts of North Africa they are kept only for milk, although trained females are common in Arabia, Australia and the Indian subcontinent. Dominant bull-camels may be aggressive in the mating season, especially if they are prevented from getting near females. If you have a mix of males and females in your caravan, this problem may arise. A mating bull can easily be recognised by his "warning flag" – a pink bladder that he blows out of his mouth, accompanied by much slobbering. If you see this sign, avoid the camel. In those countries where you must buy bulls, try to obtain 4 or 5 year olds which will rarely cause trouble.

Judging a camel's age and condition takes experience and a novice will need the help of a local. However, certain facts can be ascertained by examining the animal closely. First, make the camel kneel and inspect its back and withers. Any open galls or wounds immediately rule it out as a mount: on a long desert trek it can mean death. Let the animal stand again and look for obvious defects such as crooked legs, in-growing nails, a hobbling pace, excessive fat on the legs. Check the inside of the front legs where they meet the chest: if you find evidence of rubbing there, the camel will be weak and slow. Generally look for an animal that is well covered: no ribs showing, a fairly robust hump, bright eyes, well-formed long legs and an erect carriage of the head. Finally, have someone saddle and ride the camel: note whether it snaps, bolts or roars at its handler; lead it around and see that it walks freely; and make it kneel and stand up several times.

SADDLES, HEAD-ROPES AND HOBBLES

Unless you are simply flitting fast across short distances, try not to be seduced by romantic-looking riding saddles: the Mauretanian "butterfly" *rahla* and the Tuareg *tirka* with its long crossed arm, for instance, are designed to impress. They are much more likely to cause galls on the camel's back. If you are planning a long journey it may be better to go for the prosaic packsaddle, which is not only far more efficient for carrying heavy baggage but also makes a more comfortable ride. The pattern varies from place to place but the best all-round packsaddle I have seen is the Sudanese *hawiya* – a wooden frame supported by straw or palm-fibre pads. In Asia, the double saddle or *pakra* is in use: this is heavy but efficient for both carrying and riding and has the advantage of stirrups. It has been exported to Australia and to parts of Kenya.

In East Africa (Somalia, Djibuti, Ethiopia and Kenya) camels are not generally saddled or bridled for riding: baggage is loaded on a rudimentary saddle of matting and sticks. In Sinai and the Middle East, the best design is the two-poled *shaddad* found in most northern countries. The more primitive *rahal* or south Arabian saddle, of Wilfred Thesiger fame, is awkward and unstable – during Thesiger's expeditions in the 1940s one of his Bedouin companions fell off and smashed a leg. In some countries, such as Morocco, there is no tradition of camel riding (the Moroccans, essentially hillmen, always preferred mules). Here, camels will take riders but only as "baggage": the camel has to be led by someone on the ground.

For riding, a blanket, rug or sheepskin is useful for padding the saddle and can double as bedding. A sleeping bag won't do for saddle padding because the synthetic fibre slips about. An invaluable addition to saddle gear, invariably carried by Sudanese nomads, is a waterproof sheet, preferably of canvas. It can be used for watering or feeding the camels, for carrying firewood and other heavy material, for shade during the day, shelter against the rain and as a groundsheet at night.

On long treks avoid the traditional leather head-ropes; they are comfortable for the camel but snap easily. Buy some lengths of strong fibre rope and make your own. You can cut shorter lengths for knee-hobbles and foreleg hobbles; you will also need ropes for most packsaddles. In some countries, such as India and Pakistan, and in the western and central Sahara where nose-rings or nose-pegs are used, it may be necessary to buy special head-rope as a matter of course.

Like many animals, camels will try to intimidate their human handlers. Refuse to be intimidated. As a last resort even the fieriest animals can be controlled with a good grip on their nostrils or lips, which are very sensitive. When fixing the head-rope, however, remember to stand to one side. Camels do not spit, they vomit and if you are standing in front of an irate animal you may get the entire stomach contents in your face!

In the open desert take great care to secure your camels. No matter how well you have looked after them, they may be inclined to abscond: "Never trust a camel," the Arabs say. It is unwise to leave your camels without tying them to a tree by the head-rope or, in the absence of a tree, hobbling them by the knee. This means attaching a loop of rope around the kneeling animal's front leg, just behind the knee, which makes it more difficult for the camel to rise (Figure 31.3). All trained camels are accustomed to this process. Some ready-made hobbles have wooden pegs attached, which makes the operation much easier. If not, make sure that they are tied as tightly as possible.

The Arabs have a comical folklore figure named "The Father of the Knee-hobble" who forgets to hobble his camels at night and ends up hobbling himself in the morning after they have run away, while waiting for a thirsty death. A camel can stand fairly easily on three legs and hop a surprising distance, so most desert nomads will hobble both the animal's knees at night. There is even a special verb for this action in Arabic. Yet even this will not prevent a determined animal from crawling out of sight.

When camels are grazing, they should be hobbled with a different rope, which is looped around the lower joint of the forelegs and plaited double. The best foreleg hobbles are about a metre long and have a large knot at one end and a loop at the other. Fixing this hobble requires some practice, because it means bending down among the camel's massive legs as it stamps and fidgets, and sometimes you may have to pull the forelegs closer together so that the hobble fits. Most well-trained camels are accustomed to this, however. The foreleg hobble allows the camel to shuffle about feeding while restricting its movements but, again, beware. Camels can shuffle great distances during a night and can even – believe it or not – run with the hobble attached. A powerful camel will break a leather hobble with consummate ease: make sure that your foreleg hobbles are tightly fixed and of the strongest material – double or treble the lengths if in doubt. Many, including western travellers and hardy nomads, have died because their camels disappeared in the desert: it pays to keep watch on them all the time.

OTHER SADDLE EQUIPMENT

As camel gear varies so greatly from country to country and culture to culture, it is impossible to generalise about equipment. You need large spacious bags in which to carry your food, fuel and other gear. Many of the Tuareg and the Moors merely pack their baggage in sacks and rope them to the packsaddle. In the Sudan and northern Arabia the more efficient double-poled saddle allows one to sling everything from the saddle-horns. The Egyptian Bedouin have solved the problem of loading by developing a cylindrical basket with an amazing capacity, which stands so tall that it does not need to be lifted on to the camel's back. My instinct is always to choose the simplest, most robust and least flashy equipment: a sack is more dispensable than a hand-stitched saddlebag and probably more enduring. When packing, ensure that every item is individually wrapped inside the saddlebags or you will find nothing but a mess of bits and pieces after a few days. Fibre sacks, cheap and easily available, are ideal for this.

WATER – FOR HUMANS

The traditional water-skin has been used in the Sahara since before the camel was introduced there by the Persians in 525 BC. However, even Saharan nomads will today admit that the jerry can has its advantages. The "standard" water-skin may carry an average of 25 litres, is comfortable for the camel, easy to load, simple to repair and also keeps water cool. Its great disadvantage is that it loses water by evaporation when suspended and by osmosis when lying on the ground. A very hot wind may deplete it seriously. On an unsupported camel-trek, where water can be crucial to your survival, it is more advisable to rely on the plastic jerry can. It may be more awkward in shape and more difficult to repair but, at the end of the week or a month, it will contain the same amount of water that you put into it. Beware of the giant jerry can though – a rigid 25-litre container is murder to hump about and, if damaged, means the loss of several days' water supply. The Egyptian Bedouin, probably the last true long-distance caravaneers anywhere, use the 10-litre jerry can which is handy to carry, easily stowed away in saddle bags and, if broken, represents only a limited water loss. For a supply of cool water, use one or two small water-skins, which can be topped up daily from the jerry cans. A small military-style canteen of 2–3 litres, preferably insulated, is useful but should have a strong carrying strap. Modern plastic water bags of 2 litres with drinking tubes are ideal. You will need a large enamel bowl or giant mug for pouring and transferring water from water-skins: a funnel is useful for pouring water from jerry can to water-bottle.

The Bedouin of Egypt, who regularly cross 1500 miles of almost sterile and uninhabited desert in all seasons, reckon on 5 litres of water per person per day, including water for drinking and cooking, but not washing. This allows a reasonable margin for emergencies and unexpected rises in temperature. In the hot season the daily requirement doubles to 10 litres. If your trek involves three people travelling for 10 days between wells, therefore, your summer requirement for that period is at least 300 litres which, at a kilogram per litre, is one-and-a-half to two camel loads. If you carry this amount in water-skins you could reckon on up to a third of it being lost.

GUIDES AND CAMEL MEN

Now that desert navigation with global positioning systems (GPS) has become easy and efficient many people may think twice about taking a guide. If you decide to go it alone, first make sure that you can handle the camels. Remember, if you are on an unsupported expedition and your camels disappear or die, your life is in extreme danger. This is why, to the desert people, the camels always come first. Personally, although I have done and enjoyed treks alone, I generally prefer to take Bedouin companions even if I am doing the navigation myself. I believe, as my "mentor" Wilfred Thesiger did, that you can only really get to know the environment by travelling with people who were brought up there. If you do decide to travel with local people, however, choose carefully.

Many who present themselves as desert guides today may have only a rudimentary knowledge of camels. Even those who are official desert guides may have become so used to travelling in motor vehicles that they have grown lax. Do not fall for the "wise old Arab" syndrome. If you are perfectly confident of your navigation skills, you may prefer to take a "camel-man" rather than a guide – someone who can help you with the camels, who knows the desert, but who does not profess to know the particular route that you wish to travel. There are still some excellent camel-guides about and in my experience they are generally very honest. To the Bedouin the crime of *bowqa* or "treachery" to a travelling companion – once you have "eaten bread and salt" together as the saying goes – is an unspeakable disgrace. However, this is only an ideal and, anyway, not all desert men are true Bedouin. The Tuareg, for instance, have different standards, and once made a living out of pretending to befriend desert travellers, then attacking them.

Choosing a guide or companion is very much more difficult than choosing a camel. As the Arabs say, "You cannot know a man until you have been in the desert with him". The only advice I can give here is to ask as many questions as possible – if your man is a guide you could ask him about the route and compare the answers with the information on the map. Ask him about his habits, his likes and dislikes – particularly concerning food, his family, his personal history and his opinions on the way that you intend to run the expedition. His answers might help you to identify possible areas of conflict. In the end your choice is down to instinct. Some people, especially women, have a very good intuition about strangers. However good your guide or companion, there will inevitably be conflict sooner or later, and this will tend to revolve around the age-old question of who is the boss.

Desert guides tend to be dominant characters and, however experienced you may be, they will inevitably regard you as a stranger in a strange land. If you have a clear objective in mind, it is important not to let your guide "take over". Stick doggedly to your purpose once agreed and, while having an ear for your companion's advice, retain the right to take decisions on rate of marching, halting places and route. When hiring a guide, as when hiring animals, you should agree together on the number of days that your proposed journey is likely to take and calculate payment by the day. (To help calculate, see the next section.) If you don't agree beforehand your guide/camel-man may deliberately slow the pace to make more money. This has happened even to the most distinguished western explorers.

Obviously, if you pay a lump sum for the trek, irrespective of time, the camel-man may tend to speed things up to get it over. Once you have agreed on the number of days, you can always pay an extra bonus at the end should anything unexpected slow you down. A clear agreement with your guide will save many problems later. Agree, too, on how the guide is to return home and who is going to pay. It is accepted practice in the Sahara for the employer to provide all the food for the party. As food is a notorious bone of contention on any trek, try to ascertain what your companion will and will not eat. As an extra precaution, I have found it wise to pay half the agreed sum before the journey and half on completion and, where possible, make the payment in the presence of witnesses, preferably the local police.

MARCHING

Camels are never trotted on desert journeys because it is crucial to preserve their stamina. A walking camel covers about 5 km/h or kph (3 mph) – more or less according to the terrain. A reasonable day's trek will last from 8 to 10 hours and cover 40–50 km (25–30 miles). Marching 10 hours a day, a camel journey of 200 km

(120 miles) should take roughly 4 days, and 1000 km (600 miles) 20 days. For long journeys there will be all manner of random delays to add to your original estimate of days for the journey – certainly on trips lasting more than a month. For these you should add roughly another 1 in 3 days for watering, resting, grazing or administrative delays.

A strong camel can carry up to 300 kg for 50 km per day over a period of a month. A more practical weight for desert treks, however, is between 150 and 200 kg per camel. The number of camels that you require will be decided by circumstances but the ideal ratio is probably five camels to three humans.

Marching methods change from summer to winter. In the hot season, where noon temperatures may reach 50°C, camels will find it uncomfortable to travel in the early afternoon. Experienced cameleers will start before sunrise, halt at about 11:30 and rest till around 15:00. It is advisable to erect some kind of shelter for yourself during this period, or heat exhaustion may result. In winter, when camels shy from the intense night cold, cameleers will generally march from roughly sunrise to sunset with a short break at noon for a drink and some food. A shelter is unnecessary in these lower temperatures.

Most professional caravaneers combine walking and riding as the most efficient use of their camels. Many western novices are tempted to walk all day, either because they are nervous about mounting the camel, or out of the desire to "prove themselves", forgetting that the more physical effort they expend, the greater their water loss. This is fine if you are being re-supplied by motorised back-up but otherwise it is advisable to do as the desert people do – walk during the cooler times, ride during the hot times. In east Africa, where camels are never ridden, marches are generally shorter and the country more inhabited and more watered than in the Sahara.

CAMELS – FOOD AND WATER

It is the camel's legendary ability to go without water that, more than anything else, makes it the most efficient means of transport in deserts. In winter temperatures camels lose a tiny 1 per cent of their body weight per day and, if they find green grazing, they may go on almost indefinitely without water. I have seen camels refuse water after a hard 17-day slog across sterile desert in winter, but the distinguished Saharan explorer Theodore Monod once travelled 28 days without watering his camels. In summer, however, it is a very different story. Laden camels on the march need to be watered every 3 days in high temperatures, although they may carry on for 5. Like humans, camels are subject to habit and training, and those accustomed to well-watered regions or to being watered every day are of little use on unsupported desert expeditions. Check this when buying or hiring your camels.

Although a camel may be given a small amount of water from its own load in an emergency, it would be a fatal mistake to water camels from the caravan's own resources. Desert people never do this. Camels may drink up to 27 litres per minute and gulp down 120 litres per session. To satisfy all your camels would require a great many 10-litre jerry cans!

Although they can do without water, camels cannot do without food. They will eat many desert shrubs and grasses and, in vegetated areas such as the Sahel, their food will present few problems as long as they are given time to graze. In more arid regions, however, you must provide either hay or grain. The Tuareg salt-caravans, which still cross the Tenere Erg in Niger, carry sheaves of hay for the camels but generally hay is awkward to carry and load. Grain is easier to handle, although a camel must be trained to eat it: if necessary you should ask the previous owner if the camel had been so trained when buying it. Sorghum grain is preferred although in some areas camels eat wheat. Two to three kilograms of grain per camel per day is a good ration and should preferably be fed to the camels in cotton nose-bags to prevent them fighting over it. On a journey of 10 days without grazing, therefore, with a caravan of five camels, you will require 100–150 kg of grain – half to three-quarters of a camel-load.

WATER SOURCES

As water is crucial in the desert, your trek (assuming that it is unsupported) will naturally be from water source to water source. You can keep a length of rope and a makeshift bucket for use in shallow wells. In some places, nomads water their flocks and herds from open pools in the wet season. In rocky areas there are frequently rock pools, known in the Sahara as *gueltas*, which are hard to find without local knowledge. These pools are strictly temporary and the water is generally good. In other places, especially in the Thar Desert of India and Pakistan, nomads rely on hand-dug cisterns for a long-term supply and water from these places should be carefully boiled and sterilised for human consumption because they are teeming with guinea-worm – a very nasty parasite indeed.

If you are lucky enough to experience a rainstorm, you may find that the rainwater has formed a shallow pool from which you can fill your vessels. The Arabs call such pools "the bounty of God".

Mostly you will rely on wells for your water supply. "Hand-dug" wells as opposed to modern "deep-bore" wells are often very ancient and may vary tremendously in depth. Along the southern fringes of the Sahara the wells may be as much as 30 m deep, so unless you wish to go to the trouble of carrying a 30 m rope, not to mention a bucket, it is worth finding out in advance if the wells are currently in use by nomads whose equipment you can borrow. (Albeit 30 m of nylon parachute cord is not bulky or heavy.) Well water is generally fairly safe, although when travelling with clients I add iodine as a matter of course – one cap to a 20-litre jerry can. A last piece of advice: never pass a water source without filling all your vessels. It is too easy to tell yourself that you have enough water when you are tired or itching to press on. Never lose the chance to get water. The desert is a harsh mistress and you can never tell what tomorrow will bring. No matter how high the temperatures soar, as long as you have water you can survive.

FOOD AND COOKING

Obviously the kind of western luxuries some people take on motor journeys are impractical when travelling by camel. Food can be a major problem, especially in winter when hunger rather than thirst becomes the dominant preoccupation. I find it best to live on local rations. The nomads generally eat rice, unleavened bread, cous-cous or polenta. In the scale of efficiency, pre-packaged cous-cous scores highest; it is nourishing and very quick to cook, and requires only a small amount of water. Rice scores lowest because of the large amount of water needed. Nomads bake bread in the sand beneath the ashes of the fire and serve it with some kind of seasoning. In the Thar Desert, where water is more plentiful, the nomads make chapattis over the fire. In the eastern Sahara, *assida* – a kind of porridge – is the norm. All these can be served with a stew or sauce made of dried tomatoes, dried "ladies fingers"(okra) and onions with sun-dried meat. You can make the meat yourself by slaughtering a goat or sheep and cutting the meat into small pieces, then hanging them to dry on a bush. The drying shouldn't take more than 24 hours. If some of the meat grows maggots, simply clean them out and leave it to dry again: some nomads eat the maggots too and they seldom come to any harm! If you meet nomads on your trek you can generally buy small stocks of fresh meat to replenish your supply of sun-dried meat. Tins of corned beef and sardines are heavy but excellent supplements.

Tea and coffee, with plenty of sugar, are important luxuries in the desert and are especially valued by the desert people. Other valuable additions to the diet are peanuts, dates or biscuits, all of which require no cooking and can be eaten while you travel. Dried milk, mixed with sugar and water, makes a refreshing drink if your stomach revolts at the idea of food in 50°C heat. In many parts of the desert you will find dry firewood; in other, more arid parts, there may be none. Again, find out from local people whether or not there is likely to be fuel on your route. If not, either buy or collect firewood before entering the woodless region. Use the traditional nomad fireplace of three stones for cooking; it burns very economically. In sandy areas you can dig a three-pronged slit trench with the same effect, or carry three old tins to set up in the sand. The important factor when using this fireplace is to ensure that there is a small gap between the fuel and the base of the pot to allow the air to circulate.

Some nomads do use charcoal but it is extremely difficult to light and inefficient without a proper charcoal burner. In remote, treeless places, the nomads will use camel dung as a fuel. Fresh dung will not light, however, and they collect bone-dry stuff, which has been deposited months or years before. Even then, lighting a fire of dung and keeping it alive involves tremendous effort.

A small butane stove is a good standby and a quick method of making tea on icy winter mornings, although gas cylinders are not available in some countries and cannot, of course, be taken by air. For cooking, you need a large, robust pot with a lid and a kettle. All the Saharan nomads eat from a communal dish, even the Tuareg who use a spoon whereas the others use hands. Although Islamic peoples do not normally eat with strange women, they will often make an exception for westerners in these circumstances. To eat on your own in the Sahara is to be regarded as a very strange fish. In the Thar Desert, however, even Muslims seem to eat from separate bowls. The Rajputs on the Indian side of the border, being Hindus, will not eat or drink from the same vessel as an "unclean" Christian, nor allow a Christian to dip his or her water bottle into a Rajput cistern. You have to ask your Rajput guide to do it.

Unless water is plentiful, do not waste it in washing vessels: sand is just as effective for cleaning pots. The cooking utensils should be packed carefully in a separate bag, else the accumulated soot on their sides will quickly blacken everything.

HYGIENE, MEDICAL, CREEPY-CRAWLIES

The desert people rarely wash on long desert journeys, and unless you have water to spare washing will be your lowest priority. However, if you are in a large group, stomach problems are far more likely to result from lack of hygiene than from food poisoning. When travelling with a group, and if there is water to spare, I recommend that everyone rinses his or her hands in a communal mess-tin or bowl of water laced with disinfectant or iodine before a meal. An indispensable requirement is an excellent medical kit, including painkillers and broad-spectrum antibiotics. Travelling with clients I have found that the most common ailment is diarrhoea. Another common problem is heat exhaustion. Remember that we humans only need to lose 5 per cent of our body moisture to find ourselves in a critical condition, and it doesn't even have to be hot. The key to this problem is to drink sufficient fluids – remember that thirst, or the lack of it, is not a reliable indicator of dehydration.

I find that people's greatest fears of travelling by camel or on foot in the desert are of snakes and scorpions. Obviously the threat exists, but the fears are largely unfounded. Snakes and scorpions are generally only a danger in summer, when it becomes too hot for them to lie under their stones. Most scorpions, anyway, pack only a local toxin, which although unpleasant causes no more than a local swelling. There are two species of scorpions in the Sahara, however, that carry potentially lethal nerve toxins, and of course there are poisonous snakes such as puff adders. The chance of being bitten or stung by one of these is extremely low – after almost a quarter of a century and 20,000 miles by camel in the world's deserts, I have only ever been stung by a scorpion once. The number of snakes I have seen could almost be counted on my fingers. Two precautions that can be taken are:

1. Check out the sleeping place for snake tracks and scorpion tracks before making

camp (snake tracks are "wiggly", scorpion tracks are asymmetrical and large – almost like finger-prints – whereas beetles make a very symmetrical stitch pattern). If they are present, move.

2. Shake out boots and clothes before putting them on. If you are using water-skins, check the undersides before loading – this is a favourite place for scorpions to hang out.

A third potential hazard is the solifugid camel-spider, which has a nasty bite but is not poisonous. These things eat scorpions!

NAVIGATION

Finding your way on camels in the desert, especially when it involves searching for wells in a vast area of rock and sand, is the single most important factor in desert survival. It is unwise to delegate the responsibility for navigation entirely to your guide, even if he knows the way well. Traditional guides can be excellent but this is either because they have been going that way since childhood or because they are very skilled at using the sun. The problem with sun navigation is that at noon, when the sun is directly overhead, there are no shadows to guide you. I have known an otherwise superb guide who swore we were going south at midday when my compass told me we were heading east.

Having covered tens of thousands of miles by camel, much of it with traditional guides, I am dubious about the nomads' supposed "perfect sense of orientation" when outside the country in which they grew up, although they can be astoundingly accurate in familiar country. Often illiterate, they also have a wonderful memory for descriptions and an extremely well-developed vocabulary for geographical terms. Anyone who travels with them will notice how frequently they stop to talk to other travellers; this "nomad grapevine", coupled with keen observation and a very exact classification of natural features, is the true strength of the desert people.

Before GPS was practical for camel-borne treks, I always relied on good maps and a pair of Silva compasses for my navigation. With a watch and a compass, you can navigate by dead reckoning (DR), calculating the camel's pace at roughly 5 km/h and recording the time that you travel on each compass bearing. You must make allowances for halts, differences in terrain, sandstorms and deviations, but with a good map you should be able to pinpoint your exact position at any given spot.

Over the past few years, however, I have used a hand-held, battery-powered GPS unit. As long as the batteries last (take plenty, of course) GPS makes desert travel a great deal easier and more independent than it was in the past. It will even tell you your rate of travel and distance covered. Still, I would not venture into any desert without at least one compass as a back-up, and of course GPS is no substitute for an accurate DR record.

WHAT TO WEAR, WHAT TO BRING

It seems to me that the West has devised no better dress for travelling by camel than that worn by desert people. The long, loose-fitting shirt allows a layer of cool insulating air to circulate beneath it. The baggy trousers or loincloths worn by most desert tribes are extremely comfortable for riding. When travelling with clients I advise men, in any case, not to wear underpants while riding because these will tend to rub and cause saddle sores. The turban or headcloth, with its many layers, not only keeps the head cool but can also be used in a number of other ways, including veiling the face in a sand-storm. Personally I believe that there is no such thing as a "desert boot"; most desert people go bare-foot or wear sandals. Any kind of boot will make feet sweat in high temperatures and the perspiration softens the skin, making it more prone to blisters. In some terrains, however – especially rocky hammada, there is no substitute for the boot. In winter you will need a pullover, some kind of warm jacket and perhaps some socks. A pair of sunglasses or ski-goggles is recommended: nomads do not wear them but blue or green-eyed westerners are more susceptible to the sun.

Incidentally, the long Arab robes and baggy trousers are much more appropriate for answering nature's call in places where there is no cover. Nomads use stones to clean themselves after defecating where water is very scarce: despite the myth, sand is only used as a very last resort because it can work itself into some very awkward places! I can personally recommend the use of stones as a clean and efficient method, but if you are not prepared to do this, take a large supply of toilet paper. Remember to burn it after use, though, to preserve the environment.

For personal kit a torch is an essential, preferably two. I prefer something cheap and simple and with standard U2 batteries, because exotic bulbs and batteries are often impossible to obtain in less developed countries. A head-torch, however, is extremely useful. A Swiss Army knife or Leatherman is a must but a larger knife might be useful for slaughtering and butchering meat. You will need an axe or machete for cutting firewood. Take a good strong cup, a supply of nylon parachute cord for repairs, a large packing needle as well as a normal needle and thread, a sleeping bag and groundsheet – and perhaps a sleeping mat. Most nomads normally sleep beneath the stars, although in some regions, such as East Africa, a tent is sometimes desirable because of the threat of dangerous game.

TRAINING

Whatever country you are trekking in, travelling by camel is inevitably going to involve a great deal of walking. Cardiovascular fitness is therefore the main area to concentrate on when preparing yourself physically for a camel trek: jogging, long-distance running, cycling, swimming. Loading camels usually requires a certain amount of lifting so weight training is also appropriate. Stitching, knot tying, leather-repair work and fire lighting are among field skills that need to be mastered.

If you intend to make a long expedition – or even a short one – it is essential to spend some time getting used to the camels and their loads, and developing general riding and operating skills on short local trips before departing on the main project. When I am asked what I consider the most important factor in the success of a camel expedition, most people find the answer surprising. My strongest advice is this: learn the language. As already noted, desert nomads have a superb intelligence system and are keenly observant. They will take great pains to stop and talk to any stranger whom they meet on the way and like this they learn where rain has fallen, which wells are open and which are closed, who has passed which way and for what reason. To learn their language is always worth the effort: if you can communicate directly with your guide and the people you are travelling among, even a little, then your chances of success are enhanced by at least 50 per cent.

A FINAL WORD

Some people go into the wilderness believing that it is a dragon to be slain, a foe to be conquered. Although accepting that the desert can be dangerous, the Bedouin, who are for me the beau ideal of desert people, have never had this western attitude. They have survived by accepting that they are part of nature and by submitting themselves to its moods. Mobility and flexibility are the key to living in the desert – one of the world's most extreme, most unpredictable of environments. Be open and flexible like the Bedouin and you too will survive.

FURTHER READING

Asher, M. (1984) *In Search of the Forty Days Road*. London: Longman.
Asher, M. (1986) *A Desert Dies*. London: Viking.
Asher, M. (1986) *The Last of the Bedu*. London: Viking.
Asher, M. (1986) *Sahara* (with photographer Kazoyoshi Nomachi). London: Viking.
Asher, M. (1988) *Impossible Journey*. London: Viking.
Bulliet, R. (1975) *The Camel and the Wheel*. Cambridge, MA: Harvard University Press.
Gauthier-Pilters, H. and Dagg, A.I. (1981) *The Camel – Its Evolution, Ecology, Behaviour and Relationship to Man*. Chicago University Press.
Marozzi, J. (2001) *South from Barbary*. London: HarperCollins.
Moorhouse, G. (1974) *The Fearful Void*. London: Hodder & Stoughton.
Thesiger, W. (1959) *Arabian Sands*. London: Longman.
Wilson, T. (1983) *The Camel*. Tropical Agriculture Series. London: Macmillan.

32 TRAVELLING BY HORSE

James Greenwood

WHY ARE YOU THINKING OF USING HORSES?

You probably fall into one of two camps or a combination of both; you either specifically want to make a horse journey or the horse seems to be a good means of transport for entering or crossing the country that you wish to investigate. There is value in both. The horse offers a natural freedom to travel that can open doors that you did not know existed. Like taking a dog or child into the park, you are already more approachable. The very process of travelling with a horse opens up an extraordinary world – at an average speed of 5 km/h (kph) you get to see a lot of things and meet a lot of people.

You don't even have to ride to benefit – it is a treat to be able to cover a lot of ground without having to wear a rucksack, and horses, packed correctly, can carry heavy loads made up of rations and equipment.

Their versatility and ability to cross most terrains is matched by a level of maintenance costs that can only be admired by expeditioners using more costly means of transport such as aircraft.

Do not count on using horses to speed up your journey. Horses are very slow and require constant attention. The process of riding, using unfamiliar muscles, mixed with exercising the whole range of walking muscles when on the ground, can be very tiring. The minimum time required to prepare for a horse journey of any great length is a month. There is terrain – jungle, coastal, wetlands, rock faces – where feet outrun a horse. Looking at the accounts of the long distance rides, the average speed has been a constant 19 km/day – 11 miles a day. Of course, average daily distances are longer and horses can perform incredible feats of short-term endurance – in Argentina there are 15-day horse races of 750 km.

LOOKING AFTER YOUR HORSE

The most important thing to remember is that the horse is a naturally nomadic animal. They work on a cycle of movement, rest and replenishment. This natural cycle is the key to using horses to the best of their ability, while ensuring that they remain healthy and spirited. Your thoughts should be on sourcing the best food and water, and maintaining a natural rhythm that will get you to your destination with the minimum hassle and in maximum comfort. Emile Brager's book, *Techniques du voyage à cheval*, has everything you need to know.

Once in charge of a horse, you are in charge of another living thing, and that may be a responsibility that you don't need. Such is the lack of everyday understanding of the basic components of equestrian travel that many trips by horse fail – usually within the first 2 weeks. Considering that the last complete manual on horse transport was written in the UK by the army at the beginning of the twentieth century, this is not surprising.

For any use less than a month, hire rather than buy horses. Work with local horsemen at all times. Horsemen the world over make good company and know their environment intimately.

If you have brought horses with you, seek out and speak to those who best know local conditions ahead and who are attuned to the search for food and water. Horses need feeding and watering daily.

Horses in work are bred to travel backwards and forwards from their field, corral or stable to the fields, streets, passes or competition area – if they get ill or lame, they can have a couple of days off. You will not have that luxury to give your horses on a long journey, and you should make every effort, depending on finances, to find the best horses to buy or hire.

They should be sound, free from saddle sores, and between 5 and 15 years old. Cost and quality of horse will vary markedly depending on where you find yourself in the world.

It is absolutely essential that at least one of the members of the expedition has an extensive deep knowledge of working with horses. Detecting the earliest signs of illness or exhaustion is crucial to a smooth-running trip. A basic knowledge of horse management among all those present is a bonus. However, total novices should not have a problem riding from day one, although the person in charge of the horses should look out for the extra work that a novice rider will give a horse.

Groups should be no larger than what is sustainable – horses and people consume considerable resources, and the combination of grazing and shod hooves can denude a campsite overnight.

It is perfectly possible for a full set of modern lightweight camping equipment to be packed around the seat of a saddle with a ruthless attitude to weight. Carry only the essentials. Two horses taking it in turns to carry both kit and rider is the fastest method of travel. The second horse will adopt a herd mentality, allowing the lead horse to assess the everyday threats of the environment. Despite travelling the same distance, the follower will use a fraction of the energy of its friend.

SADDLERY

Above all, a saddle should not cause a saddle sore. Once damaged along the back, a horse will rapidly lose condition and weaken as a result of the pain, unless remedial action is taken, which is a laborious, time-consuming and difficult process. The English cavalry saddle remains in use all over the world because of its sympathetic design.

PACKING

Pack horses can be both a blessing and a curse. The ability to carry equipment and stores can make life more comfortable and fruitful, but the extra time required for saddling, packing and horse care can be frustrating. Dead weight takes much more energy out of a horse than a good rider, and is more likely to give a saddle sore. The right equipment is essential in keeping the pack steady, balanced and upright.

SHOEING

Learn the basics by spending time with a friendly farrier. It is illegal to shoe a horse in the UK without having done a 3-year apprenticeship, giving the UK the best artisan farriers in the world. As you are unlikely to have a spare 3 years, concentrate on "balance", the basis of all good shoe work. In the field, work with local farriers. The most essential tool is a pair of "nippers", and always carry spare nails.

FEEDING

Fodder is the basic ingredient that no horse can do without. Unless they have a regular intake of roughage, their digestive systems will seize up and you will have a case of colic on your hands – rice and maize straw give virtually no nutritional value but are better than nothing. The king of fodder is alfalfa, followed by hay and barley straw. Natural grazing may often be available and full use should be made of it. On the whole, horses will be tethered out to graze and the process of tethering is fraught with danger to the horse. Picking the spot to maximise the night's feeding, while ensuring that the horse will not come to any harm, can take an hour of indecision. No horse should be tethered until its knowledge of the tether is confirmed.

Barley, wheat, maize and pulses can all be given as "hard" feed. The quantities to give will depend on what the horses have been used to and what you are expecting in the way of work. If travelling for a number of days where "hard" feed will not be available, it is worth taking feed with you and gradually reducing it to "half-rations" to avoid sudden changes of diet. Grain that is in some way crushed or ground gives a greater nutritional lift than whole grain, which can simply pass through the horse's system undigested.

WATER

There is much debate about how much and how often horses should be watered. However, all agree that horses must be watered every day. It is impossible to carry sufficient water, so the availability of daily water is the main driver behind route selection. Without water, your horses will die.

CONCLUSION

Successful horse travellers of recent times have relied on a mixture of common sense, good horsemanship and an initial basic understanding of both the environment and the societies through which they were travelling. Specialist manuals and equipment are beginning to appear, particularly in France, as the commercial world catches up with an increasingly popular method of travel.

For getting close to the land and for getting close to the people of that land, I know no better way to travel.

And remember – when the going gets tough and you feel like crying, you're only pony trekking!

FURTHER INFORMATION

Further reading
Brager, E. (1995) *Techniques du voyage à cheval*. Paris: Editions Nathan.
Green, T. (1999) *Saddled with Darwin: A journey through South America*. London: Weidenfeld and
    Nicolson.
Greenwood, J. (1992) *No Guns, Big Smile*. London: Michael Joseph.
Hanbury-Tenison, R. (1985) *White Horses over France: The Carmargue to Cornwall*. London: Granada.
Hanbury-Tenison, R. (1987) *A Ride along the Great Wall*. London: Century.
Hanbury-Tenison, R. (1989) *Fragile Eden: A ride through New Zealand*. London: Century.
Hanbury-Tenison, R. (1991) *A Spanish Pilgrimage: A canter to St James*. London: Arrow.
Severin, T. (1989) *Crusader: By horse to Jerusalem*. London: Hutchinson.
Severin, T. (1991) *In Search of Genghis Khan*. London: Century Hutchinson.
Tolstoy, A. (2003) *Last Secrets of the Silk Road*. London: Profile Books.
Tschiffely, A.F. (2002) *Tschiffely's Ride*. Reprint. London: Pallas Athene.

Useful addresses and websites
Association des Cavaliers au Long Cours (CALC), La Carcarie, 30700 Montaren, France
The Long Riders Guild. Website: www.thelongridersguild.com
    An international association of equestrian explorers, formed in 1994, to represent men and women
    who have ridden more than 1000 continuous miles on a single equestrian journey.
Horse Travel Books. Website: www.horsetravelbooks.com

SECTION 7

RECORDING YOUR
EXPEDITION

33 GUIDE TO WRITING EXPEDITION REPORTS

Shane Winser and Nicholas McWilliam

At the heart of every expedition is the work that it does and the results that it produces. The expedition report is the main means of sharing these results with the world. Without a report, an expedition is relegated to the status of a holiday and personal memories.

Expeditions involve a vast number of formal and informal partnerships: sponsors, advisers, government departments, schools, universities, embassies, villages and voluntary groups. As part of your dealings with them, sharing your results is central, and again the expedition report plays a leading role.

There are, of course, many other ways of sharing an expedition's results: books, articles, films, photographs, posters, sound recordings, websites. The possibilities are vast and depend on the expedition's work and who might benefit. Other results may take shape during the expedition itself: training programmes, workshops, buildings and facilities, donated equipment, education programmes and botanical reference collections are all examples.

All expeditions and reports will of course be different. The nature of the report will depend on the aims and scope of the expedition, who it is aimed at and the means of publication. This chapter aims to show how best to produce the report. Our advice is based on reading thousands of expedition reports lodged for reference with the RGS–IBG Expedition Advisory Centre and on producing expedition reports ourselves.

THE FORM OF THE REPORT

In most cases, it is appropriate to produce one comprehensive printed report, typically well within a year of returning. In some cases, however, two or more reports may be more suitable. You may, for example, be trying to reach very different audiences: one summary report for sponsors and one full scientific report for people who could benefit from your research. If post-fieldwork identification, analysis or laboratory work is needed, your results may not be available until long after the expedition: produce as detailed a report as soon as possible after the expedition, and follow it with the full findings later.

Where English is not the first language, your results will be much more accessible and effective if you publish one version in the language of your host country and one version in English.

Web and CD versions

Many word processors and page layout programs can convert documents into HTML or into Portable Document Format (pdf) for publication on a website. This has the advantages of easy access, low-cost distribution, full colour and easy updating. However, remember that internet access may be slow and expensive from your host country and is almost certainly impossible from remote field locations. Find a site willing to host your web pages over a long period, e.g. in a university, ask whether your department would be willing to host your expedition web pages.

Producing a CD-ROM has the advantage of being easily sent by post and not needing an internet connection. HTML or pdf might be an appropriate format, because it requires only a web-browser or reader (which may be copied on to the CD too) for access, rather than any special software. If your main partners in your host country do not already have suitably equipped computers, you might be able to donate a system; discuss in advance what would be appropriate.

Despite the advantages, digital reports should be produced only as *additions* to the printed report. The printed version remains the most useful publication, easy for its recipients to refer to and available through libraries.

REPORT CONTENTS

Cover and title page

An attractive (not necessarily lavish) cover will make a good first impression and make the report more tempting to read. The title page information can be included on the front cover or as a separate first page. The following should be given.

- *Title*: the title is immediately informative if it includes the name of the institution organising the expedition, the country visited (and possibly the area) and the year of the expedition, e.g. "University of Wales Svalbard Glacier Survey 2002: Final Report".
- *Location and dates in the field*: the country, location, dates in the field and participating organisations should be given as the subtitle, if they are not already in the main title. Sometimes the locality name is given, but the country name is accidentally overlooked!

- *Aim*: if the expedition's work is not already apparent on the front cover, it might be useful to state the overall aim, summarised in a short sentence (say 10–15 words).
- *Author(s) and/or editor(s)*: these details should be included.
- *Permanent contact address, email and website*: readers wishing to find out more or planning their own expeditions to the same area may well want to get in touch with you. Use as permanent a contact address as possible.
- *Year*: you need to include the year in which the report was published.

Finally, don't forget the word "Report" (or whatever description is suitable). It is surprising how many expedition publications don't state what they actually *are* on the cover!

Contents page
A list of contents, with section titles, subtitles and page numbers, is invaluable for readers. It will also probably avoid the need for an index. The contents list is usually on a page of its own.

Abstract or summary
This is extremely useful for those who have time only to glance through the report and for anybody needing a summary for a library or other information system. The abstract should be around 200 words (one or two paragraphs), written very clearly, summarising the report with a statement of the expedition's aims, fieldwork and achievements. If plans changed (e.g. for political or security reasons), or progress was different to that expected, brief details should be given here.

It is useful (and courteous) to translate the summary into the language of the host country. Indeed, if your expedition makes findings or recommendations that may be useful to non-English speakers in your host country, consider translating the whole report.

Introduction and map
This provides background information to the expedition. How did it arise? What were the main issues involved? Who were the key partners? What was its motivation? And what were the expectations? For the reader, this puts the rest of the report in context, and is matched at the end by the conclusions.

This is also a good place to include a location map and a general map of the area. More detailed maps can go into relevant sections of the report.

Expedition members
List the expedition members, giving brief details of their relevant experience (e.g. academic status, qualifications, climbing experience) and their main roles in the

expedition. With a young persons' expedition, an indication of the age group or individual ages is helpful.

Fieldwork and research
For research expeditions, this is probably the most important part of the report. Here are the results that justify all the expedition's logistic and scientific efforts!

If you carried out several projects during the expedition, it is probably clearest to present these separately. At the end of all of them, give an integrated set of conclusions – one of the great benefits of expeditions is their ability to draw conclusions from multidisciplinary studies.

You might follow the form of a scientific paper: starting with the context and questions of your project; describing your methods in a way that would allow, in principle, the project to be repeated; presenting the results; and finishing with a discussion and conclusions. Indeed, this format can be adapted to almost any type of expedition, not just scientific ones.

Background, planning and aims
• What was the background to the project?
• Who helped you put the project together?
• What was the aim?
• Which area was chosen for study, how and why?

Methods: fieldwork and follow-up
• Give details of field methods: sampling, collecting, observing, recording, etc.
• What difficulties or useful innovations did you find?
• Did you have to modify your project or techniques in the field? Why and how?
• Did any opportunistic projects arise?
• How did you process and analyse your field data?

Results
• Present and describe the results arising from the fieldwork and analysis, with tables, graphs and pictures where suitable.
• Large sets of "raw" data can be provided on an accompanying CD-ROM or by request, with a summary printed in the report.
• Report other incidental observations, e.g. climate, bird sightings.

Discussions and conclusions
• What did the results show? Were your original questions answered?
• How do your results compare with other work in the same or other regions? How do they fit into a wider context?
• What are the key conclusions? These might relate to the results themselves, to

the methods or to other achievements of the expedition.
- Provide accurate, detailed and specific conclusions, avoiding general inferences, interpretations and recommendations (tempting though they are!).
- What were the major limitations of the project? What caveats apply to the conclusions?
- Expeditions often return with more questions and ideas than they started with: what follow-up work might be useful? Could any studies be continued, e.g. monitoring or conservation projects?

Adventurous activities

Many expeditions' achievements are adventurous, such as mountaineering, canoeing, caving or trekking, and some research expeditions will also have an adventure element. For these expeditions, the report is important both as a record of their achievements and to help others following them. Give details of the area, the environment, your aim and objectives, and the planning and achievements. The full account of the expedition's day-to-day activities and accomplishments often takes the form of a diary or log (see later).

Administration and logistics

This should give insight into the process and problems of planning and executing the project. It will be of great practical help, particularly to those undertaking similar projects in the future. Include a note of any unforeseen difficulties and how they were overcome.

Listed below (in no special order) are some headings that you might wish to include, as appropriate. Most of these require no more than one or two paragraphs:

- *Destination area*: the aim isn't to write a comprehensive guide to the area that you visited. Instead, give details that could help others but that are not found in guidebooks and previous expedition reports. The bibliography can refer to other information sources. Very detailed information (such as packing lists or financial details) can be put into appendices to avoid cluttering the main text.
- *Research materials and information sources*: which maps, aerial photographs and satellite images did you use? Where did you obtain these and other materials? Any particularly useful books, libraries, websites or advisers?
- *Training and equipment testing*: what form did these take, and where were they carried out? Were they effective?
- *Permission and permits*: how were research permits and visas obtained? How long did the process take? Was special customs clearance needed for export or import of equipment and specimens? This information can be particularly useful to future expedition planners.
- *Fund-raising*: what methods did you employ? Were they successful? What

recommendations do you have to make? Who were your key backers?

- *Finances*: how did you manage your accounts? How did you transfer money to and from your destination? The expedition accounts may not have be closed when the report is written, but still summarise the income and expenditure to date. For reports being sent to funding bodies, this is essential.
- *Insurance*: what did you insure, with whom? Was the company helpful? Was the policy sufficient? How much did it cost? What claims were made and settled?
- *Travel, transport and freighting*: how did you and your equipment reach your destination? What transport was used locally? For example, mention fuel availability, accessibility, and use of porters and guides. Give details of specialist companies, such as freighting or vehicle hire.
- *Food and accommodation*: did you buy food locally or bring it with you? Was it good? What water supply did you use? What accommodation arrangements did you have, both in transit and in the field?
- *Communications*: how did you communicate with your host-country partners pre-expedition? Did you use any field communications such as VHF radio or satellite phones? What local mail, telephone, fax and email facilities were there?
- *Specialist equipment*: in particular, mention new ideas and techniques and any equipment that the expedition may have designed. Special equipment for a particular environment should be mentioned.
- *Risks and hazards*: what potential risks did the expedition face? How were these assessed? Did they affect plans and did any problems arise? For example, insects, vehicle travel, flooding, mud slides, political instability, water quality, theft. If you did a formal risk assessment, publish it and say how you would modify it in future. Be honest about any near misses. Could they have been avoided?
- *Medical arrangements*: give details of medical personnel on the expedition, medical equipment that you took and supplies, facilities for emergency evacuation (telephone, radio, doctor, airstrip, hospital, etc.). Describe preventive medicine, inoculation procedures, and any illnesses and treatment.
- *Environmental and social impact assessment*: what pre-expedition assessment was made of its impacts? Were they monitored in the field? Did any impacts result? Were they significant? What recommendations would you make? Factors might include fuel/wood burning, construction, planting, vehicles, and cultural, political and economic effects on local communities.
- *Itinerary*: if the expedition has been working in several sites or has been a journey (walking, vehicle, boat, etc.), details can be given here, or included in a separate diary or log section (see below).
- *Photography, sound-recordings, video and film*: permission needed; any

particular problems; types of film suitable for the area; equipment used; what pictures were taken, for what purpose; where copies are held and how to obtain them.

Diary or log

For expeditions undertaking long journeys, this may form the major part of the report. For expeditions based at one site, it will probably be quite short and should be kept distinct from the research sections. Details can include: dates, distances, sites, routes, travel arrangements, and features such as bridges and petrol stations. Detailed maps are invaluable here.

Conclusion

The conclusion summarises the main text and makes clear the lasting contribution of the expedition. It answers the questions raised in the introduction and adds any new information exposed by the expedition process. For a reader in a hurry, this is likely to be the section looked at first, so make it as clear and helpful as possible, perhaps numbering the conclusions. Themes to include are: did the expedition achieve its aim? What were its critical strengths and weaknesses? What were the key results or achievements? What is their wider significance and what benefit might they bring? What recommendations would you make? What supplementary projects could be done by future expeditions?

Acknowledgements

Remember to thank everybody, individuals and organisations, who helped before, during and after the expedition.

Appendices

These are used to develop points of detail not easily placed elsewhere in the text. For example:

- *Inventory of stores and equipment*: be selective and comment on those items that require notes in the light of your experiences. Add notes on their packing, how suitable you found them, customs problems, etc. If anything you took seemed unnecessary, or if there was anything that you had wished you had taken, mention this.
- *Summary of finances*: publish a full account of income and expenditure, audited if necessary.
- *"Raw" field data*: if large amounts of data were generated in the field, they should if possible be made available in the report, but not necessarily in the main text, e.g. animal distribution observations, plant collections or stone size measurements. Summary tables and statistics can then be used in the main text.

Address list and web links
An annotated list of useful names, addresses and websites will become a great resource for those following you.

Bibliography
List the books, articles, websites, expedition reports, etc. that you used, highlighting the most helpful ones. Also list other publications and products from the expedition, as well as papers being prepared for publication by expedition members.

Distribution list
List where copies of the report have been lodged for reference by future expedition leaders and others, and where the report can be bought.

PRODUCTION AND PRINTING

The various stages of producing the report usually involve the following: initial outline of contents (as covered earlier in this chapter); writing, editing and checking text and illustrations; design and layout; production of final copy digitally or on paper; printing/reproduction; collating and binding; and distribution.

Writing the report
Above all, bear in mind who will be reading the report and the information that you are conveying to them. Unless you are producing a technical report for an expert readership, write and design it with a general readership in mind. This process not only clarifies your own thinking, but also means that far more people will read it and find it useful.

Make sure that acronyms are fully explained and use the full name for the first occurrences. Explain any specialised procedures and terms, and keep the language as simple and concise as possible. Use suitable headings and avoid long rambling paragraphs: think of one theme per paragraph. Short sentences are better than long ones.

When the text is complete, leave it for a few days before reviewing and revising it. If possible get an independent person to read through the text too, asking them to comment on the style and perhaps the content. Finally, go through the report several times again, each time looking critically at one aspect such as logical sequence, punctuation and layout.

We are assuming at this stage that you are preparing the report on a computer. You may well have contributions from team members in your host country. In this case, make sure that they have suitable facilities for producing their sections. Donating computer equipment could help tremendously with this task.

Design and layout

For a long document such as a report, finalise the text and figures as far as possible before starting on the final formatting and page layout. It's tempting to do the formatting as you go along – but this almost invariably leads to lots of time wasted by changing pagination and layouts as you make small edits to the text. It is quite possible, however, to make the main design decisions in advance: page size, margins, line spacing, heading and text styles, running heads and whether colour will be used.

Think carefully about the overall design. Is it easy to read? Do chapter headings and page headings help readers know where they are in the report and encourage them to read more? Do the design and illustrations help emphasise the important aspects of the expedition's work and expand on items in the text? Look closely at the design of other books, articles and reports; choose the best ideas!

Check carefully for consistency, both in the design and in the text itself. This applies to features such as bibliographic references, abbreviations for units, spelling of place names, chapter names, section headings and page numbers.

However you design the report, don't forget to include page numbers. Without these it is incredibly difficult to make bibliographic references to the report and to catalogue it in a library.

Choosing a printing method and printer

Deciding how the report will be printed and who will do it is best done as early as possible, perhaps even before the expedition leaves. You will need to budget in advance for the work and know what deadlines to aim for. Knowing how the report will be printed will help greatly with preparing it in a suitable format, whether on computer or on "hard copy".

The method used for reproducing the report depends on:

- the budget
- the number of copies that you need
- the facilities available to you
- the length of the report
- features such as page size, colour printing, binding, recycled paper, etc.

Write these down; discuss them with a variety of printing companies; and get estimates for the cost of the work. Ask the printer to explain where the major expenses lie and if necessary where savings may be made, e.g. by using a different page layout or smaller type size to reduce the number of pages, or printing photographs in greyscale instead of colour.

Most universities have print or reprographic services, which often offer prices well below the commercial rates. They are usually easily accessible, able to offer friendly advice and familiar with the latest reprographic technologies. Indeed, a good relationship with the university or one of your major sponsors from the start of the expedition might result in their paying for the printing and even distribution of the report. From the point of view of the university or other organisation, your high-quality report is good for them too.

Digital copying

Most printing bureaux and university printers use high-quality digital copiers for documents such as expedition reports. These copiers work in black and white and are efficient for print runs from just a few copies up to a few hundred copies. Any text, maps, diagrams and line drawings that are in black and white can be reproduced well using this method.

Typically, material to be printed can be accepted either in digital form (as computer files) or as hard copy (on paper). In principle, supplying digital files provides a more direct and efficient process. In practice, there are several potential pitfalls: close liaison with the printer is needed well in advance of printing and preferably before you start work on the report.

Here are some particular issues highlighted by one university printer to whom we spoke:

- Printer drivers: a report designed using a common laser printer driver can look different when printed with a driver for specialist copying equipment, e.g. small changes in line lengths can result in drastic changes in pagination. Be sure to obtain the right driver first from the printing company.
- Fonts: again, small differences in fonts on different computers can greatly change page layouts, so be sure that exactly the same font files are being used.
- Software and file formats: check which program and version you should use for compatibility with the printing company. Alternatively, they might accept more generic formats such as Postscript and pdf.
- Greyscales: a certain level of grey (e.g. 10 per cent) in a photograph or diagram may be easily visible when printed on a particular laser printer, but can be almost invisible when printed on other equipment. Seek advice on levels of grey to use.

As a result of these and other potential pitfalls, some printers prefer work to be given to them as hard copy. Typically, this will take the form of your own print-outs from a laser printer – the best that you can get access to and using good quality paper. You can then be certain that the page layout and diagrams, graphs, tables and photographs appear correctly. This method also allows hand-written, typed or other paper material to be incorporated easily.

Photographs and greyscales

Photographs can be reproduced well in black and white on digital copiers and most photocopiers. Ask first whether you need to carry out any screening – this is a process of converting a photograph from a continuous range of colours into a pattern of tiny black dots, the different sizes or spacing of which give the impression of grey tones. The same process of screening applies to anything else in the report that uses shades of grey (also called greyscales) rather than just black and white; graphs, diagrams, text or lines may all contain grey. Without screening, greyscales usually come out as murky patches of black and white.

If you are submitting hard copy that contains greyscales, the screening may be done by your laser printer. In this case, check with the printing company what screen settings to use in the printer driver software, e.g. screen frequency and angle. If the hard copy contains photographic prints, these will be screened by the printing company.

If you are submitting photographs in digital form, take care to choose a suitable resolution. Around 150–200 dots per inch (dpi) is suitable, where the inch refers to the size of the final output on paper rather than the size of the original picture, e.g. using 150 dpi, a photograph to be printed at a size of 3 × 4 inches should contain approximately 450 × 600 pixels. Many scanners use far higher resolutions, creating files that printing companies don't like because of their size and slowness in printing. Most scanning software allows you to specify the resolution of output files, while resolution can also be decreased to a suitable level using image editing software (a process sometimes called re-sampling).

Colour pages

Colour photocopying provides an easy way of including colour pages in the report to show maps, photographs, etc. The colour pages can be inserted at specified positions into the main document by the printing company before it is all bound together. This method can also be used to create an impressive colour front cover, in which case the colour copying should be done using as heavy a paper or card as possible. A clear acetate sheet can then be bound in at the front to protect the cover.

DIY laser printing or photocopying

If you have a very small number of reports to produce and you have easy access to a laser printer or photocopier, you could print the reports yourself. This depends on how many copies you need, how long the report is, and how much (if anything) you pay to use the laser printer or photocopier. A university department or an expedition sponsor might be willing to help by providing free facilities. Colour pages, including the cover, can be printed on an inkjet or colour laser printer and manually collated into the main document. Many offices and departments have simple comb binding or thermal binding machines, which you could use, or you might have the binding done at a print bureau or university printer.

Binding

A word on binding the report. All the previous hard work and organisation will be of little value if the report falls to pieces. This means having a good binding. There are a number of bindings. The simplest method of doing this is to use staples along one side, suitable for up to about 15 sheets, but a more professional binding will improve the appearance, durability and ease of handling. Plastic comb binding or metal wire binding is relatively cheap and allows the report to lie flat when opened. Although wire binding is the slightly more expensive of the two, it is more durable. Adhesive thermal binding and bar binding are inexpensive alternatives, although they are more liable to break. Sewn binding is used for high-quality, more expensive publications.

Offset litho printing

If you're in the happy position of having a large budget for printing or have found a friendly printer as a sponsor, offset lithographic printing might be an option. This method is good for large print runs, for high-quality black and white or colour printing, and for book-style publications.

Offset litho printing involves printing with plates, made from either hard copy (print-outs, photographs and slides) or digital files. Single-colour printing uses just one plate, whereas high-quality full-colour printing is usually achieved with four different plates, each using a different colour of ink.

Note

When computers were starting to be widely used for report production, a *Cambridge Expeditions Journal* editor commented that "computers do not save time but merely extend one's range of options!" This might be a sceptical view, but do not lose sight of the report's aim: to convey information. With this in mind, there may be times when it is best done simply.

DISTRIBUTION

For the expedition to have the most useful impact, think carefully about who has helped you and who might benefit from having the report. When that is done, the distribution requires plenty of time, money and organisation: decide on storage, packaging, sales, dispatch, and the budget for postage and packing. Be sure to record exact names and addresses of potential recipients in the host country before you leave it. Likewise, keep a careful record of to whom copies have been sent.

As well as your own list of collaborators, supporters, helpers and beneficiaries, copies should also be sent to some or all of the following:

• Legal Deposit Office, The British Library, Boston Spa, Wetherby, West

Yorkshire LS23 7BY (obligatory for UK publications: see www.bl.uk/about/
policies/legaldeposit.html)
- As this Handbook went to press, legal deposit legislation was being extended
to *electronic* publications too; check www.bl.uk for updates
- Relevant libraries, ministries, organisations and individuals in the host country
- Royal Geographical Society (with the Institute of British Geographers)
- University and other relevant specialist libraries, such as the Scott Polar
Research Institute, the Alpine Club and the British Mountaineering Council.

PLANNING THE REPORT

Planning is the key to producing a good report in good time. It is surprising how
much can be done before the expedition departs. All expeditions will of course be
different, but this hypothetical time-table shows the main stages (Table 33.1).

TABLE 33.1 SUGGESTED TIME-TABLE AND ACTION PLAN FOR PRODUCING THE EXPEDITION REPORT

January and February
- Start liaising with your partner organisations at home and in the host country
about the project, its results and useful products.

March
- The expedition's aim, team, destination and collaborators have all been arranged
by now.
- List the main products you expect from the expedition.
- Decide what sort of report will be most suitable and what it should achieve.

April
- Appoint the report editor, a key member of the team. This is perhaps *the* critical
step in planning, as everyone knows from now on who has responsibility for
everything to do with the report.
- Editor produces a time-table for the report.
- Draft table of contents; distribute to all partners for comments and agreement.
- Decide who will produce each section of the report.

May
- Draft list of recipients: who and how many.
- Give thought to the design of the report: how many pages, page size,
photographs, colour or black and white, cover design.

- Estimate the cost of printing and distribution: obtain estimates from printers, and seek sponsorship for printing.

June
- In planning your fieldwork, allocate a week at the end, before you return home, for writing the preliminary report.
- Assemble equipment for recording and report writing to take with you.

July
- On arrival, plan a place and time to work on the preliminary report and check on local copying facilities (photocopying is inexpensive in many countries, often 1 or 2p per page).

July–September
- Fieldwork: keep up to date with recording everything that might be used in the report and transfer data on to computer, e.g. daily log, species lists, GPS readings, descriptions of methods.

September
- Before leaving, write your preliminary report.
- Distribute it to all your contacts in the host country.
- Email or post the report to sponsors, helpers and other contacts at home and elsewhere – they will be delighted to receive a short, efficient report from the field.

October, after returning
- Develop and distribute your preliminary report.
- Update your website with news of the expedition's progress and the preliminary report available for download.

November and December
- Team members, at home and in the host country, write their sections.
- Prepare design and stylesheets.
- Prepare maps and photographs.
- Make final arrangement with printers.

January
- Deadline for each section.
- Editing and layout.

February
- Last sections, e.g. conclusions, abstract, bibliography, acknowledgements, finances.
- Final edits – and spell-check! Careful proofreading by editor and authors; if appropriate ask an adviser to comment.

March
- Send everything to be printed.
- Prepare packaging, labels.
- Prepare covering letters, presenting the report and possibly inviting comment, suggestions and participation for further development of the project.

April
- Stagger to the post office.
- Make arrangements for storing extra copies, sending them out if requested, and responding to feedback from the report.

May
- Almost ready for the next expedition!

EXPEDITION REPORTS AND DATABASE

The Royal Geographical Society (with the IBG) houses a unique collection of expedition reports. Over 3500 reports contain details of the achievements and research results of expeditions to almost every country of the world. The catalogue of these expedition reports, and over 7500 planned and past expeditions, are held on a database maintained by the RGS–IBG Expedition Advisory Centre. The database provides contact with a wide variety of sporting, scientific and youth expeditions from 1965 to the present day, enabling expedition leaders to share their experiences and expertise.

Expeditions and field researchers are encouraged to consult the expedition reports as early as possible in their planning because they contain much useful advice on every aspect of planning an expedition and field research project.

Please consult the Expeditions Database at www.rgs.org/expeditionreports to identify the reports that you wish see, and email eac@rgs.org for an appointment.

CONCLUSION

The expedition report collection at the Royal Geographical Society (with the IBG) is testament to the achievements of over 3500 small expeditions. Your report will be a lasting record of your work and invaluable to future expedition planners. It is well worth taking the time to get it right, because the report will reflect what you have achieved. Take care with presentation and layout. As with all aspects of the expedition, it is important to plan the report right from the initial stages. Remember that, above all, the aim of the expedition report is to share the results of your expedition clearly and usefully to as wide an audience as is possible.

FURTHER INFORMATION

Other collections of past expedition reports are held by:

The Alpine Club's Himalyan index. Website: http://himalaya.alpine-club.org.uk/
A comprehensive record of expeditions to Himalayan and Karakoram peaks over 6000 m.

British Mountaineering Council Expedition. Website: http://www.thebmc.co.uk/world.htm

BP Conservation programme. Website: www.conservation.bp.com

34 EXPEDITION PHOTOGRAPHY

Tom Ang

Photography is an indispensable part of the modern expedition. It can be a major activity in its own right – great expedition photographs can express the spirit of the adventure, define the nature of the challenge. And it can support other activities – from documentation to publicity and fund-raising. Photography should therefore be integral to every expedition's planning and execution.

The expedition photographer has three main responsibilities:

1. To make a comprehensive visual record of the entire expedition from preparation through to its return home to meet the needs for documentation, communication and sponsorship.
2. To produce visually stunning images suitable for publication, exhibition, etc. that capture the identity of the expedition.
3. To work in a manner that responsibly respects the societies, peoples, cultures and environments that the expedition encounters, to ensure that other, succeeding, photographers will not be hindered.

It is important that all members of an expedition recognise that the photographer's work is a vital and valid part of the aim and objectives, and, indeed, that photography constitutes a major part of its achievements. To meet the high standards expected today, photography has to be a full-responsibility task, i.e. not the part-time interest of the medical officer.

FILM-BASED OR DIGITAL PHOTOGRAPHY

One of the first decisions the expedition photographer has to make is whether to use digital or conventional film cameras. Modern film and camera technology now provide very high-quality images at reasonable cost whereas digital cameras can offer considerable flexibility and savings in the long run.

Photography on 35mm format film is capable of producing image quality suffi-cient for all normal use at a reasonable cost; it is familiar to most expedition members and does not require new skills; conventional cameras can be relatively robust and independent of battery sources. However, pictures taken on film-based cameras cannot be reviewed until they are processed, images cannot be used directly for transmission, film is easily damaged in transit and running costs can be high, especially compared with the image quality actually required.

Digital photography can provide substantial cost savings for documentation because no film processing is required and images are easily reviewed in the field. Images are easily annotated immediately with accession code, description and loca-tion of a find; they are easy to transmit via satellite phone to a website; images of specimens, malady, accident, etc. can be emailed to home-base consultants for iden-tification or advice.

However, the equipment is less durable and all are reliant on battery power; digital cameras are best exploited with a computer (although not a necessity); initial costs, including training, may be high; and image quality may be limited for publication use (see "Digital cameras" below).

In summary, where conditions are extremely arduous or images are needed for high-quality reproduction and where images are not needed for transmission, film-based photography is best. Where numerous images are needed for record-keeping, where conditions are not too challenging, e.g. static base camp, and there is a need to maintain websites, digital photography is most cost-effective, if not essential. Photographers on large-scale expeditions should consider using both film and digital photography.

PREPARATION

The main elements in preparation are: (1) to plan the photographic and documen-tary needs; (2) to plan for the post-expedition needs, e.g. publication, sponsorship; (3) to prepare and train on the equipment; and (4) to pack.

Photographic and documentary needs

The key task is to build the shot-list: a comprehensive listing with notes and priority gradings for all the pictures that will be needed. This requires discussions with the leader, the expedition specialists and scientific advisers, and publication editors as well as sponsors. For example, you need a picture of every person involved in the expedition; does this also mean sponsors and other backers, e.g. Mum and Dad? Will you need a photograph of every villager interviewed and, if so, will it be possible? You need pictures of the landscape: does this include geomorphological and exposed geological features as well? If so, is it necessary to include a scale, e.g. a metre rule? What kind of pictures will the sponsors need – and how many? If there is an accident, what needs to be documented?

This list should be circulated to all expedition members.

Post-expedition needs
This is the action list for the field that enables the shot-list requirements to be met, e.g. if a sponsor needs a picture of the team on the summit, you must ensure that a team photo – which clearly shows the sponsor's goods – is one of the mandatory activities in the push to the top. If you need a series of pictures showing base camp being established, the expedition leader should ensure that you are sent ahead of the main group and relieve you of other duties. Portraits of all interviewees needed? Who will take names and cross-check?

Planning for publication also means that you discuss picture needs with magazine features editors and picture editors. Ensure that you know their basic requirements, their likes and dislikes, e.g. some publications dislike ultra-wide-angle views, as well as pictures that attempt to be humorous; some dislike product placement – a trademark visible – even if it is unintended. Consider their schedules: when you return in the autumn, magazines are preparing winter issues – so will they want a summer travel feature?

Equipment
The best equipment to take on expedition is equipment that you trust and with which you are thoroughly familiar. If you are not an experienced photographer you will need to train and practise. This means learning to use the camera blindfold, so that you know which way to turn the dial for a shorter shutter-time with your eyes shut, where the button for holding auto-exposure is, which way to turn for longer focal lengths, and so on. Practise using the camera without film, taking vertical as well as horizontal format shots, so that you can operate it rapidly, without thinking and can release the shutter without shaking the camera.

Two months before departure, thoroughly test the equipment (simulating expedition conditions if possible particularly for polar work: test in a walk-in freezer) with film, and all the different combinations, for example of lens and camera, with and without flash, using different exposure modes and settings, etc., to ensure that it is all working fine. If you discover any problems, you have 6 weeks to repair and re-test the equipment.

Confirm that your equipment is covered by the expedition insurance or, if your own, inform your insurer of the expedition and the nature of the risks to obtain written assurance of continued cover.

Packing
Rehearse packing your gear, and then take it out for a trial run. It is so easy to forget the most obvious things, e.g. space for survival gear such as water, space-blanket, food. And will you be able to trek for 15 km at 3000 m with that load? What will you

take as cabin luggage and what will go in the hold? Will your carrier accept your big camera bag as carry-on? Can you persuade team members to take some gear or film?

OBTAINING SPONSORSHIP

Film distributors may offer special discounts for expeditions with scientific or humanitarian aims: enquire at their head office in the country where you are working. You may be able to obtain discounts on purchases of equipment and film processing, but forget about asking for the loan of cameras or lenses unless you have a solid or high-profile reputation. A website about the expedition – showing a portfolio of your pictures – can help to establish credibility.

SELECTING PHOTOGRAPHIC EQUIPMENT

Modern equipment can produce excellent results for all but the highest-quality half- to full-page magazine reproduction. Inexpensive modern cameras and lenses can produce image quality better than the best of older equipment. Therefore poor image quality usually results from poor technique. In general, the more you pay the better the quality, responsiveness, reliability and versatility that you obtain. The best-known makes, i.e. Canon, Nikon, Minolta, Pentax, Contax, Leica, all produce cameras that will meet the needs of most expeditions. It is not necessary to use mechanical cameras for extremely arduous conditions because modern auto-focus SLR cameras can be highly reliable.

The 35mm SLR camera

For highest image quality, precise and rapid operation, e.g. with exact exposure control and motor-drive plus auto-focusing, as well as versatility, i.e. its ability to cover every subject matter, the 35mm SLR is a beauty. But it is also heavy and bulky, and requires costly lenses to exploit its potential fully. If you do not fit more than one lens or do not need an SLR for specific uses, e.g. close-up or long-distance photography, a high-quality auto-focus compact camera is preferable.

Almost all modern SLR cameras are electro-mechanical, needing a battery to operate. This is not necessarily a problem for the modern expedition, which also runs radios, global positioning systems (GPS) and laptops. Wholly mechanical cameras, e.g. Nikon FM3, Olympus OM-4 and Leica R6, are high quality but costly and even second-hand examples of cameras such as Nikon FM2, Nikon F2, Canon F-1 (all of which are recommended) are not cheap; they offer good reliability, but they are slow to operate. On balance, the facilities given by modern auto-focus SLRs, e.g. rapid and precise focusing, motor-driven film-winding and excellent exposure control for available and flash light, far outweigh their battery dependence.

The 35mm range-finder cameras
These cameras offer convenience and compactness, but only the most costly, e.g. Leica, Contax, Nikon and Konica, can be relied on for publication-quality images. Auto-focus compact cameras are best when the expedition's photographic needs are modest, e.g. for pictures of people and general scenics intended only for reports or website.

APS compact cameras
Auto-focus compact cameras using APS format film are very compact and inexpensive. They are useful for informal shots, e.g. expedition parties, the grinning border-guard.

SELECTING FILM
Colour transparency (slide) film continues to be preferred for publication and is easily duplicated for slide presentations. However, it is also the hardest for the inexperienced to control and it is unreliable under expedition conditions (see "Precautions during the expedition"). All modern makes, such as Fuji, Kodak and Agfa, give excellent results: which you use is a matter of personal or a picture editor's taste. Films with high contrast and rich colour, e.g. Fujichrome Velvia, are more difficult to expose correctly than those lower in contrast, e.g. Kodak EliteChrome. Prefer "amateur" to "professional" film unless you have refrigeration or work in cold environments because amateur types have a longer shelf-life. For general use, choose ISO 100/21° speed film and for specific, low-light conditions use an ISO 400/27° speed film.

Black-and-white film is relatively robust under expedition conditions but requires more handling at the printing stage. Some publications, e.g. newspapers, will use black and white but many will much prefer colour transparencies.

Colour-negative film, i.e. for making colour prints, is disliked by publications. But it has advantages where publication is a low priority: it is relatively robust, is tolerant of inaccurate exposure control and high contrast, can be processed by street-corner laboratories throughout the world and is inexpensive. Colour-negative film is a good choice for informal shots and personal records of the expedition.

PACKING LIST
It is easy to pack too much and extremely difficult to pack too little. One approach is to select a camera bag that you can carry easily, and then see what you can get into it.

Camera bag or case
For base camp or porterage in arduous conditions use a water-/dust-proof case.

Choose plastics and resin cases, e.g. Pelican, which are almost indestructible (and available in orange), over aluminium cases. Use two or more smaller cases rather than one large one (to spread risk and load). When travelling on foot use a backpack-style camera bag, e.g. from Lowepro, Tamrac or Tenba. These distribute weight well, are comfortable to carry long distances but are not so convenient as shoulder bags for access to equipment. A sternum strap that stops the camera from swinging side to side is useful for trekking.

Camera outfit

Take two camera bodies to use in the field and, ideally, leave a third at base camp. The highest-power motor drives are seldom needed. Use at least two lenses: a wide-angle zoom, e.g. 28–70mm able to focus close-up, and a long zoom, e.g. 80–200mm. For a third lens, add a 50mm lens either relatively high speed, e.g. f/2, or a macro-lens. Avoid zoom lenses with very large range, e.g. 35–350mm unless they are top class. Avoid lenses shorter than 24 mm unless you know what you're doing.

If you take only one lens, use a high-quality wide-ranging zoom, e.g. Nikon 24–120mm, Canon 28–135mm (this has a image stabiliser to reduce camera shake). Picture quality depends on the lens that you use, so spend until it hurts. Supply all lenses with ultraviolet filters (minus ultraviolet, they look colourless), lens hoods and lens caps.

For close-up photography, the easiest option is a close-up lens that screws on to your main lens: it is cheap and improves image quality. The best but most costly option is a macro-lens with extension ring(s).

Flash is useful – not so much for dark conditions but for the very contrasty situations typical of many expedition environments. Take a flash unit that offers a tilting, rotating head with a fully automatic exposure system that is compatible with the camera.

Take NiMH (nickel metal hydride) rechargeable batteries for flash and camera motor drives. These are efficient and easily recharged. Take battery chargers appropriate to your expedition: from the mains, through vehicle lighter socket or solar power.

Take a tripod that can be set down low – camera about 15 cm from the ground – and up to at least 1.6 m. Carbon-fibre models are relatively lightweight but costly. A ball-and-socket head is lightweight and compact but less easy to set than a three-way pan-and-tilt head. Lightweight tripods can be made more rigid and steady by hanging a heavy weight, e.g. camera bag, from the centre-plate – where the legs meet – so that the bag just touches the ground.

Also take the following: a rubber puffer to blow dust away plus wet-wipe lens-cleaning tissues; jeweller's screwdrivers to fasten loose screws; permanent marker pens to label cassettes and CDs; film extractor to remove film tongues from the cassette; white 15cm rule as a scale; neutral grey sunglasses (to avoid distorting your colour vision); and spare lens caps.

Film

Budget on at least four rolls per day in the field where photography is a high priority and one roll per day otherwise. Alternatively, count up the shots in the shots-list and multiply by 10 to estimate the total number of shots that you can expect to shoot; divide the result by 35 for the number of rolls. Plan to return with some unused film – or you'll be in for some very nervous final days. A professional who manages to hit one top-class shot per roll is, well, on a roll – and is an extremely happy bunny. Therefore shoot far more than you think you'll need: it is better to regret "wasting" film than to regret not taking a shot. For longer expeditions with re-supply drops, bid for film to be included in the drops and, if possible, have exposed film taken to the home base for processing.

DIGITAL PHOTOGRAPHY

Digital cameras store images on small memory cards and provide instant review of pictures made. Those offering image resolutions of 1–2 megapixels provide image quality sufficient for websites and small-size reproductions, e.g. in reports. Sensor sizes of 4 megapixels or greater are sufficient for good quality reproduction, e.g. magazines to about A5 size. As with film-based cameras, quality follows price. Prefer cameras from well-known makes such as Canon, Nikon, Olympus, Sony, Kodak, Minolta, Fuji, Pentax, Casio or Epson. Prefer cameras using CompactFlash cards to others, i.e. not SmartMedia or MemoryStick. A computer is not necessary: special self-powered portable hard disks, e.g. NixVue or DigitalWallet, can download and store thousands of images off camera memory cards. Modern laptops, e.g. Macintosh iBook, offer convenient reviewing and cataloguing as well as a built-in CD burner, providing ideal back-ups for images.

Ensure that there are at least two ways to transfer images from camera to computer: a reader for the memory cards plus direct download from the camera. Take at least one spare cable of every type required as well as spare memory cards and batteries. Take a CD containing all software in case any need re-installation. Small, portable, battery-powered printers are available, which enable prints to be made in the field but running costs, e.g. consumables, can be high.

BASIC CAMERA TECHNIQUE

Exposure

With SLR cameras, use spot- or selective-area metering on auto-exposure to pick out the key tone, e.g. face, sky of a scene; then hold that exposure and re-compose the shot. Modern "intelligent" or evaluative exposure systems are effective and achieve better success rates than an inexperienced photographer. For key pictures, take several exposures and bracket them, i.e. make one shot that is deliberately underexposed, one that is overexposed plus the camera's "correct" exposure. Expose carefully and bracket generously when shooting transparency film in tricky lighting conditions, e.g. high contrast, against the light or at night. You cannot get correct exposure without making some mistakes.

Focusing

Focus carefully and deliberately or allow the auto-focus time to do its work. Take special care when working close up, i.e. less than 0.5 m distant or with a long focal length lens. If using an SLR check the viewfinder image carefully. Squeeze the shutter button to avoid shaking the camera. At the telephoto end of a zoom, i.e. 200 mm and working close up, use a support whenever possible: best but most inconvenient is a tripod; a nearby tree or wall is convenient but seldom in the right place.

Portraiture

The best way to take a photograph is to give something of yourself. To take a portrait of "someone" is intrusive; to take a portrait of a friend is fun. Therefore, talk to people, spend time with them, respect them. Fifteen seconds of sign language and smiles can open an entire family to you, whereas one distant portrait snatched with a telephoto confirms you as a stranger at best and gets you pelted with rocks at worst. Portraits are most effective at a lens focal length of between 50 mm and 135 mm, with the face or part of the face filling the frame, i.e. from well within their personal space, and the focus on the nearest eye. Diffused lighting from one side, e.g. from an open door, with a plain dark background, e.g. plain walls, often gives effective portraits. Measure exposure from the lighter side of the face.

Landscape and topography

Every landscape has its most pleasing lighting – usually, but not always, with low, raking light soon after dawn or before sundown. But note that, in tropical latitudes, sunrise and sunset last only a few minutes. In mountainous regions, light can reach certain valleys only relatively late in the morning. Use interest in the foreground, e.g. flowers, shapely rock, to lead to the background. Take portrait (vertical) format pictures as well as landscape (horizontal) format. Don't be afraid of including the sun in the picture. Landscapes can be taken with telephoto lenses as well as wide-angle ones: a long view detailing a distant mountain village is more effective than a vista of the valley showing nothing in particular.

Use flash in bright light

Under the equatorial sun, shadows are very heavy and highlights burn out: modern flash units are effective at putting some light into shadows, greatly improving pictures. Don't be afraid to use flash in the midday sun: everyone will think you're mad, but you know better.

Basic composition

For non-documentary work, let the picture situation guide you; listen to the inner voice: go close, let it blur, hold the camera askew, shoot the unobvious – whatever. Do not follow any rules or you'll miss the spirit of the place. Get close or you'll miss the action – it is always better to be too close than too far. Above all, don't think: just be guided by what you can see in the viewfinder.

PRECAUTIONS DURING THE EXPEDITION

Keep all your equipment covered when not in use. Keep lenses covered front and rear when not in use. Keep an ultraviolet filter on all lenses at all times. Clean all your equipment regularly; especially blow dust off the front and rear of lenses and from the eyepiece of cameras. Above all, look after yourself: one sick camera is a nuisance but one sick photographer disables *all* the equipment. And avoid taking unnecessary risks, e.g. climbing a dodgy tree limb for a better view: it is not worth it.

In transit

Keep all film and digital cameras and key items in the hand baggage. Never leave unexposed or unprocessed film in hold luggage. Check formalities with the embassy's trade attaché if you plan to take in more equipment than a normal tourist would: obtain official letters of support. It may be necessary to prove, through carnet procedures, that the equipment is a "temporary import", i.e. you will take home everything you bring in. This can be costly and extremely troublesome if equipment is lost in the field.

Cold conditions

Keep camera and battery packs, e.g. for motor drive or flash, warm, i.e. close to your body, until you need to use the equipment. In extreme cold, take care not to place your face to the camera or touch metal parts, e.g. the tripod, with bare hands. Make those difficult-to-reach settings before going out. When returning from the cold into the shelter, enclose the equipment in a bag or case and allow to reach near room temperature before exposing to warm conditions; in the humid conditions of a hut, a difference of only a few degrees is sufficient to condense water on your camera.

Hot, dry, sandy conditions

Use ample supplies of plastic bags to isolate against sand and dust. Many kinds of zip- or pinch-fastened plastic bags of varying durability and cost are available. A bin-bag closed with rubber bands is cheap and effective; one inside another is even better. As a bag becomes dusty, do not hesitate to dispose of it (make someone in the team happy) to use a new one. Where extremely hot and in a static expedition, consider keeping film in a deep hole in the ground: temperature will be lower than above ground. Keep film and equipment away from high temperatures such as inside cars or tents in full sun. Take particularly good care of transparency film. Be aware that a windowsill in the shade in the morning may receive full sun later in the day.

Hot, humid conditions
Store film and unused items in air-tight food containers with sachets of silica gel to absorb moisture and discourage moulds from partying on your film or equipment. Refrigerate film if possible, but remember to leave in containers to warm up before loading film to prevent condensation on the film. Process exposed film as soon as possible. Keep film and equipment away from high temperatures, e.g. inside unsheltered tents or from direct sun. Take particularly good care of transparency film. Remember that a car dashboard in the shade in the morning may receive full sun later in the day.

Wet conditions
For rafting, kayaking, monsoon rains as well as snorkel diving, cameras may be protected from water in specially designed Ziplok plastic cases. These provide protection, ranging from water splashes to scuba-diving depths, according to design and costs. Alternatively, use splash-proof or underwater cameras: the range runs from inexpensive point-and-shoot models to those suitable for scuba-diving depths.

LEGAL CONSIDERATIONS

Before the expedition
It should be explicitly agreed from the outset that the copyright in photographs should remain with the photographer. The role of the expedition in working with the photographer and making the venture possible is best recognised with a licensing agreement between photographer and expedition, whereby the expedition is entitled to use freely the photographs for reports and the website. In addition the photographer agrees to pay the expedition a royalty – usually 50 per cent – of net earnings (earnings less direct costs of sales) from the photography, e.g. publication fees from magazines, sales of prints. All expedition members should also sign a model release to the photographer allowing use of the pictures in which they feature.

During the expedition
Avoid photography close to military or government areas – anywhere a national flag is flying should be avoided. Do not photograph military convoys, naval ships or military aircraft. Stop immediately if ordered by any person in uniform: no picture is worth your life or detention. Stop immediately if your subject appears distressed or embarrassed. Approach photography of nudity and of minors in the nude with

caution; know the local laws. And ensure that, if you sustain stereotypes, e.g. "noble savage", "innocent native", you do so consciously. This is important when you publish expedition results in the host country and because websites are globally accessible.

ON YOUR RETURN

Process your film in relatively small batches to avoid disastrous loss and allow review of the results for adjustments to be made to the processing of the remaining rolls. Caption and mark up every roll immediately. Mount the best and keep them very, very safe: cool, dry, dark and secure. Make duplicates of your best shots; scan them if you can, but never, ever, project your best originals.

Publications

Immediately re-establish contact with publications that you contacted before departure: tell them you're back: "Great trip, great pictures – will show you some soon; is that OK?" Most publications will accept scanned images for review; some will insist on the originals for publication. Before you send irreplaceable material, obtain confirmation that the recipient will accept liability for loss or damage to originals. Then send a delivery note enclosed with the pictures which states clearly the fee for loss or damage and that the recipient accepts this liability on accepting the consignment. If you do not obtain that confirmation, you must weigh very carefully whether you will risk irreplaceable material to a stranger. In general, use your material with imagination and determination, and you can easily be published, earning useful money.

Caption everything clearly, present all work neatly and, whenever possible, support the pictures with a well-written article.

Picture libraries

The market for travel photography is saturated, over-supplied and highly competitive. All but the most skilled, well-equipped and utterly determined should look elsewhere. If there is a strong news or magazine feature angle, e.g. discovery of an unknown tribe, picture of a live Dodo – or that ilk – there may be syndication possibilities with news picture agencies.

FINAL WORDS OF ADVICE

Stay fit, healthy and awake: photography demands all your faculties working at top notch. Go the extra distance: all good vantage points are where the expedition group is not. If in doubt, press the button: you cannot go back. Always have your camera within reach: you will be given only one chance. Enjoy and accept what life offers your camera: life is what is worth sharing through your photography.

35 WRITING FOR MAGAZINES

Miranda Haines

When you return from your expedition, your brain will be awash with wonderful stories of discovery, adventure, frustration, science and culture. But how might you convey these to a wider audience? The secret to having your findings published in a magazine is to follow the basic guidelines written below and then add a little of your own individual imagination, initiative and expertise. Remember that there are exceptions to every rule.

The difficulty is working out which magazine will be interested in what and when. *Geographical*, the official magazine of the Royal Geographical Society (with the Institute of British Geographers or IBG), is probably one of the most likely magazines to be interested in printing an article on your expedition. Stories of discovery, geography, history, science, people, culture and environment are all top priority subjects for *Geographical* to feature. And what better place to have your findings presented over beautifully designed pages with stunning photography? However, if you look on the magazine shelves in a big newsagent you will find an amazing array of publications with a global or outdoors focus. Equally, looking in the *Writers' and Artists' Yearbook*'s magazine listings will turn up unknown titles that may be ideal for the story you have in mind.

Magazines are the holy grail of press coverage for your expedition because you may have noticed that very few national newspapers publish stories from unknown expeditions. Certainly, it will not be an easy "sell" to most publications, because space for these types of features is extremely limited, with most of the emphasis on the travel pages being consumer-oriented pieces. If the publication cannot sell advertising on the back of your expedition piece and you are not a famous explorer, the story has to be extremely strong, as well as being well written and original.

There is no doubt that, wherever your feature finds a home, the result should be a fantastic boost for the expedition, your own portfolio, the sponsors and the next venture that you want to do. Do not expect to earn a fortune for your masterpiece, however.

SELLING YOUR PIECE

Mind the gap: dealing with the media

Strangely, and most editors moan about this, it has been very difficult to find reliable professional expeditions to write a good, relevant, interesting piece that stands out from the pile. Undoubtedly, both sides are missing something here. It is always worth remembering that we both want the same thing: a good and visually exiting result in print that readers will enjoy and come back for more. So how do we get to this point?

On the face of it, the gulf of misunderstanding between expeditions and publishing folk could not be wider. We sit in our safe offices with smug looks on our faces, avoiding your telephone calls, drinking lattes, and worrying about small things like spelling and printers. You travel all over the world, with the latest kit, boasting about how great your expedition was, what amazing things you saw and how stupid we would be not to print your story and photographs.

Quality control

First of all you must answer this question truthfully: can I write well? If you find this hard to judge yourself, ask your colleagues and friends. If they enjoy a 300-word piece that you have written from the field or even about the planning stage, the chances are that you can write. If the answer is no, you should consider either finding a good writer to come on the expedition or simply concentrate on getting your story across to another journalist or in-house staff writer who can write it up for you. In other words, you can be your best public relations officer.

Before you contact a publication, have a think about what you are trying to achieve. Remember, by making it as easy as possible for journalists, who are lazy (I mean busy), you are aiding the chances of your story being snapped up.

The press release

Basic principles of writing press releases are often not applied and as a result end up straight in the bin. You have a few seconds to get an editor's rapt attention – or at least intrigue them to want to know more. Maximise this opportunity by including the following:

- date it
- heading: always summarise the news, "hot item" or angle in a bold single statement at the top
- story first: explain in a few sentences the central theme of the feature
- hard-and-fast facts are of utmost importance
- vital pieces of information are often hidden among the hyperbole – lose it
- bullet points can help list facts/achievements/figures
- quote yourself, locals, colleagues and patrons

- include humorous/quirky element – something that you would retell in the pub
- mention whether images are available
- include further sources of information: contact details, maps, diagrams, website, etc.
- remember to include your contact details at the bottom.

A sample press release is given in Appendix 7.

Selling features

So you can write but despise the PR role. Let's go about pitching your story idea to the editor. First of all, remember that there are often many commissioning editors on each publication, e.g. certain sections are commissioned by different staff members – with the main features usually in the editor or feature editor's realm. Research each title, paying close attention to the masthead and telephone the editorial assistant to corroborate findings if you have any queries.

Once you have identified the title and the editor in question, ask yourself three questions:

1. Why would the editor be interested in my piece?
2. Why would he or she publish it now?
3. Does my piece suit the style and content of the publication to which I am selling?

These are all key questions that the editor has to answer – he or she has a publisher and readership to answer to and if they cannot be answered easily you do not have a chance of publication. Space is always so tight that the story has to fight its way to the top.

Presenting the story

I would always recommend writing a proposal *before* writing the whole piece. This will help with pre-selling the feature before the expedition and save an awful lot of time should the publication reject the feature outright. Also, each publication will require a different style and a different angle depending on its readership. You do not want to have to rewrite the piece each time you make a submission to an editor.

Do be prepared for rejection letters. Until you become quite experienced at selling your pieces, this will happen regularly. Simply identify another magazine, another approach and try again with another letter. In time you will find that you will succeed almost every time, but this will take a good knowledge of the market and relationship building with magazine staff over time. You do not have to know anyone in publishing to get published, but of course, once you do get to know people, this will always ease the passage of your proposal in the future.

If you are an unknown writer, you may be asked to file your story "on spec", which means that the editor is not bound to pay you if he or she does not publish the piece. Always follow up your letter a week or so later with emails or telephone calls. An editor is extremely busy and usually quite grateful for a reminder call to prompt a response.

If you have sold your story proposal – and it is always worth looking at the types of stories that are making it into print – you will next need to agree a word count, deadline and fee. Most magazines have set fees per 1000 words that are paid on publication, but it is always worth asking if that is their final offer, especially if it does not match your expectation.

Don't give up!

Of course if you have failed to sell a feature before leaving, you may find that you have an even better story when you return, and at this point it may be worth writing up the feature in full. This should be useful for your own reporting and lecturing that you may do on your return, and the piece may just capture the editor's imagination in a way that your proposal was unable to.

Photographs

For magazines in particular, photographs are just as important as the words. So often, a popular feature idea falls at the second hurdle because of poor quality pictures that the art director has to reject or pay vast sums of money to track down relevant images to supplement your piece. There seems to be a mismatch of expectation here. Art directors require top-quality pictures and journalists think fuzzy prints of a sunset are fabulous (see Chapter 34).

It is really worth having a good expedition photographer because good pictures in a jungle or desert are extremely difficult to achieve and, if you can offer them with your feature idea, this increases everybody's prospects of a happy result.

Websites

A good website regularly updated proves that your expedition is serious. This will help in gaining sponsorship (logos can be proudly displayed here), picking up interest from around the world, and letting the media know everything about your plans.

In addition to housing maps, facts and information, and a gallery of (high-resolution – 300 dpi – and low-res quick-loading) images, the site should be updated remotely by you or your team during the expedition. This gives people a reason to return to the site time and again to check how you are progressing.

Design and easy navigation of such a website are paramount, but it is not something that professionals alone can create. Always think of the casual user, who will not bother waiting around for intricate graphics or large files to load. A choice of text-only might be a good idea for people with slower modems.

CONCLUSION

If you do follow these common-sense guidelines and catch a publication's imagination and enthusiasm for your expedition, all there is left to do is file your copy on time, in a professional manner.

- Most editors like to receive hard copy and an electronic version if possible.
- Keep the writing style simple, succinct and factually correct.
- At the same time, readers need to laugh, cry, and be amazed and fascinated.

If you can do all this efficiently, you will have a good commercial publishing outlet for the future because the editor will trust you and your clippings portfolio will help demonstrate a proven accomplishment. Mission complete.

FURTHER READING

Writers' and Artists' Yearbook (annual), A & C Black, PO Box 19, St Neots, Cambridge PE19 8SF. Website: www.acblack.com
Gives listings of media contacts.

Geographical, Unit 11, 124–128 Barlby Road, London W10 6BL. Tel: + 44 20 8960 6400, fax + 44 20 8960 6004, email: magazine@geographical.co.uk, website: www.geographical.co.uk
The *Geographical* magazine is owned by the RGS–IBG and published under licence by Think Publishing. This lively, colourful, monthly magazine presents geography in its broadest sense, with exciting and beautifully illustrated articles on people, places, adventure, travel, history, technology, science and environmental issues.

36 SOME TIPS ON LECTURING

Richard Snailham

One of the peripheral but important aspects of going on expeditions is talking about them afterwards. It can be both pleasurable and profitable. If the expedition has sprung from some institution (school, college, university, exploration society or whatever), the leader and his or her team may feel obliged to offer a presentation to the other staff and members of their institution. This will normally be a team effort rather than a solo performance by the leader.

The points that follow are, however, addressed mainly to the expedition leader or member who has it in mind to give lectures to audiences single-handedly: not just "duty" talks, but talks that might raise a bit of money in the often-critical post-expedition phase. Although it naturally forms part of this phase, bear in mind that you can, in a limited way, use lecturing as a money-raiser before the expedition. This is more easily done if you have slides or film from some reconnaissance expedition or previous visit to the expedition area.

GETTING THE LECTURE TOGETHER

Aspiring lecturers must make sure before the expedition that they are going to have access to a large enough range of good images with which to illustrate the lecture. If they are confident of their own photographic work there will be no problem. Otherwise, they must arrange to have copies of the best slides that the other members have taken. Most leaders get their members' written agreement on this before the expedition.

An hour's talk should be accompanied by between 40 and 100 slides. Don't over-ice the cake by showing more than 100 unless (1) you are a very good speaker who can hold an audience in comfort for over an hour, or (2) you don't intend to speak on all the slides but merely present them rapidly as a pictorial "essay". They must, of course, all be of excellent quality. Discard all poorly exposed, badly focused slides unless they are of exceptional interest. As a general rule never be tempted to show more than one slide of the same thing. Endless views of the same glacier from slightly different angles can pall. In general, pictures of people are more interesting than landscapes and some shots of local people should be shown even if they were not directly involved with the expedition. You should try to find a slide to illustrate every aspect of the expedition, from the journey out to the culmination. When you have chosen a set, make high-quality duplicates and get them protectively mounted as soon as possible. Never project your originals if you can possibly help it.

Have plenty of good anecdotes: the funny story that people chuckled over during the expedition, or on the way back, is the sort of thing to remember. Amusing little things about the way you lived, the way people behaved or what they said go down best. After the first few talks you will soon find what is well received, and you will jettison some stories and introduce and refine others as time goes on.

Don't write out your lecture as a script. At all costs it must not be read out. Speak from the heart. If you need an aide-memoire have a list of all the slides with a brief caption and, if necessary, a codeword that will remind you of some anecdote or observation that you wish to make while the slide is on the screen. At first you may find that the slides act as your notes and jog your memory as they appear on the screen. After giving the lecture a few times you will begin to know what is coming up and be able to lead up to the next slide with a few anticipatory remarks. In this way the talk will begin to flow. Glance only fleetingly at the screen to check what is there. Address your remarks to the audience, never to the screen.

FINDING THE AUDIENCE

This is the most difficult bit. There seem to be more speakers with a story to tell than there are audiences to listen to them. And yet the lecture, that joy of our Victorian forebears, is not dead – despite television – and in the winter months it flourishes.

For the young prospective speaker, with no ambitions at this stage to get on the national networks, I suggest the following audiences: Townswomen's Guilds (TWGs); Women's Institutes (WIs); Rotary Clubs; Round Table; National House-wives' Register (NHR); young wives' clubs; wine circles; university exploration societies; parent–teachers associations (PTAs); preparatory schools; and old people's day centres. There are many other potential audiences, but these are tried and true, and all of them proliferate at a local level.

There are no firm rules about getting in touch with these bodies in your own locality. Personal contacts are the best. The phone book is useful. TWGs and WIs are organised in County Federations. The WI County Federation HQ (find it in the phone book) will give you the name of the woman who produces, in booklet form, an annual panel of speakers for her county's institutes. If you can get on to this panel (there is sometimes an audition), your name, the subject of your talk and your fee will all be circulated to the institutes in your county and you should be off to a good start. TWGs are also grouped in County Federations; their Edgbaston HQ will give you the name of your County Federation Secretary who may be prepared to send a circular to all the guild secretaries in your area. Most of the women's institutions listed above meet monthly and some of them all year round, except for August. They often have outside speakers. Travel and exploration are popular subjects.

Schools make excellent audiences. All schools can be approached. I only singled out prep schools because they are the most receptive and pay better. They, like public schools, favour Saturday evening, whereas state schools expect you to talk during the morning and for little or no fee. Wine circles are the most hospitable: you may have to arrange to be collected at the end. Old people's day centres are run by local government and offer a set rate for talks. The Expedition Advisory Centre maintains a register of lecturers on expedition topics, which is circulated to enquirers, especially school and university exploration societies. Ask them for a form to be added to the list.

THE TECHNICAL SIDE

University exploration societies will put you into a smart lecture theatre with all the most modern facilities. You need only to bring your slides. All the others will, as a general rule, offer only a darkened room. You should have a portable projector and an extension lead, and remember to ask the organiser to provide a screen or a white wall.

If your projector has a remote control, so much the better. Make sure that its flex is long enough to enable you to change the slides yourself from in front of the audience. If offered the choice of a projectionist or self-operation, and other things being equal, opt for self-operation. Volunteer projectionists, especially in schools, have a way of dropping magazines, putting slides in upside down, and not catching your prearranged signals for "the next one, please". A pointer torch with a tiny illuminated arrow is a useful bit of kit, as is a spare projector bulb.

THE LECTURE ITSELF

Remember to pay tribute, however briefly, to those who made the expedition possible. Mention your major sponsors – this is one of the few chances that you have of repaying them with some publicity. Show slides of their product in use, if this is appropriate.

The beginning should be arresting, with some unusual fact or figure. The end should be fairly resounding and point hopefully to the future. The middle bit is quite important, too. I wish you luck!

37 SOUND RECORDING AND RADIO BROADCASTS FOR EXPEDITIONS

Neil Walker

Few people would ever go travelling without packing a camera. It would be unthinkable to travel to the most beautiful place in the world and not be able to capture images to remind yourself later in life. And yet, if you just take a camera, you get only half of the picture. Without a camera you are blind, but without an audio recorder you are deaf.

Figure 37.1 *Damian Welch, winner of the RGS–IBG Radio 4 Journey of a Lifetime Award carrying out an interview on Tokelau, a South Pacific atoll threatened by sea level rise (© Damian Welch)*

At this stage I hear you say that the answer must be to take a camcorder instead because that way you capture sound and images. Perhaps. But the art of capturing moving images is vastly different from stills photography. And, frankly, the sound recordings picked up by most camcorders are only a slight improvement on silent movies.

The aim of this chapter is not to persuade you to take an audio recorder in preference to a camera or camcorder, but to convince you that you can capture a huge chunk of your personal experience if you do take one.

For many years I have helped beginners to make programmes about their travels for the BBC. They received only a basic training about what to record but were astonished on their return by the high quality of their own recordings and how vividly they rebuilt memories of certain times and places. To quote a well-worn media cliché: "the pictures are better on radio."

Before discussing the cost and what to buy, I would like to take you on a journey. This will be a fictional trip but my guess is that it will not be very different from what you are planning.

It starts before you even leave home when you confide to your audio recorder what problems you are having raising the finance, who is giving you grief, and what are the major things that you still have to sort out before you can leave. It's like a diary and the sort of things you record now are soon forgotten once the trip is under way. But these pre-trip problems and your thoughts at this time are just as important as what happens later and will no doubt produce a wry smile once the adventure is all over. These thoughts and emotions are easy to record on audio, but could you take a photograph or try to get them on videotape? I doubt it.

Once your expedition is under way, your audio recorder takes on a new role. Frequently, it becomes a friend and confidante, especially for those travelling alone. At any time in almost any place you can switch on your recorder and describe your mood and the people and places that you are encountering. It is so subtle that it fails to attract the attention caused by holding up a camera. And if your descriptions are good, the images conveyed to others when they hear your recordings later will be as vivid as any colour photograph, but with the added value of your opinions and emotions expressed at the same time. Not only that, but you will very often be able to hear the very thing you are describing.

As you proceed on your journey you will encounter other people. An audio recorder gives you an opportunity to record the voices of others. These may be fellow expedition members who are willing to share their thoughts and experiences, or they may be strangers whom you may wish to know more about. A microphone and recorder give you a golden opportunity to ask many questions that might seem prying or impertinent in other circumstances, and it does not matter a jot whether the sun is shining or you are in the pitch black.

So, below is a recap on what you should record on this journey.

Figure 37.2
Expedition member doing a personal voice piece (© Neil Walker)

DIARY PIECES

Quite simply, this is you talking to your machine as if it's a telephone, with your best friend on the other end. It should happen spontaneously and not only when you are happy. Nor should you record at the same time each day. Travelling is an emotional experience and you must record the lows as well as the highs.

INTERVIEWS

An interview is a conversation between two people – it is not two people giggling into a microphone pretending they are disc jockeys. Such things may seem funny at the time but invariably it sounds naff later. An interview is a fact-finding mission, so treat it as such. Ask questions, and then shut up long enough to get a full answer. Presenting yourself to others in a formal way and with serious intent gives you a great opportunity to be taken on guided tours of people's homes, factories, monuments, etc. because most people love to be interviewed. It makes them feel important. Out of politeness, make a point of listening very carefully to their answers. They often give clues to things that they want you to ask them about.

Figure 37.3 *An interview situation with microphone slightly closer to interviewee, but standing close together. This is ideal (© Neil Walker)*

Figure 37.4 *Interview too far apart. The microphone is intimidating in the gun-like position. Not ideal (© Neil Walker)*

ACTUALITY

Actuality is the sound of anything but without narration. By recording the sounds of the people and places you visit, you will paint a panoramic image more colourful than anything you could describe or photograph. Never hurry when doing this. Stand still (if you can) and simply let the sounds wash into the machine for several minutes. Check your watch to make sure that you have at least 3 minutes' worth of each of these stimulating sounds. These might be the clamour of a railway station, the crying of a woman beggar, geese honking as they fly overhead, the drone of prayers from a mosque, feet scrabbling on rocks, someone shrieking in panic – all these are essential for reconstructing your journey, and will be a useful soundtrack to a photographic slide show.

If you are now inspired enough to consider taking along audio-recording equipment, let's consider what to do next.

WHAT TO BUY

Technology is changing so fast that it is impossible to make firm recommendations. In the last 10 years the equipment used for collecting broadcast-quality material has moved from reel-to-reel quarter-inch tape, to tape cassette recorders, to digital cassette (DAT) recorders, and lately to minidisk or portable CD recorders. Even as I write, the BBC is experimenting with portable machines that record on to silicon chips. The chips are then inserted into a computer for editing. However, it seems likely that minidisk recorders will be with us for a few more years and these offer high-quality recordings at a modest cost. They are reasonably robust, simple to operate, smaller than the average Walkman and run on minimal power for long periods of time. The disks are also light, cheap, not easily damaged and store masses of material. You can also edit your recordings as you go, thus maximising the space available while minimising on weight. The downside of minidisks is that non-professional machines have tiny buttons and so are fiddly to use, especially if you are wearing gloves. High humidity and cold weather rapidly drain the batteries, and it is easy to record accidentally over your previous material.

If you decide to use cassette recorders (tape) you should be aware that they are prone to give you hissy recordings. Even professional machines will give high-quality recordings only if you are intimately familiar with how to set the recording levels. If funds are extremely limited you could choose the most basic machine of the lot: the dictaphone. For simplicity and cheapness it is hard to beat, although the sound reproduction will be poor. This can be improved by purchasing a decent microphone. With this machine you can easily make diary recordings and even do interviews, but actuality recordings will be severely limited. Whichever machine you choose, read the manual thoroughly and experiment as much as you can before setting off.

Figure 37.5
*Mono, lapel and stereo microphones
(© Neil Walker)*

Figure 37.6
*Lapel microphone on the end of a
bamboo cane being used to record birds
in a tree (© Neil Walker)*

Many people make the mistake of buying cheap microphones or decide to use the microphone that is built into their recorder. This is not recommended. You can get reasonable recordings with a good microphone and a poor recorder, but not the other way round. Mono microphones are easiest to use but tend to be more expensive than stereo microphones because there is less demand for them. Stereo microphones suffer more from hand noise and wind blast. Even gentle breezes generate blast interference unless you protect the sensitive tip with a windsock. In very windy conditions you will need to add further layers of protection. Radio journalists tend to use mono microphones for speech recordings and stereo for actuality. Unless you are recording for broadcast, my recommendation would be to seek out a good quality mono microphone.

To find the best, take your audio recorder into a shop and make several recordings using different microphones, then listen back to them (preferably at home on a hi fi) before choosing. Most in-built microphones are useless because they also record the whirring of the motor. A handy tip is to buy a tie-clip microphone as a back-up. These are very cheap from any electrical store and their quality is surprisingly good. They are incredibly light, will run for ever on tiny batteries, and can be adapted to form a stick microphone. You do this by simply taping it on to the end of a stick which can be a few centimetres long for face-to-face interviewing, or 3 metres long if you need to get close to people in a group situation (or want to record a bird up a tree): see Figure 37.6.

HOW TO GET THE BEST RECORDINGS

As with all things, time and experience can turn you from an amateur into a gifted recordist. How good you become at recording depends on how much effort you put into it. Just like a camera, you can whip it out when something obvious comes your way and click on to record. But as with photography, such snapshots are unlikely to impress anyone. Making good recordings is an attitude of mind. It is not that you need to be constantly doing it, but you do need to be aware and ready to pounce every time a recording moment comes along. That will happen only if you think about it a lot in the early days and train yourself to be aware of the sounds and events going on around you.

Having alerted yourself to a "recording moment", the next task is to decide how to capture it. No good cameraman would simply point and shoot. He would pick his subject, decide where to stand, which angle and what exposure, and then try several versions to get the best.

Sound is similar. Try recording a speech from the back of a crowded room and you will hear nothing worthwhile. You have to pluck up courage, march to the front and hold your microphone as close to the speaker as you can possibly get without offending him or her (it's not as hard as you think and few people object if you are courteous). Alternatively, be creative and, if the speaker is using a public address system, hold your microphone right next to a loudspeaker. This principle of getting close to your subject is the most vital, yet for many it is the hardest to overcome. Natural shyness prevents most people from entering into close proximity with others, but if you do not do this you will be wasting your time. The principle of close proximity must be adopted for virtually every situation. If recording live music, get right up close to the band. If recording in a market you need to make two recordings. One of the general background sounds, and a separate recording of individual stall-holders. Stand right next to them when they shout their wares. If you explain what you are doing they will happily comply, and often put on an extra show.

It is no good being half-hearted if you want good recordings. At the very worst your victim can only say no and ask you to go away. In my experience, they rarely do. When you do your diary pieces you should work close to the microphone, about 30–45 cm from your mouth. You need to be close to create that intimate sound associated with a diary piece. In interviews, manoeuvre yourself and your participant so that you are close together and the microphone is positioned equidistant between you, at about chest level, so that it is out of direct eyeline with your subject. His or her answers are more important to hear than your questions so be more concerned with getting the microphone close to him or her. Point the microphone straight up (as Figure 37.3) and not at your interviewee, unless you are forced to interview at full arm's length. A test recording followed by playback will tell you if you have got your positioning correct. Even with your friends you should be firm and direct. If you let people muck about, you have lost control and will only ever record rubbish. If you aren't going to try to make a good job of this, don't waste time and money taking an audio recorder on your travels.

RECORDING FOR BROADCAST

This should not be your prime concern. Millions of people go travelling and very few of them have experiences worthy of crafting into a radio programme. It is true that many people have interesting or amusing experiences, but they rarely have enough of them in one trip to fill a 30-minute documentary. As a rough guide, it is normal to record approximately 100 minutes of material for every one minute that is used in the programme, so you can see how dedicated you have to be in order to meet the demanding requirements of broadcast organisations. It is therefore my heartfelt recommendation that you record your expedition for your benefit and no one else's. This should make it fun to do and fun to listen to.

If these dire warnings have not put you off wanting to make a radio programme, consider the following. Is your trip really going to be full of interesting happenings that will translate to radio? Is there a good story to be told – one that hasn't been heard before? Are you sure that you will have the time to devote to gathering the recordings, or will the radio work get in the way? Will the other members of the team be cooperative, especially when under severe stress? If you can satisfy yourself on all these points and you are convinced that there is a good story to be told, the next step is to approach a broadcast organisation with your idea. Let me say straight away that you should not regard this as any kind of money-making venture. If you are offered any money at all, it will not be a fortune. However, an involvement with media organisations can boost your fund-raising possibilities elsewhere.

Before approaching any broadcaster you should be aware of the kinds of programmes that they transmit. Commercial radio, which is heavily music based, is unlikely to want to carry large chunks of speech, but is likely to be interested in an expedition that involves lots of music. Scan through the schedules of each broadcaster, then approach directly the producers of those programmes most likely to be interested. Initially letters or emails are always preferred to phone calls, and remember to supply as many contact numbers and addresses as you can. You should make these approaches at least 6 months before departure and, in the case of some networks, 12 months ahead. Budget allocations and schedules can be determined as much as 18 months ahead by some BBC departments, so you do yourself no favours by leaving it until the last minute. You can also approach independent broadcasting companies. There are several hundred of these in Britain, each of which sell programmes to radio stations or networks. Most specialise in certain kinds of programmes. A telephone call enquiring whether a network buys from freelancers will give a yes or no. If the answer is "yes", ask for a list of the names and addresses of the companies with which they work. The system of commissioning programmes is different in each country, but the method of approach is broadly the same. If you are fortunate enough to get a commission, the organisation will either send their own crew or provide you with recording equipment. Good luck.

Journey of a Lifetime Award

The Royal Geographical Society (with the Institute of British Geographers or IBG) and BBC Radio 4 award an annual travel bursary for an individual to undertake their "Journey of a Lifetime". This substantial award, worth up to £4000, is offered to someone who plans to undertake a journey that will inspire an interest in peoples and places, and who would like the opportunity to communicate their experiences through the medium of radio broadcasting. BBC Radio 4 is keen to discover new broadcasting talent among those with a genuine curiosity for the world around them. The closing date for applications is usually in October. Full criteria for this award and details of past winners can be found on the website: www.rgs.org/grants.

38 VIDEO AND FILM-MAKING FOR EXPEDITIONS

Phil Coates

The advent of the mini-digital video camera has changed the concept of expedition filming forever. The traditional 16mm film documentary, however much one might regret its passing, has already become a rare and costly luxury.

JIM CURRAN, SEPU KANGRI EXPEDITION, 1998

Now, as never before, modern lightweight digital camcorders enable anyone to record an expedition. Mini DV digital camcorders, both the consumer and professional models, have proven themselves in all conditions and all environments. They have become the standard for remote newsgathering and cost-effective documentaries across the globe. Unless you are working on a professionally funded, high-quality film or television production, film is simply too expensive and complex a proposition for expedition filming. We are now well and truly in the digital age.

This chapter therefore focuses on the use of these digital camcorders and how to get the best from them on location.

In TV speak, producing and delivering a "film" is split into three easily distinguishable parts; these constitute the production process:

1. Pre-production: story development, research, treatment and script writing.
2. Production: location shooting and sound recording, producing the goods.
3. Post-production: editing, adding music, sound and possible narration.

IN THE BEGINNING: PRE-PRODUCTION

The moment that you decide to take a video camera with you on an expedition you are entering into the communications business. If you are unable to share your ideas and stories effectively, it is simply not worth trying to record your expedition or attempting to make a film. Without this most basic skill, your endeavours will invariably result in disaster.

Ask yourself what you are trying to achieve, what story you are trying to tell and how you are planning to tell it. Remember that the single most important piece of equipment for any expedition film-maker is the pencil. The more times you write out your ideas, the more times you can look at them objectively, rethink them and rework them. This shaping and developing is a very organic and dynamic process. The more experienced you become at this process, the more skilled you will become at "seeing" your story and your shots in your mind's eye. For those of you with a word processor the limits are endless!

Many first-time film-makers say: "I'm going to see what happens and record the expedition as it goes along." Invariably they return home with a bunch of disjointed video shots that will never edit together and don't make any sense.

Remember that the simplest ideas are always the best; people like watching people, "interesting people"! A short writing or journalism course might not be a bad idea either.

TABLE 38.1 COMMUNICATING YOUR EXPEDITION: REFLECTIVE TO INTERACTIVE COMMUNICATIONS

Communication style	Level/activity	Impact on expedition
Reflective	Observational and authored narrative film	Need to buy in from team
	Shoot footage, return home to edit and possible broadcast	Could deliver as a one-person band
Active	On-location editing	Need dedicated team members, two- to three-person job and good resources
	Sending news reports Producing video diaries	
Interactive	Location broadcasting	Full-on team at home and on location
	Web cams Email chat Video conferencing	Serious back-up
Event	Full interactivity, "media circus" Live links into news programmes	Big team Big back-up Big "bucks"

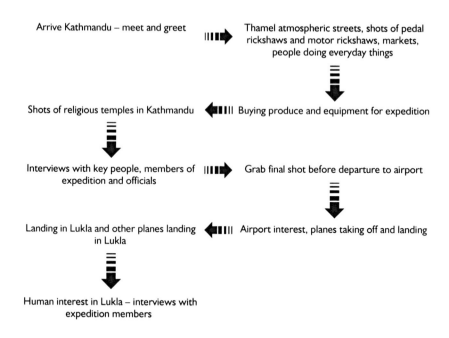

Figure 38.1 *Idea and production flow chart: filming expedition to the Khumbu region of Nepal*

The treatment

In professional television productions, the organic process of writing up your idea is called development; this is where you'll be doing extra research, finding out about where you are going and the things you will be seeing. Once you are happy with the idea, you need to write a first-stage document. This is known as "the treatment". You should need no more than one side of A4 to do this. If writing prose isn't your thing, try to construct a schematic flow chart, so that you can sketch out the idea. This can describe shots and sequences that you may wish to get on location. This initial concept will soon develop into your production flow chart (Figure 38.1). The good thing about doing it this way is that you can easily cross something out or add something as you go along.

Once you have asked yourself these simple questions, you should be able to fill in the blanks quite easily:

- What is my story?
- How am I going to tell it?
- Who are the main characters?
- What or who is of interest?

- Where do we want to go?
- How are we going to get there?
- What are we going to see en route?

It's a good idea to watch programmes that you really enjoy and analyse them. Watch them over and over again and try to understand how they are shot, or simply note down the shots that you really liked. However, don't choose a big budget drama, because you will be more than likely working with a very limited budget and have limited resources at your disposal. Reviewing and analysing travelogues and adventure travel documentaries is a great way to begin to learn the craft of expedition film-making.

If in doubt, KISS – keep it simple, stupid

If it is stupidly simple back home, it should still be relatively simple when you are on location with all the inherent complications of expedition logistics in a remote environment. Remember, you should be able to operate the equipment and shoot your "film" when you're tired, hungry or feeling too hot or too cold, and when both your cameras and your own batteries are running low.

Involving your team

It is important that you work closely with your expedition team from the outset and share with them what you are trying to achieve and the story that you want to tell. The effects of filming on your fellow expedition members can be enormous and may have a great impact on group dynamics. It is very easy to find oneself in a tense and awkward situation where animosity builds.

Are you going along as a constructive observer, or are you planning to make a film or programme where the subject happens to be an expedition? The former is not that difficult to achieve with a group of friends on an expedition, the latter is a full-blown production that will fundamentally change the nature and dynamics of your relationships with other expedition members.

When shooting a "fly on the wall" documentary or observational film, time, effort and energy need to be spent getting to know your subjects. You will need to build an excellent trust and rapport with the people before you "shove" a camera in their face. This is particularly important if you are going into an area where there are cultural sensitivities about filming, or if you are trying to make any anthropological observations. The better you know your subjects the better you will become at predicting what they might do next.

Managing the expectations of others is all important. You should always try and exceed people's expectations. Promising to deliver a broadcastable piece of television is a serious and "professional" (paid or unpaid) undertaking that requires commitment from yourself and a team of dedicated people. Be honest and realistic with yourself and others about what you can achieve with the time and resources available to you.

THE PRODUCTION PROCESS

With your treatment and understanding of the production process, you can begin to write your production flow chart (see Figure 38.1). This describes what you need to do and the shots and sequences that you will need to film. Think of it as a "road map" for your production journey. It will show you where you want to go, how you are going to get there and hopefully what you're going to see en route. Refer back to this as often as possible and note down any changes as you go along.

Choosing your camcorder

Choosing the right camcorder for you and your expedition is a very important process. All too often, camcorder selection and procurement are left literally to the last minute, in the hope that one might be lent or given free. A camcorder is a complex piece of kit and film-making is a significantly more complex process than taking stills. Film-making is all about managing the moment, managing resources and being pragmatic about compromise. Don't compromise when choosing your camcorder.

Always get your hands on your camcorder as early as possible, and remember that, if you're going to visit a cold region, try out the camcorder with gloves. Make sure that the buttons are easy to see and use; the feel and positioning of the buttons should be intuitive and the rocking motion on most zoom buttons should be fluid and smooth from wide to telephoto and back again. Make sure that the closures around the recording mechanism (where you put the tape in) are good and secure, if there are gaps and you can see daylight, dust and water will easily get in.

The pros and cons

If your budget allows, always try to go for a professional camcorder. These are usually sold through broadcast or professional dealers and are not generally available on the high street. The additional investment is easily worth it because not only do you tend to get more rugged camcorders, but they will also have better sound-recording capabilities and functionality.

Chips with everything, singles or threesomes?

Image quality on small lightweight mini DV camcorders can differ dramatically depending on how many chips they have. The chips referred to here are "charged coupled devices" or CCDs. Small palm-sized camcorders tend to have only one chip; the slightly larger camcorders that you have to hold with two hands or rest on your shoulder have three chips.

Without getting too technical, the chip or CCD is responsible for turning all that the lens can "see" into electrical impulses. Three CCDs are better than one and larger CCDs are better at doing the job than smaller ones. If you are looking to develop your video filming skills or to produce any material that you will be showing to an audience whether by TV, web or lecture, go for a three-CCD camcorder. Single-chip camcorders are best for video diaries, undercover filming and when you're on the move, say on a cycling expedition, or for a summit dash on a big mountain; they are light, very portable and can be easily stowed in a warm pocket or secure pouch.

Doing it digital

If you are looking to tap into the world wide web or undertake a spot of home editing on a Mac or PC, going digital with a digital DV or DVCAM camcorder is the only solution. Modern digital camcorders can now be connected to other devices through a single cable called iLink (i333) or Firewire (Apple trademark). You can now literally plug and play!

Budgets

If you're on a seriously limited budget, there are fortunately a large number of Hi 8 and early digital video kits available on the second-hand market. Check camera dealers and the classified sections of video magazines. In the UK, the national photographic retailer, Jessop's, offers an excellent national search for second-hand equipment in their many branches. However, older camcorders may use NiCd (nickel–cadmium) batteries. Always check all of the batteries on a second-hand camcorder and view a freshly recorded tape on a TV or monitor before purchasing.

Many people looking to take a video camcorder on expedition think that it is a way of making money: "for sure we can get paid to make a film, there are loads of channels out there that'll want to buy our stuff", they say. When you go through the true economics, it is simply not the case. If you're lucky, you may be able to cover your costs.

The purist's way to fund your expedition film is to take on an extra job, do some overtime or live a frugal and monastic existence, living on the three Bs – "beans, bread and beer" – for a period before your expedition, and save the cash that you need. Remember that, if you start pursuing commercial sponsorship, you will need to be able to offer the sponsor something of value in return. The more serious the investment from the sponsor, the more serious the value of sponsor remuneration is needed.

A basic "catch 22" situation can easily develop where the film is used as a device to secure funds for the expedition, and the expedition cannot go ahead if the film is not made. Budgets quickly spiral. If you are going to accept commercial sponsorship, it is wise to sign a contract checked by your solicitors. Know what is expected of you and your expedition, regard this as a business proposition and deliver your obligations accordingly. Be wary of so-called professional third party organisations that will broker deals for you.

Without doubt you should understand what you are taking on; do not let your ambitions, promises or desires run away with you. Remember these points: are you over-committing yourself and your team? Are you and your team media friendly and media savvy? What is your comfort zone?

In the field

If you're both physically and mentally prepared for your expedition, coping with the extra demands of filming shouldn't be too much of a problem. Try not to be over-ambitious when filming, and work well within your own comfort zone; this will allow you to get the shots that you have been looking for from the start. It is so easy to let yourself go, with dire consequences, e.g. when filming in the cold and taking off your gloves it is so easy to suffer cold damage to your fingers. If there is no one on your expedition to watch out for you, always set aside some time each day to make sure that you're fit and well. No matter how well your kit is, if you're not working properly, you don't have a film.

Managing the kit

Irrespective of the environment or location you are visiting when filming, you should "always use protection". Make sure that your equipment stays in good working order. The big hazards to watch out for are: extremes of temperature; water, especially salt water; and dust. Using a protective cover or housing will help in all of these cases and always make sure that you are not taking the camcorder from a cold to a hot environment quickly because condensation can easily build up.

- Heat: try to keep your camcorder out of hot vehicles; use a reflective insulation jacket.
- Cold: insulate your camcorder from the cold and keep the batteries warm. Heat packs can be useful; however, remember that they do not work well at altitude!
- Water: plastic bags and rubber bands help with light rain. "Ewe Marine" splash and waterproof housings work well up to shallow depths; then you need a full underwater marine housing. Silica gel (available at chemists and camera shops) is great for absorbing moisture; always take plenty.
- Dust: this is just as damaging as water, so undertake the same precautions and measures to stop it getting inside the camcorder in the first place.

Be fully familiar with all the display readings on your camcorder because this can help when trying to diagnose what is wrong and what you need to do. The more familiar you are with your kit, the better you able will be to do any maintenance and running repairs. As with stills photography, make sure that you understand the fundamental basics of composition. Always keep the camcorder as steady as possible.

Always remember to check what you've shot against your original written ideas, treatment, script or flow chart; in this way you will be able to work out how you're progressing with your shoot.

On expedition many things occupy your mind. Always carry a small pocket notebook with you when filming; it's then easy to write your thoughts down as they occur to you and review them later.

Getting the shot

Next time you're at a tourist destination, watch people "videoing" the scene. Suddenly you find people turning their camcorders into Hoovers, trying to suck up "all the action", waving their camcorder from side to side, vacuuming the sky. Getting the shot is all about preparation and observation; it's an art and a craft. If you are used to composing shots for still photography, you are well on your way to shooting sequences with your video camcorder. The important thing is to shoot a variety of close-ups, mid-shots and long shots with what are called cut aways (shots that you can use in the edit to cut a sequence together). Consult a book on creative production to help you with this.

The basic idea is to keep your shot for at least 10 and preferably a minimum of 12 seconds. Don't zoom in and out and pan the camcorder too fast. The zoom is there to "frame" the shot; panning and zoom can be used to good effect to show a sense of scale only if you know how to put them to good effect! Panning from left to right seems to be inherently more pleasing, maybe because it follows the direction that we westerners read. To improve your shots, practise panning using a fluid head tripod and using the rocker button for zooming from wide to telephoto and back again.

Steady as she goes

Just as important as getting the shot is keeping it steady. You can do this by using a variety of devices: tripod, monopod, shoulder support or brace, Handyman stabiliser, beanbag or cord loop. A tripod is a must. There are many excellent lightweight, fluid head pan and tilt tripods available. Choose one that suits you and your pocket. A monopod is a good option because it eliminates vertical movement; shoulder braces do the job but can be rather cumbersome on expeditions. A stabilisation system is an excellent yet rather expensive solution to the problem of stability, the great advantage being that you can hold the camcorder on the top of the unit and "carry" it as if it was floating through the air – perfect for those shots of someone walking through a busy street or on a mountain path.

Beanbags are great if you are using the camcorder low down or want to strap it to a car or plonk it on a wall. With the cord loop, you place a loop around your boot that will tighten by itself, and attach the other end to the base of the camcorder. If you have adjusted the length of the cord correctly you should be able to create tension in the cord by pulling up the camcorder and that tension should hold it steady! A stable camera significantly improves the quality of the resulting image; most camcorders now have image stability systems built in. For technical reasons, optical steady shot systems are always better than electronic steady shot systems, which tend to degrade the image quality.

Capturing sound

All too often, video sound quality is sacrificed for "the shot" or simply forgotten about altogether. Sound recording is one of the most under-regarded, yet one of the most important components of your film. Your eye is very forgiving; you can endure and even "enjoy" poorly lit, slightly out of focus, "atmospheric"-looking footage of people on expedition. When it comes to sound quality and the listening experience, your ears are almost totally unforgiving. Poor sound quality basically says a poor film. Always try to use a microphone other than the one built into the camcorder. The onboard camcorder microphone is in the wrong place, and can pick up camera noise, and is not designed for interviews or recording sound in "detail". Be aware of any background noise, use headphones to check or "monitor" the quality levels of the audio. Try to use good quality headphones; however, if your production is constrained by budget, Walkman-type headphones are adequate. Avoid fully encased headphones when on expedition, because you will not be able to hear anything going on around you, especially important when working in a mountain environment, on a boat or near a busy road.

To record crisp clean sound for an interview or a specific activity, you need a directional microphone that you can point at the subject. There are two types of these directional stick mikes: condenser or dynamic. Condenser microphones need power; usually they get it from the camera, a "phantom" power supply or, alternatively, they need a separate battery. Dynamic microphones don't need extra power and tend to work better in extreme humidity and really bad weather. You can also use a lapel microphone for interviews. These are easy to hide and can be positioned very near to the face of the person and so pick up their spoken words clearly.

Professional microphones have a special type of secure socket called an XLR (eXtra Long Run or eXtra Low Resistance) audio connector, designed for public address work. To get an XLR connector to fit a consumer camcorder, you need an adapter. A popular unit for allowing you to use professional XLR microphones on a consumer camcorder is a BeechTek box. This little device allows you to use two professional microphone sources and put them through a standard "phono" (Walkman-size headphone connector) input on your camcorder.

Whatever type of microphone you choose, you should always use a wind gag or "fluffy dog"; the number one company that produces these is called Rycote.

When you need to get your microphone in close to your subject, you can use either a "pistol grip" to attach to your stick microphone or a boom or "fish pole". A telescopic trekking or ski pole with a small universal clamp or even gaffer tape can easily substitute for a proper boom. If using a boom the person holding the pole should wear good headphones, so that they can monitor the "quality" of the sound, making sure that they are pointing the microphone in the right place and that the sound doesn't "spike" or go off the recording scale. Take extra cable for both your microphone and headphones, because the person recording the sound can sometimes be a few metres away from the person with the camcorder.

And then there was light

Modern camcorders have an amazing ability to record quality images in very low light conditions. In general, small lightweight camcorders are better in these conditions than in very bright sunlight. High-contrast bright light is bad, low-contrast soft even light is good. This is the result of the way that the imaging chips or CCDs of mini DV camcorders convert the light energy into electrical impulses.

When in very bright light try diffusing the light as much as possible by using a neutral density filter; this limits the amount of light entering through the lens. Good quality camcorders usually have them built in. Try to use whatever available light you have to illuminate a scene. Video lights can prove to be very useful, but beware of them creating a harsh look on your subject when the lights are fitted to your camcorder.

Power

Without power, your electronic camcorder simply will not work. Without a working camcorder you simply will not have a film. Good energy management is vital.

Batteries mainly fall into two types: primary, single use, non-rechargeable cells and secondary, which are rechargeable and multi-use. Primary batteries, especially lithium cells, are excellent for expedition filming; they hold their charge longer than secondary cells and work well in very cold temperatures.

The most common rechargeable batteries being used in older camcorders and battery packs are NiCd. NiCd batteries develop a memory effect if not drained as low as possible before recharging, e.g. if you use it 15 minutes, then charge it over and over again, after a while you have a 15-minute battery. Next comes nickel–metal hydride (NiMH) batteries, which give you longer usage between charges and can be charged whenever you like with no danger of developing a memory effect. Finally there are lithium ion (Li-Ion) batteries; these have the advantages of NiMH batteries as well as being thinner and lighter than the others. Unfortunately they are usually much more expensive. Lead acid batteries must be fully recharged after each use, just like a car battery. Leaving them uncharged for long periods of time can ruin them. Even while being stored, lead acid batteries must be occasionally charged.

Another consideration with batteries is to have them either mounted on, or within, the camera or off camera. On camera works best because you don't have any leads or cables to a separate battery pack or battery belt, which can easily become tangled in other equipment, especially climbing gear. Last but not least, remember that cold kills batteries. In such conditions, halve or even quarter the standard continuous recording times stated on the battery by the manufacturer. Always remember that you are using valuable battery power when zooming the lens in and out while framing your shots and also by just leaving the camcorder on standby mode.

Taking charge
With primary cells there are the environmental considerations of disposal. Where possible they should be brought back to your home country. Many less developed countries just do not have the infrastructure to dispose of your batteries in an appropriate way. When charging batteries you have many options: wind and water turbines, solar panels and liquid fuel generators. Modern solar panels packed with voltaic cells are super-efficient and you can now charge batteries in most conditions. Always try to take sufficient fully charged secondary or sufficient primary cells for your recording needs; total reliance on recharging a couple of batteries is risky.

Taking stock
The question of how much tape to take is just the same as the question photographers ask about how much film to take. You should be able to answer your question by checking how much you can afford to buy, how much you can afford to carry, how much power you have and what type of film you are making. The minimum should be 10 hours of tape for a 1-hour film, or preferably a half-hour film! Keep your tape in a waterproof, airtight container inside a larger Peli Case, and always mark the tapes that you have shot

THE CUTTING ROOM FLOOR

With digital production and the relatively low cost of computer processing power, home editing on a Mac or PC has become a serious reality. There are many cost-effective solutions available on the market; however, remember that the kit doesn't do the storytelling and great editing hardware and software don't necessarily lead to a great edit. If you can afford it, use a system that has "DV compression"; this simply means that the type of digital information that it is processing is the same as the DV digital information recorded in your camcorder. Firewire or iLink enables you to transfer DV footage directly from your camcorder to your editing computer. Try to avoid all the "special" effects supplied with the system; editing is all about pace and the craft of storytelling, not about how many dazzling effects you can string together within a short sequence!

Line up, line up

On modern computer-based editing systems you will see the words non-linear. Non-linear editing means simply editing not in a line. Today you can start building your story at any point you wish: start at the end and work backwards, start in the middle and work outwards. Imagine non-linear editing as playing with a pack of cards. You can create any sequence that you wish from the cards that you have in front of you; if you don't like it you can always shuffle the cards and start all over again! However to ensure good storytelling many editors still work from the beginning.

Bringing it home

Most expedition films are rarely completed, and of the limited few that are, many are never broadcast. The expedition films that do hit our screens are mostly broadcast 6, 12 or even 18 months after the expedition is over. This hardly makes the expedition newsworthy and for many sponsors the time lapse is just too great. By harnessing the amazing power of satellite phone systems and a couple of boxes of relatively easy-to-use hardware and software, live or "near live" transmissions can now be performed from almost anywhere on earth.

Telling the world

In recent years the media business has changed beyond all recognition. To most viewers, things look virtually the same; however, we are now living in a truly multi-channel and multimedia world. The good news for those interested in expeditions is that there are now channels dedicated to the subjects. The potential distribution outlets for professionally made films featuring subjects that are in demand are numerous; however, trying to sell a completed film in this way doesn't guarantee any results. It is a very risky process, so you should try to obtain funding or a "commission" up front.

The potential outlets include:

- broadcast: terrestrial, cable and satellite channels or web streaming
- sell-through on video or DV: specialist distributors; specialist retailers; direct mail.

The following are other ideas worth considering.

VNRs

Video news releases (VNRs) are short (i.e. under 3 minute) pieces about a given subject, which are given to news organisations, broadcasters or "magazine" shows that may be interested in your story. Through this process you may be able to secure exposure and be able effectively to promote your expedition and fulfil any sponsor "requirements" without making a full-blown film! You can also use these video sequences on your website.

Going local
It shouldn't be too difficult to get involved with your local regional television station or news provider. Almost all have an early evening magazine programme that combines news, features and chat. Try to secure a feature slot in their show before and after your expedition. If you have the resources at your disposal you could even do a live link. Securing local airtime and exposure is how many well-accomplished film-makers started out; start small, with a big vision.

Epks
Electronic press kits (EPKs) are not that dissimilar to VNRs; however, they tend to be less "newsy" and may promote a sponsor or organisation a little more. If a VNR is news, then an EPK is more PR (public relations). Again they should be short and well edited, supplied on at least a Beta SP tape, and be accompanied by still photographs on CD-ROM or DVD, an article and accompanying notes.

Web-wise video and location broadcasting
Satellite phones provide the widest range of options for expedition communications. Their ability to provide both mobility and continuous communication links in the remotest of locations makes them an extremely valuable, if not essential, piece of equipment for anyone operating in areas beyond the reach of regular communications.

The most commonly used satellite system for simple voice and data communication is the INMARSAT Mini-M system. Mini-M phones are both extremely compact and weigh about 2 kg and so provide high-quality mobility with 98 per cent land mass coverage. The system does not work above 70° of latitude, north or south, so it is not a useful form of communication for Antarctic and high Arctic expeditions.

Mini-M phones provide high quality and relatively inexpensive international voice, fax and email transmission at current rates for around $US2 per minute. However, as data connections are only up to 2.4 kbps, large data communication and internet access is not really practical. To send and receive relatively large data files such as stills and very large files such as video, the INMARSAT M4 system is required. At around three times the price of the standard Mini-M system, it does seem rather expensive, but the system offers high-speed data services with connections up to 64 kbps.

The central components that make field video and stills communications viable over satellite phones is the ability of modern compression systems to reduce file sizes and then squeeze images and audio down the system without any real loss in quality. Compression makes the data smaller, and so less expensive to store; it also makes it much easier to send around networks and systems.

The actual size of the data file to be transmitted depends on the required quality of the resulting footage: the higher the quality, the higher the video encoding rate

(Mbps) and the larger the file that is recorded, therefore, the longer the transmission time. Normal "broadcast quality" material is around 2.5 Mbps MPEG, which, using a single M4 terminal at 64 kbps, would take 40 minutes to transmit 60 seconds of video footage. Obviously, this has not only huge time considerations, but also huge cost implications when, at current prices, the rate is about $US7/minute, e.g. at $US7/minute a 2-minute VNR would take 1 hour 20 minutes to transmit and cost around $US560.

The greatest problem of using satellite communications is political rather than technical. Some countries are very sensitive about satellite communications and tend to ban the use of such equipment and communications in their country or territory.

Encoding, crunching the numbers

To get your DV film on to the web, you have to make it available in a web-friendly format. This is where encoding comes in. The three most popular video streaming media players are Apple QuickTime, Microsoft Windows Media Player and Real Networks Real player.

Always try to provide links on your website to media plug-ins for the different media format players and, most importantly, check that all the links work!

Consider a web version when you're shooting an expedition film, a web version being one that is more suitable to watch on a computer screen when the video is being streamed over the web.

Web video top tips

When shooting a presenter or interviewing someone close up, get them to wear neutral solid colours; this helps with the compression of the image. Try to avoid stripes or patterns and really strong colours such as reds and yellows.

When viewing encoded video that will be used for web streaming, view it on a computer screen, and find out for yourself what works and what doesn't.

Remember that web video is viewed at only a small size, so long static shots work best. Use long close-ups where possible. Quick cuts don't compress well, so concentrate on video content, not video effects.

Research and test the variety of Internet Service Providers (ISPs) available to you and then test again the one that you finally choose.

Where possible use a dedicated media server for your video material and make sure that their capacity is sufficient for your needs, or understand the limits and capabilities of the server and work within them. If a large company sponsors your expedition, you could always develop a subsite of their corporate site and ask if their IT team can do the work.

Road test your connections beforehand by uploading material to the site and server before you depart on your travels.

As connection speeds to the internet increase, more creative and complex productions will be deliverable via web streaming; until then, like all good things, keep it simple.

No matter what type of communications you plan to undertake on location, you will need a dedicated, hard-working team back at base to make sure that your reports are coming in "loud and clear" and that all your hi-tech gadgetry works.

AND FINALLY ...

If you decide to take up the challenge, remember, even when you've got the kit, you are still a world away from producing a well-crafted, witty, intelligent and engaging film! As with most things it is "one per cent inspiration and ninety nine per cent perspiration".

FURTHER INFORMATION

Lists for everything
An easy and convenient way to manage and keep a track of everything is to produce lists.

List 1: kit list – the basics
Camcorder
Wide-angle lens adaptor
Filters, polarising, neutral density, skylight
Filter wrench

Synthetic lens cloth in pouch
Lens cleaner
Fluid head pan-and-tilt tripod
Steadying foot cord
Beanbag
Headphones
Lavalier microphone
Shotgun microphone
Microphone cables
Microphone boom or pole
Tapes
Spare batteries
Mains charger
12-volt charger
White balance sheet
Notebook
Pens and pencil
Carry case
Camcorder pouch

List 2: spares and repairs
Spare filters
Super-Glue
Gaffer tape, lots of it!
Watchmaker's screwdriver that fits the
  screws on your kit
Multitool, Leatherman or similar
Clear plastic bags
Silica gel
Cable ties, assorted sizes
Elastic bands, various lengths and widths
Jubilee clips, assorted sizes
Bulldog clips, assorted sizes

Useful addresses: contact details

Camera Care Systems (case manufacturer and supplier), Fotolynx Ltd. Tel: +44 117 963 5263, fax: +44 117 963 6362, email: info@ccscentre.co.uk, website: www.ccscentre.co.uk

CKE Distribution (professional equipment dealer). Tel: +44 1274 533996, fax: +44 1274 533997, email: info@cke.co.uk, website: www.cke.co.uk

CP Cases (case manufacturer and supplier). CP Cases Ltd (London). Tel. +44 20 8568 1881, fax: +44 20 8568 1141, email: info@cpcases.com, website: www.cpcases.com

Digital Reproductions Limited (video duplicator and tape supplier). Tel: +44 1274 688068, fax: +44 1274 688071, email: sales@digital-reproductions.co.uk, website: www.digital-reproductions.co.uk

Inmarsat (satellite communications organisation). Tel: +44 207 728 1504, fax: +44 207 728 1179, email: information@inmarsat.com, website: www.inmarsat.com

Integrated Communications Solutions (satellite communications supplier). Tel: +44 1844 260560, fax: + 44 1844 339091, email: info@icomms.com, website: www.icomms.com

Kendal Mountain Film Festival (adventure film festival). Tel: +44 1539 725760, fax: +44 1539 734457, email: info@mountainfilm.co.uk, website: www.mountainfilm.co.uk

Optex (professional equipment dealer). Tel: +44 20 8441 2199, fax: +44 20 8449 3646, email: info@optexin.com, websites: www.optexint.com, www.optexdirect.com

Peli Cases (case manufacturer and supplier). Tel: +44 161 832 5335, fax: +44 161 833 4488, email: (UK sales): sales@pelicases.co.uk, (UK support): support@pelicases.co.uk, website: www.pelicase.co.uk

Prokit (professional equipment dealer). Tel: +44 20 8995 4664, fax: +44 20 8995 4656, email: enquiries@prokit.co.uk, website: www.prokit.co.uk

Rycote (equipment manufacturer). Tel: +44 1453 759338, fax: +44 1453 764249, email: info@rycote.com, website: www.rycote.com

Solar Century (solar panel manufacturer and supplier). Tel: + 44 20 7803 0100, fax: + 44 20 7803 0101, email: enquiries@solarcentury.co.uk, website: www.solarcentury.co.uk

Sony Broadcast (professional equipment manufacturer). Tel: + 44 1932 816000, fax: + 44 1932 817014, website: www.sonybiz.net

Total Audio Solutions (professional audio equipment dealer). Tel: +44 1527 880051 (24 hours), fax: +44 1527 880052, email: sales@totalaudio.co.uk, website:www.totalaudio.co.uk

West Herts Media Centre (runs adventure filmmaking courses and hires equipment). Tel: + 44 1923 681602, email: mediacentre@westherts.ac.uk, website: www.mediacentre.westherts.ac.uk

39 WRITING AND PUBLISHING SCIENTIFIC PAPERS

Adrian Barnett

WHY YOU SHOULD PUBLISH

Expeditions often get to parts that other fieldworkers do not reach. The data that they gather are often of immense value, either as information snap-shots in their own right or as the basis for future work. Yet, although most RGS–IBG-approved expeditions produce a report, the number that also produce contributions to scientific journals are very low indeed. Given the quality of some of the work done by expeditions, the main reasons for this must be ignorance and fear – ignorance of the process by which to get the information published and fear that it wouldn't be good enough anyway.

Be confident. Pre-fieldwork people: there's no reason why your data shouldn't be as good as anyone else's; just make sure that you collect them correctly. Post-field-work people: remember your literature search? Would you have been pleased to come across a paper containing the data that you now have? If the answer is "yes", then publish.

However, getting your results published can be quite a time-consuming and, sometimes, tedious process. But it *is* a worthwhile exercise. There is a great inequality between the amount of information available about, say, ecology in Europe and North America and that available for most other countries in the world. This is all the sadder, considering that this northern bias in knowledge is in inverse proportion to the distribution of global biodiversity, e.g. between January 1989 and September 2000 the *Zoological Record* listed 2798 papers on British mammals. The UK has 50 species of known mammal. In the same period, there were 1231 papers on mammals in Brazil (400 species), 170 in Colombia (363 species) and 190 for Costa Rica (206 species). With 122 papers, the UK's common woodmouse (*Apodemus sylvaticus*) had almost as much written on it in this time as did all the mammals of Colombia! This pattern occurs not just in mammals; it is present in almost any subject at which you care to look.

ISN'T A REPORT ENOUGH, THEN?

Reports are valuable sources of information, but should be regarded as only the first stage of information processing. Reports are great for bringing everyone's data together, but not that good at disseminating the collated information. This is mainly because, generally, even the best-funded expedition can afford to produce only a few dozen copies of its report. Even when they are deposited in all the appropriate places, they are not always easy to find. As they are not refereed, people may be uncertain how much value to place on their contents or conclusions. Publication of your results in a journal gives them more credibility and makes them more accessible to a wider audience; for example, in November 1993 the *Journal of Zoology [London]* had 1143 subscribers, and in 1992 the *Journal of Ecology, Journal of Animal Ecology* and *Journal of Applied Ecology* had, respectively, 3600, 3200 and 3000 subscribers. The *Geographical Journal* reaches 10,000 people worldwide in 173 countries, of whom 1700 are subscribers. As many of these subscribers are libraries, the potential readership is obviously even larger than these numbers suggest.

SOME COMMON MISCONCEPTIONS ABOUT PUBLISHING SCIENTIFIC PAPERS

The following are not true:

- You have to pay the journal to publish your paper.
- The journal will pay you to write the paper.
- You have to have a degree or PhD before you are allowed to publish.
- You have to be attached to a university faculty before your work will get published.
- You can publish only discoveries of truly earth-shattering greatness.
- No paper will be accepted unless it has statistics in it somewhere.
- Only long and complex papers get published.
- You need to publish all your results at once and in the same journal.
- You need to put in everything you did and everything you found in the same paper.

WHERE SHOULD YOU PUBLISH

Choosing your journal

Don't set your sights too high. Your first paper will probably not get into *Nature* or *Science*. From your literature searches you will know what journals cover your field. Try one of those. To increase your chances of publication try one that is not top-flight. There is no shame in this; it is just the pragmatic realisation that some journals may be more open to publishing work by undergraduates than others. Any academics you know will probably be more than happy to tell you which journals these are.

Publishing in the host country
Journals published in less developed countries are sometimes held in low esteem because some do not have as rigorous a system of peer review as others elsewhere, they just don't look as good or they get printed on lower quality paper. Don't let this put you off. It should be seen as a duty and a mark of courtesy to publish in a journal from your host country. Even better, publish in the language of that country: not everyone reads English. In countries where financial resources are often limited, academic journals printed in Europe and North America are often too expensive to be widely available. Nationally printed ones can bridge that gap.

Publishing elsewhere
Many host country organisations are likely to have exchange programmes to help defray their costs; funding constraints may also limit print runs. This means that some host country journals don't get circulated as widely as others in the same field. Accordingly, you may also wish to publish a second paper elsewhere. This seems to be acceptable, providing the language is different, the texts and data-sets are not so totally alike that you can justifiably be accused of autoplagiarism, the title is not identical and the authors appear in a different order.

Multi-language summaries
When publishing outside the host country you should endeavour, whenever possible, to include a summary in the language of the host country. If no one in your team feels up to this, arrange it with someone in the host country before you leave. The British Council may also be able to help.

HOW YOU SHOULD PUBLISH
The publication process

Submission
Journals do not work like popular magazines. You do not ring up the editor and ask if they would like something on a specific subject. You have to write the manuscript and prepare everything fully. Then, *using registered mail,* submit three copies and wait for a response.

What happens next?
Precise details will depend on the journal. Generally you will get an acknowledgement of receipt from the editor. She or he will keep one copy and send one each to

two anonymous referees (generally acknowledged experts in that particular field) who will then review the manuscript. They return their copies to the editor with their comments and, once she or he has noted them, the modified manuscripts are passed on to you. This is known as peer reviewing.

How long does all this take?
This rarely takes less than 2 months. But it can sometimes take much longer, especially if one of the referees is away. Just keep tabs on the process and drop a note to the editor every few months to check on how things are coming along. Don't get neurotic about it and remember that editors have a lot of papers to deal with.

Providing space for referees' comments between lines is why you are asked to double space the text that you submit. Providing space for other comments is why you are asked to give it wide margins.

What then?
Modify the text in line with the referees' recommendations and resubmit. Don't get upset if your text comes back covered in blue pencil. It doesn't mean that he or she thinks that it is bad – it's just that people are trying to be helpful (after all, referees aren't paid – it's a service provided *gratis* for the general benefit). Make the suggested modifications, but also use this as an opportunity to put in any new references or angles that you have found (editors like it when papers are as up to date as possible – it makes their journal look good, although don't make the modification process to such extremes that it's almost a new paper or people will start wondering why you originally submitted a draft and wasted their time).

If you do not follow a referee's recommendations either to the letter or at all, explain why in a letter accompanying the modified manuscript, but don't be stroppy or cheeky if you do so. Sometimes the recommendations from the two referees will be in direct contradiction to each other. If so, say in your letter to the editor which you have chosen and why.

Once you have sent it back, the modified version will go round the system again. It may even have to go round twice. The manuscript may then need to be approved by a controlling committee. Eventually you should get a note saying that your paper has been accepted for publication. The paper will be set by a printer, and then you get galley or page proofs.

Printing and proofs
The gap between acceptance of the paper and receiving an envelope full of reprints because your paper is published can be up to a year, or longer. You should receive a set of page or galley proofs with a deadline date for their return – make sure that you meet this deadline. If you are likely to be away when the proofs are due, ensure that you delegate the checking and return of the proofs to someone else.

Scrutinise the proof
The proof is a typeset version of the text. Barring accidents, what is there is what will appear in print unless you change it. Look at the text very closely. Check spelling, typography, line and paragraph order, etc. If making modifications, you can either write your instructions in (but do so very clearly) or use printer's marks. Some journals will send you a set of printer's marks with the original reviewer's comments (this is helpful because the reviewers will have used such notations on your text); other journals will not do this and you will have to talk to a librarian to locate a sourcebook.

When you get a proof, work on it as soon as you can. Your paper is going to appear in the next issue and the editor is waiting. If you do not send any modifications you might have quickly, the editor will assume that everything is OK and will publish it as it is. With the proofs, an offprint order form will often be included. Make sure that this is completed and returned.

Nowadays, when many manuscripts are submitted electronically, it is rare for such errors as typos and chunks of text being out of order to creep into a text once it is in the journal's hands. However, some journals are still typeset by hand (especially in tropical countries). Check such proofs really thoroughly (especially for homonyms, but also the tables).

All this may all sound silly, tedious and trivial. But remember, if the text has errors, people might start to think that you are equally sloppy with your data collection and interpretation, and begin to mistrust your paper.

Do papers ever get rejected?
Yes, sometimes it comes straight back from the editor, other times the reviewers recommend its rejection. But normally you will be given some reason why. Don't despair, just re-jig the thing and send it to a different journal. Unless it's really rubbish, it will find a home eventually.

Safeguards
Keep copies of your disk, computer file and/or the manuscript. Things do get lost in the post or in the editing system. For the same reason, and to guard against computer viruses/crashes, keep hard copies of the manuscript at all stages of editorial modification.

Preparing for publication

How many papers?
How finely do you slice the information in the data-set? Obviously it depends on what you've got. But a common split is one general paper based on the data in the whole report and one on each of the main research projects. Each of these might be

subdivided into short papers on, say, reproduction, diet and community ecology of a particular group such as birds. You might also want a special subject paper on, for example, the conservation importance of the area that you've visited. This is not all done so that you can get a massive list of publications, but because many people find it much easier to work this way. Doing papers as small discrete units can make things much more manageable; the text has much less of a tendency to sprawl, you can set sharper targets, and so tend to get much less bogged down and depressed while researching and writing the papers.

How long should a short paper be?
Short papers are generally less than six pages, including references, and are variously known as "notes", "research reports", "reports from the field", etc. – depending on the journal. Single observations of interesting behaviours or geographical phenomena are often published as half-page notes.

Stepping back
When you have finished the final version and are happy with it, stick it in a drawer for a week and then look at it again. It's amazing how many blemishes you can then see on your formerly pristine piece.

Pre-submission peer review
Once you have done the first draft, send a copy of the manuscript to other expedition members. Stress that it is a draft and ask for comments. Once you have received these, pass a modified version to an academic adviser for further comment. Once all this has been done, you are ready to submit.

When you should publish
Publish as soon as possible after you get back and have had a chance to write the report. If you leave it, other things will crop up and you are likely to forget some of the subtleties of your work.

Planning for publishing

Pre-expedition work
Make sure that you are really familiar with the relevant literature before you go. This should mean that you can spot and follow up interesting things when you are in the field, and it will also reduce the time that you have to spend on literature searches when you get back.

Budgeting for paper production
Papers cost time and money to produce. Allow for this when drawing up your budget

and planning post-fieldwork tasks. As a guesstimate, say that each paper is going to cost £50 in postage, photocopying and library services and take maybe 10 days of solid work to get it written up and ready for a pre-submission peer review.

Budgeting for data analysis and specimen identification
You will also have to allow time to analyse your data and budget, and time to check field specimens that you may have to have identified. Museums don't have the staff, time or money nowadays to do it for you and are likely to charge for their time.

Literature searches
These are absolutely essential for the production of a worthwhile paper or notes. For the biological sciences *Zoological Record, Ecological Abstracts, Ecology Abstracts, Excerpta Botanica, Forestry Abstracts* and *Abstracts in Anthropology* provide the best starting points.

There are other compilations (e.g. *Biological Abstracts, Citations Index, Current Contents*), but these are generally held to be of less use in primary literature searches.

Some of the groups for which a lot of work gets published have a specialist publication that appears regularly at short intervals (every 1–2 months, e.g. *Current Primate References* for primates). This can help you find the most up-to-date publications on a specific species in that group.

Organising data in the field
Make sure that you take your data in a way that will be easy to analyse later. Write up your notes daily. Use abbreviations as little as possible. Keep regular notes. Trust nothing to memory. Organise your notes with the clarity that would allow someone unfamiliar with them to extract and analyse the data with ease.

Protecting the field data
Your data should be regarded as the most important thing you get from the expedition. Always make copies of your data and deposit them in a safe place; one set left in town and one set sent home is a good way.

Make a duplicate copy of your data while you are still in the field (copying them up every night is best). When you are travelling give one copy of your notes to someone else. Treat your data book as the most valuable thing after your passport and do not let it out of your sight.

How to decide who is on the list of authors and in what order
It is important to agree on this. Different people on the team can have very different reasons for thinking that they have earned a right to be listed as an author, even as first author. Did they do most of the literature search? Did they do most of the data analysis? Did they collect most (or all) of the data in the field? Did they identify all

the specimens? Be certain that everyone knows whether or not they will be authors and agree on the order of the authors for each paper.

A system used by Bristol University Zoological Department provides a good rule of thumb. Divide the workload into four parts: the original idea, the carrying out (fieldwork and logistics), the data analysis and writing up. If an individual has been involved in two or more of these parts they are on the list of authors. If they were involved in one part, place them in the acknowledgements section. The order in the list of authors is a fine mixture of reward and realpolitik.

If you have worked with host country counterparts, it is only fair to include them as authors. First authorship by a national of the host country may speed acceptance of a paper in a host country journal.

What's in a name?
Hopefully, your papers will be read for many years to come. So, even if you have a nickname now and are happy with it, you may not want to be known as "Biffa" Smith in 30 years' time. Best to use your legal name.

People may want to contact you – either for reprints or to discuss your results. This may not happen for several years after the paper is published. You should therefore try to include a contact address that will last and/or can be relied upon to forward things. If you give the name of an institution, (1) ask them first and (2) make sure that you keep them abreast of subsequent changes in address.

Another approach is to c/o the paper, with an academic supervisor's address and name – or bring them on to the author list and use them as the contact point. An email address is also a useful contact point, providing you keep it up, of course.

Some notes on style

Conventions
Each journal has its own unique publishing conventions. You are expected to follow them. They govern everything from the way in which references are to be cited in text – e.g. (Smith, 1978, 1979; Jones, 1990; Apfelbaum, Schmidt and Gaynor, 1991) or (Apfelbaum *et al.* 1991; Jones 1990; Smith 1979 1978) or any of many other variants – to the use of "%", "&", ":" and ";" and the way references appear in the bibliography, to how to do tables and figures. Journals regularly publish "notes to contributors", usually on the inside back cover (check the most recent one – they sometimes change).

Follow the instructions to the letter when doing the original manuscript. The instructions for authors don't always cover every eventuality (how to cite two different authors with the same name, for example). So, if in doubt check recent papers in the journal of your choice until you find an example of what you are worried about. Or email the editor. A journal will not re-format your text to fit in with its conventions. You must ensure that your text agrees with their way of doing things otherwise it is very likely to get sent back. Doing so can be the most niggardly and tedious of tasks, but it saves you doing it all over again later.

Title

Keep it short, simple and informative. Make it the kind of title that you could see in a literature search and have a good idea of the content of the paper and its relevance, without actually seeing it. If possible, include the country and location within it. If you focus on a particular species, include its Latin name in the title. "Random observations on South American mammals" (a real title!) is probably my favourite example of how not to do it.

American English

Remember that if you are going to submit to a journal published in the USA that you will have to do so in American English ("color", "behavior", "program", etc.). Doing this will irritate your spell-checker only slightly less than not doing it will irritate an American editor.

Written style

Remember that many of the people who read your paper may not have English as their first language. Try to write with clarity and precision. Avoid ambiguity (remember how annoyed you feel when you aren't sure exactly what an author means).

Descriptive prose has little place in a scientific paper. Save it for the popular pieces that you should also be writing.

Try to avoid needless jargon or tortured pseudo-scientific burbling ("the methodological parameters were operationalised"). You are, theoretically at least, supposed to be informing people rather than impressing them.

Use short sentences. Try very hard not, unless it cannot possibly be avoided, to write sentences that, in an effort to cover all possible points of view in the compass of the same sentence, tend to ramble on rather a lot, sometimes a great deal, indeed occasionally an inordinate amount, and may contain so many subclauses that, unless one keeps a very careful eye on what is happening, which naturally is not always possible, even under the best of circumstances, despite one's training for such eventualities, one does rather tend to lose the thread of what is actually being said.

Footnotes

These seem to be generally disliked by both readers and editors. They are best avoided wherever possible. Their only *raison d'être* seems to be if you have a very big new piece of information to put into a galley proof and it's so big that it won't easily fit into the existing text without tearing it apart. If notes have to be added they are best at the end of the manuscript.

Quotes
Never use a quote unless you have first authenticated it. When you do use it you should include the page number as part of the citation (e.g. "The sea is salty" [Smith, 1979, p. 43]). If you are quoting from a book, make sure that your reference list mentions which edition of the book was used.

If you quote from a foreign language and provide a translation, you should say so in the citation (e.g. Schmidt, 1970 – my translation). Check carefully the current rules concerning lengths of quotation and permissions.

Dividing the paper up into sections
Short papers are often exempt from such requirements and the whole text just appears as one uninterrupted piece. However, for longer papers, most journals ask that you divide your text up into: Introduction, Site description, Methods, Results, Discussion, Conclusions (the names may vary). It can take a lot of concentration to ensure that sentences do not bleed ideas from one section to the next. Such occurrences can spoil the clarity of your piece. Avoiding them is one of the reasons for stepping back now and again, and for letting others have a look at the text before you send it off.

Some notes on figures and references

Authenticating references
Always authenticate your references. Never quote a reference in your paper that you have taken out of another paper and not seen: first, because you have to be sure that the cited author actually said what you're being told he said and, second, to be certain that your source author actually cited the reference correctly in the first place. Errors creep in to reference lists all the time. Do not perpetrate or perpetuate them. Duff references make life harder for others and reflect badly on you. In some cases you can look really silly (because the citing of completely fictitious references does occur now and again).

If you can't avoid it, at least do the decent things and say "not seen" – either in the text or in the references (depending on the journal). Only do this for really obscure or ancient stuff, however. Otherwise referees and readers will, not unreasonably, ask "why?"

Abbreviating references
If the journal offers you the choice, try to cite the source title in full. It can be somewhat annoying to waste time trying to find something that has been over-abbreviated down to "Ag. Fd. Pn. Wd.", only to find that a non-standard abbreviation has been used (especially if its title is not in English, making educated guesses somewhat harder). There is no internationally agreed standard for abbreviating journals. Leland Alkire's *Periodical Title Abbreviations* can be of help. On the web, there is also "All that JAS", a site from the University of Iowa that lists a large number of standard

sources of journal abbreviations in many fields including biosciences.

Accents
You must add your own in both text and references. If your computer or printer won't do it for you, then doing it by hand is acceptable. Editors are unlikely to do this for you. It's an extra, fiddly thing but without it your work looks sloppy and it may offend people of the host country whose language you are mangling.

Figures
Some journals refer to maps as figures, others refer to them as maps. Check before submitting. Graphs and anything else that isn't a "table" is usually a "figure". Generally, it is a waste of time to do your own paste-up job and put your figure in your own text. Few editors accept this. Figures should be submitted at the same time as the text and the rest of the paper, mounted on card (size A4 generally), and be in their final form. Always make sure that, on the back of the figure, you clearly write the author's name, address and the full title of the paper. This minimises the chance of the artwork being lost or mislaid. Indicate in pencil in the margins of the text where you would like the figures included.

Maps
Generally a map is a helpful inclusion. But a badly drawn map can be more confusing than none at all. Keep them as simple and as uncluttered as possible. Don't put in anything that you've not referred to in the text, unless absolutely necessary. You cannot photocopy someone else's map and stick it in with your paper. It is rare for journals to take the trouble to touch-up or re-draw a badly done map. Computers now allow clear maps to be produced with relative ease.

Try to use standardised spellings for your place names. Many countries have officially recognised gazetteers that list place names. As an adjunct to your map(s), try to include in the text of your paper the code of any large-scale maps to the area that are available and include the grid reference of a nearby town or recognisable feature. Don't forget the scale and orientation and give your source for maps/aerial photos either in the text or at the foot of the map itself.

Tables and appendices

What to put in
Tables should not be used to pad your paper. They should contain numerical data or lists that it would be unwieldy to include in the main body of the text. Make sure that they are self-explanatory. Expand all abbreviations and numerise column headings in a key. Give the table a title that describes exactly what it is designed to show. Ideally, the data in the table should be comprehensible to someone should they read just the table alone.

A note on a dry-season collection of butterflies from South Western Guinea, West-Africa
By: **Madeleine L. Prangley*, Adrian A. Barnett* and Cheik Oumar Diallo****
*formal address
**formal address
email address [senior author only]
Address for correspondence: [normally that of the first author, repeat even if the same as above]
Contains: three pages of text (excluding this one), and two pages of references. Plus acknowledgements, two figures and one table.
Running header: Butterflies from Guinea [this will appear at the top of each page of the paper and helps maintain continuity during printing]
Key words: Butterfly, community ecology, deforestation, Guinea, Monsoon forest. [normally a maximum of 10 words and written in alphabetical order, not in any order of importance they might have within the paper]
Submitted as a note to the *Journal of African Butterfly Studies* on 11 March 2002.

Figure 39.1 *Format for manuscript cover page*

What to leave out
Leave out anything that is trivial and not germane to the text and to the points made in it. The material in a particular table should amplify a particular point not smother it. Try to avoid really long columns of figures; try to break the table up into subsections if this looks as if it is going to happen.

Appendices
Journals differ, but generally appendices are reserved for large papers and are places where lists of raw data are presented. Avoid them.

Presenting the manuscript
Whether you are submitting by post, email or both, the manuscript ("typescript", if you want to be really pedantic) and its entourage of references, tables, figures, etc. should have a covering page which follows the format given in Figure 39.1.

All of this should be accompanied by a short letter to the editor of the journal saying that you wish to submit this note/paper for publication. Some people like to put in a couple of paragraphs about the paper and what it is trying to achieve. To find out whether this is going to help or not, take advice from anyone who knows the editor of the journal to which you are submitting.

WHAT TO DO ONCE YOU'VE GOT THE PAPER PUBLISHED

Correction of misprints – asking for an erratum
Now and again things go wrong at the printers and what gets printed isn't what you

and the editor agreed on. Mistakes like this most often take the form of references missing from the bibliography or a missing figure or paragraph. In such circumstances you may contact the editor and politely request that an erratum be printed. Say exactly what you want to be printed. Editors normally leave space for this eventuality and your erratum should be out within the next couple of issues. Don't bother to request an erratum for minor errors of typography or spelling. If you didn't catch them in the galley proof (or they weren't there then), unless they really spoil the meaning, they are probably too trivial to bother about. You don't normally get reprints of your erratum, so photocopy it and make sure copies go out with your reprints. When citing your paper, accompany it (in parentheses) with the location of the erratum.

Ordering reprints
Most journals will give you a certain number of reprints free of charge. If you want more you may have to pay for them. What you do about this depends on the number of free ones that the publisher will give you and how confident you are that people will write to you asking for a reprint. Generally, you can only get extra copies printed off at the time the journal is going to press.

What to do about reprint requests
Convention says that you should supply reprints (free of charge) to anyone who asks for them for the benefit of wider scientific knowledge.

If you have an erratum for a paper, keep a list of who asked for reprints and send them a copy of the erratum when it comes out.

It is common to make a pdf file of your published paper and send that instead of a paper reprint. Check first; some people simply prefer a paper copy.

Information dispersal
Send out reprints (or photocopies) to all those in the acknowledgements and to anyone else whom you think might be interested (e.g. government departments, national and university libraries in the host country). This is common courtesy but also helps to get the information spread as widely as possible.

MISCELLANEOUS

Acknowledgements
Remember to thank the government officials and academics in the host country as well as the sponsors and people in the UK – but put host country people first, at the top of the list. Otherwise, you can create a bad impression and mess things up for later expeditions. You can't thank everybody in every paper, so rotate some of the less vital ones through the acknowledgements of various papers.

Data – singular or plural?

This is used both ways. Plural is correct, so it's "data are presented" rather than "data is presented". (The singular is "datum".) But check recent editions of the journal of choice for guidance.

Numbers

It is common practice to write out the numbers from one through to ten and use numerals thereafter. The only exception is when numbers start a sentence – then they are written out in full (e.g. "Twenty-nine palms ..."). Journals differ on whether they write 1000 or 1,000 or 10,000 or 10 000 – check the "Advice to contributors" in the journal.

Internet information

The internet is great, but it's not peer reviewed. It's best to be sure of any facts that you get off the net. Try to check them with an authority before slipping them into print. That's how errors get perpetuated.

Citing websites

Web addresses change frequently. Try citing just enough to get people to the home-page. They can move around from there. Also, as information changes all the time, you should give the date of your visit as part of the bibliographical citation. Note: not all journals are happy about websites as citations. You may have to use a "personal communication" citation.

Citing films

Sometimes a critical piece of information will come from a TV programme. If it is simply a fact, you should be able to track down the researcher and get their original source. But, if it is the visual event itself (the display of some rare bird, say), it is valid to cite the TV programme. How this is done depends on the journal and is rarely given in the guidelines. Generally you do it something like this: *The Lure of the Gerbil* (BBC/NHU, 2002). Then you ring up the production company and find the name of the producer, the date of first showing and their library code for the programme. That way any one who wants to should be able to get a copy at a later date.

Product information

If you specify a particular product in your methods, do include an address where the manufacturer can be contacted. Don't bother with the phone number, but you may cite a web page (either in the text or as part of the acknowledgements or in a small appendix – journals seem to vary on this, not really having sorted themselves out about it yet).

Things in parentheses

Make sure you know what (pers. comm.), (in litt.), (ibid.) and all those other odd abbreviations mean before you start using them yourself. For a personal communication, check with the person first that they are happy with you doing this. For an unpublished document, again try to ensure (if the document's not too ancient) that it is OK to use it.

FURTHER READING

Alkire, L.G. (ed.) (1989) *Periodical Title Abbreviations*. Volume 1: *By Abbreviation*. Volume 2: *Titles to Abbreviations*. Detroit: Gale Research Inc.

Barrass, R. (1986) *Scientists must Write*. London: Chapman and Hall.

Booth, V. (1993) *Communications in Science: Writing a scientific paper and speaking at scientific meetings*. Cambridge: Cambridge University Press.

Cooper, B.M. (1964) *Writing Technical Reports*. Harmondsworth: Pelican Books.

Council of Biology Editors, Style Manual Committee (1995) *CBE Style Manual: A guide for authors, editors and publishers in the biological sciences*. Cambridge: Cambridge University Press.

Davis, M. and Fry. G. (1996) *Scientific Papers and Presentations*. London: Academic Press.

Day, A. (1989) *How to Write and Publish a Scientific Paper*, 3rd edn. Cambridge: Cambridge University Press.

Day, A. (1993) *Scientific English: A guide for scientists and other professionals*. Phoenix: Oryx Press.

Godfrey, J.W. and Parr, G. (1960) *The Technical Writer*. London: Chapman and Hall.

Hans, E., Bliefert, C. and Russey, W.E. (1989) *The Art of Scientific Writing from Student Reports to Professional Publications in Chemistry and Related Fields*. Weinheim: BVHC Verlagsgesellschaft.

Huth, E.J. (1994) *Scientific Style and Format: The CBE manual for authors, editors and publishers*, 6th edn. Cambridge: Cambridge University Press.

Mitchell, J. (1974) *How to Write Reports*. London: Fontana/Collins.

O'Connor, M. (1991) *Writing Successfully in Science*. London: HarperCollins Academic.

Paradis, J.G. and Zimmerman, M.L. (1997). *The MIT Guide to Science and Engineering Communication*. Cambridge, MA: MIT Press.

Partridge, E. (1973) *Usage and Abusage: A guide to good English*. London: Penguin.

Pechenik, J. and Lamb, B.C. (1994) *How to Write about Biology*. London: HarperCollins.

Perelman, L.C., Paradis, J. and Barrett, E. (1997) *The Mayfield Handbook of Technical and Scientific Writing*. Mayfield Publishing Co.

Royal Society (1974) *General Notes on the Preparation of Scientific Papers*, revised edn. London: The Royal Society.

Shortland, M. and Gregory, J. (1991) *Communicating Science*. Harlow: Longman.

Zehr, J. (1993) *Creating Environmental Publications: A guide to writing and designing for interpreters and environmental educators*. Stevens Point, Wisconsin: Foundation Press, University of Wisconsin.

APPENDICES

1 A SAMPLE PLANNING CHECKLIST

Project aim
Main aim
Supporting objectives
Overall time frame
Size of team
Realistic budget/cost per person
The end products

Researching the opportunities
Library and website research
RGS–IBG Map Room and Expeditions
 Database
Past expedition reports and journals
Maps and guidebooks
Key references – bibliography
Ideas from past projects
University contacts (international)

Science programme
Pure research or applied
Methodology
Sampling framework
Linking the disciplines (earth, life and
 social sciences)
Surveying/monitoring
GI science and mapping
Habitat surveys
People-oriented research

Specialist equipment
Laboratory requirements
Training opportunities for young
 scientists
Publishing plan

Contact with your hosts
Embassy of High Commission in UK
UK diplomatic missions
Government departments
British Council offices
Local non-governmental organisations
 (NGOs)
Museum or herbarium
Research institutes
Local field centres
Protected areas (national park offices)
University and other networks
Links through schools

UK administration
Establishing a planning office
Communications
Reliable email facility
Website
Meeting rooms
Filing cabinets/storage
Documentation:

– passports with visas
– immunisation certificates
– permission and political clearance
– insurance documents
– international driving licences
– permits: collecting and climbing
– letters of support from host bodies
– customs clearance
– maps and aerial photographs
– bibliography

Team building
UK and host country members
Roles within the team
Choice of key disciplines/skills
Inclusive approach
Applications and selection
Early planning meetings
Joining guidelines
Costs and joining fee
Training plans for all
Delegation of responsibilities
Personal details
Medical checks
Next of kin
Communication:
– newsletters
– meetings

Field logistics
Accommodation
Transport
Catering
Equipment and stores (see below)
Fuel requirements
Water supplies
Communication
Navigation
Reconnaissance visit (see Appendix 2)

Budget and finance
Budget (initial drafts)
Appointing a treasurer
Open bank account
Managing the finances
Loan facilities
Bank branch or corresponding bank in
    host country
Travellers' cheques
Credit cards
Bank transfers
Letter of credit
Auditing the accounts

Fund-raising
Target sum
From your own organisation/university
Trusts and other grant-giving
    organisations
Appeal to commerce
Sponsorship opportunities
Fund-raising events
Working to raise funds

Public relations
Project brand
Key message to public
Image and branding
Project logo/crest
Brochure
Launch plans
Press release/conference
Media coverage

Travel

Air
Budget flights, concessionary fares
Advance booking
Excess baggage
Deadline for payment

451

Air freight
Implications of stop-overs

Land
Mode of transport
Own versus public
Vehicle preparation
Spare parts/accessories
Route maps
International carnets de passage (with
  bank guarantee)
Insurance cover for all drivers
Mechanic, maintenance training
Advance ferry bookings

Stores and equipment
Acquisition
Procurement
Storage
Testing
Qualities
Packing, containers
Value and insurance
Spares available in host country
Purchase in host country
Equipment lists:
  – secretariat
  – packing containers
  – scientific
  – surveying/mapping/GIS
  – navigation
  – communication
  – general base equipment
  – field and camp equipment
  – catering/kitchen
  – rations
  – replenishable stores
  – workshop stores
  – laboratory stores/chemicals
  – medical
  – specialist – mountaineering,
    diving, caving, kayaking, etc.

  – personal
  – transport
  – photographic and film
  – documents

Packing
Immediate storage
Packing/box sizes
Packing requirements in the field
Full contents list
Labelling/weighing
Value for consignment for insurance
  and freight
Final list for customs – export/import

Freight
Shipping dates
Available ships
Passage payment
Discounts, deposits
Guesstimate of freight – weight/bulk?
Container size
Documentation
Bills of loading
Customs clearance
Insurance during passage
Agents in the UK and in-country
Delivery to docks
Duplicate lists

Insurance
Medical and life insurance
Casualty/evacuation
Life
Personal injury
Third party indemnity
Equipment and transport
Cancellation of project
Personal belongings
Cash
Specimens

Health and safety
Recommended guidelines
Legal requirements
Risk assessment
Qualification and experience of team
Training needs
Standard operating procedures
Code of conduct
Next of kind details

Medical
Medical risk assessment
Medical officer
First-aid training
First aid and medical kits
In-country support
Casualty evacuation plans
Communication with insurance
    company
Medical questionnaire for members

Recording the project
Film or video
Still photographs
Copyright agreement
Tape recordings
Artist in residence
Daily log and diaries
Project recorder for the project
Central data (e.g. met. readings)

Sharing the results
The report (see below)
Project website
Scientific papers
Educational publications
Popular press/magazines
Tape recordings – CD/tape
Television/radio
Public lectures
Photographic exhibitions
Posters
Making electronic images available

Post-project administration
Settle bills and close the bank account
Cataloguing the photographs
Sponsor's reception
Lecture programme
Thank-you letters
Report to host country and sponsors
Insurance claims
Lecture programme
Maintaining links in host country

Final reports
Appointment of editor(s)
Author(s)
Length, binding and format
Photographs
Maps/drawings
Number, print run
Cost
Distribution list
Despatch/postage

2 A SAMPLE OF A RECONNAISSANCE CHECKLIST

A planning visit to the location of your expedition is a good investment and will help with your planning and risk assessment. During a visit collect as much detailed information as you can, even though you may not think it necessary at the time. When you are back in the UK, your colleagues will value the detail. Establishing good local contacts will make your planning much easier.

Research
Government ministries
British Embassy or High Commission
British Council
Research institutes
University and schools
Tourist agencies
Other non-government organisations
Aid organisations
Library and museum
UK companies

Administration
Research permits
Government departments
Local government
Immigration (visa)
Police
Customs and Excise
Health authorities (hospitals)
Survey and maps
Aerial photographs

Emergency assistance (casualty evacuation)
Financial advisers
Insurance (in-country)
Bank – transfer of funds and exchange rate
British companies
Couriers

Accommodation
Hotels/B&Bs
Rest houses
Tented facilities
Hostels/clubs
Friends' houses

Science programme
National research programmes
Links to academic community
Development agencies
Support bodies (e.g. Met Office)
Scientific supplies

Technical support
GIS back-up

Education programme
Government links
Local education schemes
Local schools
Lecturing plans

Media
Key newspapers
Other publishers
Television networks
Radio stations
Press agencies
Press releases

Transport
Aeroplane/helicopter
International, internal
Charter, military
Bus, taxi
Car hire
Boat
Porters
Mule/camels
Trekking guides

Workshops (nearby)
Electronics
Computer
Engineering
Welding
Garages
Carpentry
Marine
Photographic

Communications
Email and website links
Mobile phone coverage

Radio network and frequencies
Post office facilities
Telecommunications (buy telephone
  directory)
Local radio station
Pigeon, cleft stick

Supplies
Timber
Roofing
Pipes
Stationery
Chemicals (for lab)
Medical
Consumables, e.g. batteries, candles,
  etc.
Hardware

Rations
Fruit
Vegetables
Meat
Rice/pasta
Tinned food
Drinks (alcoholic)
Soft drinks

Water
Collection
Purification
Sterilisation
Storage

Fuel
Methylated spirits
Petrol/diesel
Kerosene (paraffin)
Calor gas/butane
Camping gaz
Coal/charcoal
Hexamine blocks

Medical, health, safety
National medical advisers
Inoculation requirements
Map of nearby hospitals
Medical back-up
Flying doctor service
Private ambulances
Nearest airport (for ambulance)
Local medical resources
Back-up in emergency
Decompression chamber

Field equipment
Tents/campsheets
Camp beds/hammocks
Mosquito netting
Cooking equipment
Rucksacks/kitbags
Local clothing/boots
Climbing equipment
Scientific supplies
Lamps (pressure and wick)
Photographic/film
Containers
Data loggers
Surveying companies

Main field base
Laboratory facilities
Water supply
Water pumps/showers
Power – generators
Solar power
Battery chargers
Refrigerators
Lighting
Computer area

Liaison
Home agent
Ex-patriate community
Local government and officials
Local chief/headperson

3 AN EXAMPLE OF A CRITICAL PATH PLAN: FOR A SCHOOL EXPEDITION TO ICELAND

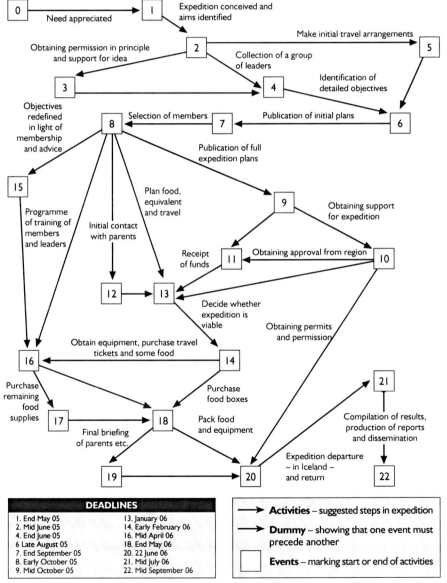

DEADLINES	
1. End May 05	13. January 06
2. Mid June 05	14. Early February 06
4. End June 05	16. Mid April 06
6 Late August 05	18. End May 06
7. End September 05	20. 22 June 06
8. Early October 05	21. Mid July 06
9. Mid October 05	22. Mid September 06

➤ **Activities** – suggested steps in expedition

➤ **Dummy** – showing that one event must precede another

☐ **Events** – marking start or end of activities

This critical path was originally designed and used by the East Lothian Schools North-West Iceland Expedition

4 A SAMPLE COUNTDOWN: PRIORITIES MONTH BY MONTH

This expedition probably has approximately ten members, will be in the field for 3 months and is freighting approximately 2 tons of stores and provisions.

PRE-EXPEDITION

July:
12 months before
expedition

Start collecting general information on: expedition area/country, flights, equipment/rations, insurance, politics, freighting/ shipping, transport/travel costs
Start research on specific items listed in Appendix 1
Obtain permission in principle from the main sponsor (school or university)
Register with EAC, to be on their mailing list

August:
11 months before
expedition

Prepare an office space/HQ to cope with the administration (telephone, typewriter/computer, files, etc.)
Design headed paper/crest
Possible reconnaissance visit (in Appendix 2)
Visit bank manager (open account)
Briefing weekend for key expedition personnel: medical officer, scientific directors, quartermaster, catering officer, mechanic, etc.

September:
10 months before
expedition

Approach grant-giving bodies for their application forms (see Appendix 9)
Produce a preliminary handout (or reconnaissance report) for members, sponsors and supporters
Start researching the correct procedure for obtaining permission for your expedition (visa requirements, collecting permits, scientific permissions)

First draft of the budget

October:
9 months before
expedition

Print headed notepaper (for examples, see page Appendix 6)
Start negotiating on flights, freighting, equipment, rations, insurance, etc.
Start finalising on team
Start approaching grant-giving bodies (deadlines usually October–March)
Book for EAC "Planning a Small Expedition Seminar" in November

November:
8 months before
expedition

Attend EAC seminars
Finalise team objectives of the expedition
Allocate responsibilities to members
Draft brochure for comment
Second draft budget
Start administration of personnel (personal details, etc.)
Check passports
Prepare timetable for inoculations
Book courses for any specialist training required

December:
7 months before
expedition

Final budget (more detailed)
Production of expedition brochure
Ration requirements finalised
First training weekend for whole expedition
Possible reconnaissance visit (see Appendix 2)
Equipment lists drafted and typed
Apply for visas (may be required earlier)

January:
6 months before
expedition

Sponsorship drive for support, equipment loan and funds
Official approach to host country for permission (some countries need earlier notification)
Notification to Foreign and Commonwealth Office and British Embassy in host country
Start collecting and packing equipment (containers?)

February:
5 months before
expedition

All equipment to have arrived for packing
Packing weekends (weighing, numbering, listing)
Prepare manifest (list) for insurance, freight and members
Send manifest with costings to insure equipment from date of packing until date of return to UK. Settle insurance premium before freight departure

March:	Send freight off minus deliver to port. Liaise with customs
4 months before	Arrange customs and storage in host country
expedition	Book flights (deposit down)
	Discuss baggage allowance
	Possible launch for sponsors and relations

April:	Tidy up expedition paperwork. Copy key files to take
3 months before	Training weekends (get boots worn-in, etc.)
expedition	Medical checks by expedition doctor
	Finalise financial arrangements in host country (open account)
	Final briefing reports for contacts in host country British
	Embassy and Foreign Office
	Review final situation and amend plans accordingly

May:	Contingency (e.g. chase permits if not already received)
2 months before	Press release and possible press conference
expedition	Contact editors about possible articles

June:	Final briefing of parents (circular)
1 month before	Newsletter to sponsors
expedition	Organise/brief home agent
	Establish communication lines
	Prepare "in-flight" dossiers for members

IN THE FIELD

July:	Advance party departs
month 1	Field accounts opened
in the field	Establish regular communication with UK home agent
	Main party departs
	Reception for supporters in host country

August:	Send newsletter/postcard to sponsors (take out ready-typed
month 2	address labels)
in the field	Send mid-expedition report to home agent for friends and
	relatives
	Take photographs and/or test products (this will take longer than
	you imagine, and many expeditions fail to do this well)

September: Thank-you reception for hosts – main party returns
month 3 in   Medical check for all members
the field   Rear party returns
                Thank-you letters to host country
                Field accounts closed

POST-EXPEDITION

October: Leader to prepare preliminary report
1 month after   Sponsors thanked and informed (photographs sent)
expedition   Approach local and national media for possible articles/interviews
                Press release about expedition achievements
                Review financial situation

November: Collect equipment from port (samples to scientific leaders)
2 months after   Members' weekend to collect data for report, look at expedition
expedition   photographs, repair and return equipment and discuss overall
                expedition plan
                Prepare final report outline (see Chapter 33)

December: Slide presentation of the expedition to the main sponsors and
3 months after   supporters and members
expedition

January: Collect chapters/scientific reports for final report
4 months after   Further photographs and articles (if promised) to sponsors
expedition

February: Prepare first draft of report. Ask graphic artist friends to work on
5 months after   layout and cover. Select photographs
expedition   Estimate costs

March: Final report (draft) circulated to contributors for comment
6 months after
expedition

April: Choose appropriate printing technique and prepare copy for
7 months   printers
after expedition   Final report to printers

May:	Circulate final report – including copies to the five copyright
8 months	libraries and RGS
after expedition	Let EAC know your forwarding address
	Close accounts

June:	Start planning next expedition
9 months	
after expedition	

5 A PROPOSED BUDGET FOR THE LOTOGIPI COMMUNITY FOREST PROJECT

The Lotogipi Community Forest Project is a small 10-week geographical survey looking at community use of forest resources in Central Africa. It has three local researchers and three visiting scientists from the UK. The group aims to carry out the project within a budget of £3000 per member, namely £18,000.

PROPOSED EXPENDITURE

PRE-EXPEDITION EXPENSES (£)

			Max.	Min.	Factor*
Administration	Office secretariat	Adjustable	650	–	A
	Meetings/workshops	Adjustable	300	–	A
	Email/telephone	Adjustable	400	200	A
	Printing	Adjustable	500	200	AB
	Publicity	Adjustable	200	100	A
	Launch	Adjustable	250	–	AB
Equipment	Field	Adjustable	600	200	BC
	Boats/engines	Adjustable	3000	600	BC
	Medical	Adjustable	600	60	BC
	Scientific	Fixed	600	600	BC
	Photographic	Adjustable	350	–	BC
	Radios	Adjustable	1500	–	BC
	Packing	Adjustable	500	–	BC
	Maps	Fixed	100	100	BC
Insurance		Adj./fixed	1650	800	

Training	Adjustable	300	200	
Freight	Fixed	2300	–	D
Flights (3)	Fixed	2100	1500	BE
Planning visit (1)	Fixed	1000	700	E

FIELD EXPENSES (£)

		Max.	Min.	Factor*
Living costs (70 days × 6 × £10 to £5/day)	Adjustable	4200	2100	C
Vehicle hire	Adjustable	2500	–	BC
Fuel allowances	Adjustable	1500	–	BC
Travel	Adjustable	1500	500	
Miscellaneous		1000	500	
Customs duty Fixed		1650	–	F

POST-EXPEDITION EXPENSES (£)

Administration	Adjustable	300	100	A
Report	Adjustable	800	–	AB
Photography	Adjustable	600	200	AB
Sponsors reception	Adjustable	400	200	A
Subtotal		£31,350	£8860	
Plus 10% contingency		£34,485	£9746	

*Factors: The six major factors affecting possible reduction in expenditure (your list will be longer!)

A: initial sponsorship by an organisation, school, college society or institution can reduce or eliminate these costs depending on their degree of involvement.

B: an appeal to commerce and industry for the donation of stores and field equipment.

C: loan of equipment from the government, other organisations and scientific bodies.

D: sponsorship from shipping or freighting companies.

E: ability to secure flights at a substantial discount (charter, student, APEX).

F: ability to gain permission to enter the country without having to pay customs duty on equipment.

EXPECTED EXPEDITION INCOME

BEFORE DEPARTURE (£)

		Max.	Min.
Personal contributions	Local researchers (£300 to £100 each)	900	300
	UK members (£800 to £500 each)	2400	1500
Grant from sponsoring organisation		4000	1000
Grants from scientific organisations		4000	500
Grants from trusts and other grant-giving organisations		3000	500
Donations from commercial companies		3000	1000
Income from film/photography		1000	–
Advertising offers (on naming boat, etc.)		500	
Minor fund-raising projects, e.g. raffle, etc.		2000	500

ON RETURN OF EXPEDITION (£)

	Max.	Min.
Sale of equipment	2000	800
Lecture programme at £50/lecture	1000	400
Articles in magazines	500	200
Totals	£24,300	£6700

OTHER ITEMS NOT INCLUDED IN THE ABOVE BUDGET

The above list is just a start. Presents, storage costs, immigration costs, port costs, reception and/or thank-you party in host country, lecturing, slide duplicating, mailing reports/postcards to sponsors from the field, extra costs of living in towns, hire of local guides and the rate of inflation could all be added. Look at budgets of previous projects to make sure that all the items are included in your budget. Share it with others to ask them if anything is missing.

WILL YOU ACHIEVE YOUR TARGET?

On this model, the maximum possible that can be raised is £24,300 and the absolute minimum cost is £9746. This makes the proposed final budget of £18,000 a realistic and attainable target for this particular project. Make sure that you do a similar calculation for your own project. Keep a working copy to hand at all times.

Good luck!

6 POSSIBLE HEADINGS TO INCLUDE IN AN EXPEDITION BROCHURE

The brochure will act as a prospectus and the impression it gives can make or break your expedition. It should look professional but does not need to be particularly lavish if care is taken with the design and layout. If you are mailing overseas, you might consider having a smaller, lighter version produced. The Expedition Advisory Centre keeps examples of brochures from previous expeditions for your reference and inspiration.

COVER

This is your branding. Make it work. Be innovative. Try to design a logo that is distinctively you, making it clear with whom you are collaborating here in the field.

INTRODUCTION

This is the most important part and should be no more than half a page. It must clearly state who you are, what you are going to do, when and why in a very few sentences. Most will read this and then flick through the remaining pages.

MAP OF THE AREA

Drawn neatly with an indication of scale to give an idea of field site location and operating area. This can be annotated and include photographs.

THE PROJECT

A more detailed description of your objectives, why they are exciting, important and unique, and how they may help local people, the world of science or society in general. If relevant, give details of how your project came about.

MEMBERSHIP

A breakdown of all the team (UK and local) taking part, their qualifications and ages, and their role on the project, highlighting previous expedition or research experience. Make the CVs short but interesting and not just a list of exam results. Include a recent photograph of the team or each individual if you can.

SAFETY, LOGISTICS AND TIME-TABLE

A brief guide to your operating procedures, logistic plans, travel plans and time-table. Your route should be marked on your map.

BUDGET

List main headings to give an idea of what money is needed and how money is to be spent. Itemise major expenditure and try to make it reflect the special needs of your project. Also show how you plan to raise funds and list any sources to date.

SPONSORSHIP OPPORTUNITIES

Those reading the brochure might say to themselves – "How can I help?" Make this clear, with details of what you offer individuals and corporate sponsors.

BIBLIOGRAPHY

List key references for the area to be visited and work to be done in order to illustrate that you have done your homework.

ACKNOWLEDGEMENTS

List those who have helped you to date, key scientific advisers, initial sponsors and the printer if he has printed the brochure free of charge.

Contact names and address, telephone, fax, email, website for future correspondence both in UK and in the field.

7 A SAMPLE PRESS RELEASE

This press release was issued to a number of publications before the start of the expedition, and subsequently generated coverage in 12 magazines and newspapers. Note the quote from the subject in paragraph five, which was written so that it could be easily be manipulated by the subeditor to read "... Glenn told [name of magazine]."

To   [name of editor]

From   [name of person sending the press release]

Date   12/10/01

Re   Press release, for immediate release [if sending in advance, write "... for release on (date)]

Disabled Explorer to Kayak off Coast of Antarctica

Disabled explorer Glenn Shaw is all set to fulfil a lifetime ambition – to kayak off the coast of Antarctica. By anyone's standards an ambitious project, the single detail that makes this venture all the more remarkable is that, when he is not in his kayak, Glenn is confined to a wheelchair (or "snowmobile", as Glenn calls his custom-built vehicle).

Glenn suffers from a medical condition known as "brittle bones". He knows that the slightest knock can result in a broken limb. A simple fall might kill him outright. But Glenn is an adventurer, and treats his physical handicap as simply one more difficulty to be overcome.

Glenn is drawing up his expedition plans with the assistance of former British Antarctic Survey Base Commander, Paul Rose. His support vessel is being operated by seasoned Antarctican Greg Mortimer, the first Australian to climb Everest and K2.

Glenn will be setting out for the frozen continent in December. He hopes to take to the (rather chilly) waters around Antarctica at Christmas. Glenn is also planning to camp on the Antarctic Peninsula during the course of the expedition. He will be posting regular updates at www.glennshaw.com before and during the voyage.

"I've wanted to kayak among the penguins for years," Glenn enthuses. "Now, thanks to a Winston Churchill Memorial Trust Fellowship, I have been able to turn my dream into reality."

Glenn's previous adventures have included a trek in the Himalaya, four attempts to cross Canada's Continental Divide using huskies, and a solo canoe voyage through British Columbia and Alberta. A Fellow of the Royal Geographical Society (with IBG), Glenn is a past recipient of a ski scholarship from the American National Sports Centre for the Disabled. He has also worked with Brunel University's Research and Development Team to test and develop kayaking equipment for fellow disabled explorers.

[End]

For more information or an interview, call [telephone number], or email: [email address]

For details of the Winston Churchill Memorial Trust: tel: +44 20 7584 9315, website: www.wcmt.org.uk

---

8 GRANT-GIVING ORGANISATIONS FOR EXPEDITIONS

Adrian Ashby-Smith Memorial Trust
c/o Mr Jan Ivan-Duke, 39 Sutherland Drive, Newcastle-under-Lyme, Staffordshire ST5 3NZ
Preference to those under 40 years of age who are taking part in their first expedition. There are three categories: member of scientific/exploratory expedition, handicapped member of expedition, member from an under-privileged background.

African Bird Club Awards
African Bird Club, c/o BirdLife International, Wellbrook Court, Girton Road, Cambridge CB3 0NA
Website: www.africanbirdclub.org
Supports small conservation projects in Africa, including the ABC Expedition Award, an annual award of £1000

Albert Reckitt Charitable Trust
Southwark Towers, 32 London Bridge Street, London SE1 9SY
Must be supported by a university. Grants not available to individuals or schools. Average grant: £750.

Andrew Croft Memorial Fund
c/o Mrs James Korner, The River House, 52 Strand-on-the-Green, London W4 3PD
Provides grants for young people under the age of 30 to participate in expeditions, particularly research expeditions to the Arctic and for the benefit of Arctic communities. Average grant: £500.

Andy Fanshawe Memorial Trust
Email: afmt_adm@hotmail.com
If you are under 26 and lack funding for a climbing expedition or other outdoor initiative, you could qualify for a grant from the Andy Fanshawe Memorial Trust. The Trust could also help fund your place on a training course provided that your primary reason for attending it is for personal development rather than professional qualifications. Maximum grant: £500. Deadline: 20 March.

Augustine Courtauld Trust
Red House, Colchester Road, Halstead, Essex CO9 2DZ. Website: www.augustinecourtauldtrust.org
Assists expeditions to the Arctic or Antarctic. Average grant: £800.

BP Conservation Programme Awards
c/o BirdLife International, Wellbrook Court, Girton Road, Cambridge CB3 0NA. Tel: +44 1223 277318,
fax: +44 1223 277200, email: bp-conservation-programme@birdlife.org.uk,
website: www.bp.com/conservation/
Open to teams (not individuals) from all over the world planning high priority conservation research projects of a global priority. Successful teams should have a majority of student participants, demonstrate local collaboration, and have clearance from the host government and local institutes. Grants: £3000–£10,000. Deadline: 31 October.

Bat Conservation International – Student Scholarship Program
Student Scholarship Program, PO Box 162603, Austin, TX 78716-2603, USA. Website: www.batcon.org
Supports research to help document bats' roosting and feeding habitat requirements, their ecological or economic roles or their conservation needs. Average grant: $US2500.

Bird Exploration Fund
c/o The Walter Rothschild Zoological Museum, Akeman Street, Tring, Hertfordshire HP23 6AP.
Email: r.prys-jones@nhm.ac.uk
For ornithological projects leading to the enhancement of the Natural History Museum bird collection; open to individuals and groups working in the UK and abroad. Grants: £100–500. Deadline: 1 April.

British Canoe Union
John Dudderidge House, Adbolton Lane, West Bridgford, Notts NG2 5AS. Website: www.bcu.org.uk
Grants for international canoeing expeditions only; normally first descents and/or very challenging river or sea trips in wilderness areas of the world; considerable canoeing experience and previous expeditions a prerequisite. An "Expeditions for Youth Fund" is also administered by the BCU, which provides small grants, typically £50, to assist young people to undertake canoeing expeditions. Applications for Youth Grants may be made at any time. Grants: £200–500.

British Cave Research Association Research Fund
Bill Tolfree, BCRA Research Fund, 6 Ledsgrove, Ipplepen, Newton Abbott, Devon, TQ12 5QY.
Email: research-fund@bcra.org.uk, website: www.bcra.org.uk
Grants are made for specific scientific projects in any field of speleology. Grants: £50–500.

British Ecological Society
Education and Careers Committee, Blades Court, Deodar Road, Putney, London SW15 2NU.
Tel: +44 20 8871 9797, email: info@britishecologicalsociety.org, website: www.britishecologicalsociety.org
Groups (minimum of three) whose projects are intended to widen the ecological experience of participants overseas rather than just produce publishable data. For sixth forms only an expedition within the British Isles will be considered. Deadline 31 December. Average grant: up to £2000.

British Mountaineering Council
177–179 Burton Road, Manchester M20 2BB. Tel: +44 161 445 4747, email: office@thebmc.co.uk,
website: www.thebmc.co.uk
Screened at the same time as the Mount Everest Foundation. New Peaks, new routes, British firsts or innovative-style ascents in the greater mountain ranges. Average grant: £1500.

EXPEDITION HANDBOOK

British Sub-Aqua Jubilee Trust
Telford's Quay, South Pier Road, Ellesmere Port, South Wirral CH65 4FL. Tel: +44 151 350 6200,
website: www.bsac.com/services/jubilee/jtrust.htm
Awards made to individuals for diving-related projects or expeditions, so long as the work done involves aqua-lung diving. Both amateurs and professionals may apply because applications are judged on individual merit. Grants may be used for scientific or non-scientific projects so long as the work done involves diving on the aqua-lung. Grants: £100–1000 (exceptionally £1000).

The Captain Scott Society
c/o United Services Mess, Wharton Street, Cardiff CF1 2AG
Spirit of Adventure Award for individual or expedition displaying similar "spirit of adventure" so nobly demonstrated by Captain Scott and the British Antarctic Expedition of 1910. Average grant: £1000.
Sir Vivian Fuchs Young Adventure Award for young persons aged between 11 and 19, displaying the above criteria. Average grant: £250.

Carnegie Trust for the Universities of Scotland
Cameron House, Abbey Park Place, Dunfermline, Fife KY12 7PZ. Tel: +44 1383 622148,
website: www.carnegie-trust.org
For Scottish university undergraduates on supervised field research expeditions, approved and supported by the university and accompanied throughout by a member of staff. Grants are made only to expeditions; no individual applications will be considered. Average grant: £1800 (max. £2000).

Charles A. and Anne Morrow Lindbergh Foundation
2150 Third Avenue North, Suite 310, Anoka, MN 55303-2200, USA. Website: www.lindberghfoundation.org
Grants to individuals whose proposed project represents a significant contribution towards the balance of technological progress and preservation of the natural/human environment. Grants: up to $US10,580.
Deadline: 11 June.

Commonwealth Youth Exchange Council (CYEC)
7 Lion Yard, Tremadoc Road, Clapham, London SW4 7NQ. Tel: +44 20 7498 6151, fax: +44 20 7622 4365,
email: mail@cyec.demon.co.uk, website: www.cyec.org
CYEC supports visits only by groups of young people aged 16–25 going to a Commonwealth country. Trips should involve joint activity and meaningful contact with young people of the overseas country, and hosting a group on a reciprocal visit to Britain, helping to educate the wider community about the partner country and the Commonwealth. Grants: up to 35 per cent of the international travel costs.

Dudley Stamp Memorial Trust
c/o The Royal Society, 6 Carlton House Terrace, London SW1Y 5AG. Website: www.royalsoc.ac.uk
Grants are available to young geographers to assist them in research or study travel leading to the advancement of geography and to international cooperation in the study of the subject. Awards are given to applicants under the age of 30. Grants are not given to assist expeditions. Average grant: £4500 awarded among a number of projects.

Earth and Space Foundation
Email: Foundation@earthandspace.org, website: www.earthandspace.org
Grants to expeditions requiring the use of space-derived or -dependent technology, e.g. using remote sensing data, satellite communications, and expeditions undertaking fieldwork in the interests of space exploration, and space-related activities. Average grant: £250.

472

Edinburgh Trust No. 2
The Duke of Edinburgh's Office, Buckingham Palace, London SW1A 1AA
All expeditions (but must have the backing of a recognised society). Average grant: £3000.

Edward Wilson Fund
c/o Scott Polar Research Institute, Lensfield Road, Cambridge CB2 1ER
Expeditions to polar regions. Grants: £200–600.

Eric Hosking Trust
Pages Green House, Wetheringsett, Stowmarket, Suffolk IP14 5QA. Tel: +44 1728 861113
Bursaries for ornithological research through writing, photography or painting. Average grant: up to £500. Deadline: September.

Explorers Club Exploration Fund
46 East 70th Street, New York, NY 10021, USA. Website: www.explorers.org
For scientific field research and exploration. Grants: not more than $US1200. Deadline: 31 January.

Explorers Club Youth Activity Fund
46 East 70th Street, New York, NY 10021, USA. Tel: +1 212 628 8383, fax: +1 212 288 4449,
email: youth@explorers.org, website: www.explorers.org
For high school and college undergraduate students to enable them to participate in field research in the natural sciences under the supervision of a qualified scientist. Average grant: $US1000. Deadline 31 January.

Fauna and Flora International (100 per cent fund)
Great Eastern House, Tenison Road, Cambridge CB1 2DT. Tel: +44 1223 571000,
website: www.fauna-flora.org
Projects must directly help endangered species of flora or fauna. Not suitable for undergraduate projects. Average grant: £2000 (max. £5000).

Frederick Soddy Trust/Geographical Association Expedition Grants
3 Woodgate Meadow, Plumpton Green, Lewes, East Sussex BN7 3BD
For the study of a human community and life in a particular area. Grants to groups only. School and similar expeditions particularly welcomed. Grants: £300–500.

French Huguenot Church of London Charitable Trust
The Clerk to the Trustees, 1 Dean Farrar Street, London SW1H 0DY
For those under 25 years of age for individual projects at home or abroad, preferably in connection with France. Grants will not be given for the sole purpose of learning a foreign language. Preference is for those who are or whose parents are members of the French Protestant Church of London, and to people of French Protestant descent. Maximum grant: £300. Deadline: 28 March.

Fuchs Foundation
c/o British Antarctic Survey, High Cross, Madingley Road, Cambridge CB3 0ET
To help young people who would otherwise be precluded for reasons of family background or financial status to undertake organised adventurous outdoor activity. Age normally 14–18. Individual applications only. Applications from institutes of higher education expeditions are not considered. Average grant: £500.

Ghar Parau Foundation
Secretary: David Judson, Hurst Barn, Castlemorton, Malvern, Worcs WR13 6LS
Email: d.judson@bcra.org.uk, website: www.bcra.org.uk
For original exploration, photography and survey of caves; scientific/speleological studies in caves, cave areas or associated features, preferably in little-known, little-studied or remote areas. Evidence of experience, ability and research required. Grants: £100–5000 dependent on application and competition.

Gilchrist Educational Trust
Mary Trevelyan Hall, 10 York Terrace East, London NW1 4PT
For British expeditions, involving three or more members, undertaking scientific fieldwork. Any necessary permissions must have been obtained. Involvement of local counterparts is an advantage. Average grant: £900.

Gino Watkins Memorial Fund
c/o Scott Polar Research Institute, Lensfield Road, Cambridge CB2 1ER.
Website: www.spri.cam.uk/about/funding/ginowatkins/
For expeditions to polar regions only. Grants: up to £1000.

Gordon Foundation
PO Box 214, Cobham, Surrey KT11 2WG. Email: Gordon.Foundation@btinternet.com
To support under-30s in performing arts, particularly music, drama or design and/or to allow them to engage in educational travel that involves physical challenge or endeavour.

IUCN Small Grants for Wetlands Programme (SWP)
Netherlands committee for IUCN, Small Grants for Wetlands Programme, Plantage, Middenlaan 2B, 1018DD Amsterdam, The Netherlands. Email: henri.roggeri@nciucn.nl, website: www.wetlands.nl
Small grants for local non-governmental organisations (NGOs) and community-based organisations implementing projects to promote the conservation of wetlands. Applicants must be from priority countries and follow specific criteria.

Jim Bishop Memorial Trust
c/o Young Explorers' Trust, 1 Kensington Gore, London SW7 2AR. Website: www.theyet.org
Open to individuals under 19 years taking part in adventurous activities at home or abroad. Average grant: £50. Deadline: 1 February.

John Jarrold Trust
B. Thompson, Messrs Jarrold & Sons, Whitefriars, Norwich NR3 1SH. Tel: +44 1603 660211
The Trust supports a wide range of organisations including churches, medical, arts, environment/ conservation, welfare and overseas aid. It prefers to support specific projects, rather than contribute to general funding. Educational purposes that should be supported by the state will not be helped by the trust. Grants: about a third of grants are for £1000 or more. Smaller grants are generally £100–500.

APPENDIX 8: GRANT-GIVING ORGANISATIONS FOR EXPEDITIONS

L.S.B. Leakey Trust
Dr Peter Andrews, Department of Palaeontology, The Natural History Museum, Cromwell Road, London SW7 5DB. Email: palaeosecretary@nhm.ac.uk
To support studies relating to human evolution. Priority will be given to research into environments, archaeology and human palaeontology of the Miocene, Pliocene and Pleistocene, into the behaviour of the Great Apes and other Old World primate species, and into the ecology and adaptations of living hunter–gatherer peoples. Grants: not exceeding £300. Deadline: 15 April.

The Merlin Trust
The Dower House, Boughton House, Kettering NN14 1BJ. Website: www.merlin-trust.org.uk
Grants for young (preferably between the ages of 18 and 35) horticulturists, to extend their knowledge of plants (in the wild or in gardens). Suitable projects may include visiting gardens in different parts of this country or abroad, or travelling to see wild plants in their native habitats anywhere in the world. Grants: up to £750.

Mount Everest Foundation
W.H. Ruthven, Gowrie, Cardwell Close, Warton, Preston PR4 1SH. Email: bill.ruthven@ukgateway.net, website: www.mef.org.uk
For British and New Zealand expeditions proposing mountaineering exploration or research in high mountain regions. Also awards Alison Chadwick Memorial Grant "to further British and Polish women's mountaineering in the greater ranges". Grants: £200–1700.

Mountaineering Council of Scotland
The Old Granary, West Mill Street, Perth PH1 5QP. Tel: +44 738 638227, fax: +44 1738 442095, website: www.mountaineering-scotland.org.uk
Expeditions whose members have a strong connection with the Scottish mountains and Scottish mountaineering and an objective of excellence and adventure; in greater ranges, pure rock or ice in less remote areas that are worthy of international recognition. Membership of the Council required. Grants: £100–1000. Maximum budget of £2000.

National Geographic Society
1145 17th Street NW, Washington DC 20036-4688, USA. Website: www.nationalgeographic.com
Committee for Research and Exploration
For post-doctoral projects undertaking scientific field research and exploration. All proposed projects must have both a geographical dimension and relevance to other scientific fields and be of broad scientific interest. The Committee is currently emphasising multidisciplinary projects that address environmental issues, e.g. loss of biodiversity and habitat, effects of human population pressures. Grants: $US15,000–20,000.

Expeditions Council
For exploration, adventure, and related technologies that provide new information about areas either largely or completely unknown, to cover field costs. The programme is editorially driven and projects must have the potential for a compelling written and visual record. New Explorers' grants are also awarded each year to talented and emerging explorers who offer future potential. Grants: $US5000–35,000.

Conservation Trust
Grants awarded to projects that contribute significantly to the preservation and sustainable use of the Earth's biological, cultural and historical resources. Applicants are not expected to have advanced degrees (PhD or equivalent); however, they must provide a record of prior research or conservation action. Grants: $US15,000–20,000.

Nick Estcourt Award
c/o The Secretary, 24 Grange Road, Bowdon, Altricham, Cheshire WA14 3EE
For expeditions attempting an objective of mountaineering significance. This might be a previously unclimbed face, ridge or summit, or a repeat of an existing route in more challenging style or conditions. Applications are considered solely in terms of their mountaineering merit; other objectives are not taken into account. Average grant: £1000.

Paul Vander-Molen Foundation
Michael Coyne, 92 Belgrave House, Wanstead, London E11 3QP
To provide opportunities for people with disabilities to enjoy and participate in adventurous activities. Open to individuals, schools, clubs, expeditions, etc. Grants: £200–1000.

People's Trust for Endangered Species
Unit 15, Cloisters House, Cloisters Business Centre, 8 Battersea Park Road, London SW8 4BG
Email: enquiries@ptes.org, website: www.ptes.org
One of the main roles of this charity is to provide financial support and encouragement for education projects in the field of conservation including those undertaken in the summer months by students in higher education. The project must have as one of its principal aims the conservation of an individual endangered species and must be organised by the students themselves.

Percy Sladen Memorial Trust
c/o The Linnean Society, Burlington House, Piccadilly, London W1V 0LQ
For field research in natural history outside the UK. Monies are given for specific research topics and not to undergraduate expeditions, for completion of degrees or visits to institutions. Average grant: £400.

Polartec Challenge
Email: brownr@maldenmills.com, website: www.polartec.com
The Polartec Challenge is an international grant programme designed to encourage the spirit and practice of outdoor adventure. Applications are evaluated on the basis of their credibility, originality, responsibility and ability to stand as a role model to outdoor enthusiasts worldwide. Grants: $US2000–8000.

Rainforest Alliance – Kleinhans Fellowship
Website: www.ra.org/programs/
For research into the development of new markets for non-timber forest products in Latin America or the expansion of existing markets. The successful applicant will have a Masters degree in forestry, ecology, environmental science or an appropriate related field. Doctoral candidates or post-doctoral researchers are preferred. Grant: $US15,000.

Reg Gilbert International Youth Friendship Trust
The Appeals' Secretary, Rathlyn, Blatchbridge, Frome, Somerset BA11 5EE. Email: GIFT@care4free.net, website: www.GIFT.care4free.net
Applicants must be aged 14–25 and have an outstanding project involving international friendship mainly through homestays. Successful applicants are those deeply involved in the normal, everyday life of the host family/community, and largely out of contact with their group except in an emergency. Grants: £120–300.

Reserve Forces Ulysses Trust
Directorate Reserve Forces & Cadets, Applications Secretary, Room 711A, MOD St Giles Court, St Giles High Street, Northumberland Avenue, London WC2H 8LD
Units of the Reserve Forces, university officer training corps (OTC) and cadets of all three services are eligible. The Trustees wish to encourage and support unit expeditions worldwide. Average grant: £900.

RGS–IBG/British Airways Travel Bursaries
Grants Officer, Royal Geographical Society (with the Institute of British Geographers), 1 Kensington Gore, London SW7 2AR. Email: grants@rgs.org
Bursaries for researchers who wish to undertake tourism-related fieldwork or research outside the UK. For individuals at postgraduate level or an established researcher under 35 years of age. Applicants must be registered with a UK higher education institute. One month minimum fieldwork period. Average grant: a free return flight (British Airways flights only), under certain conditions.

RGS–IBG Expedition Research Grants
Expeditions with scientific objectives related to geography and exploration normally outside Europe. Undergraduate level and above (not mountaineering, sporting or university departmental expeditions). Host country participation where possible. Grants: £750–3000.

RGS–IBG Henrietta Hutton Research Grants
Grants Officer, Royal Geographical Society (with IBG), 1 Kensington Gore, London SW7 2AR. Tel: +44 20 7591 3073, fax: +44 20 7591 3031, email: grants@rgs.org, website: www.rgs.org
Two grants for female students under 25 years of age who intend to undertake field research overseas as an individual or as part of a multidisciplinary team. Field research must be of more than 4 weeks' duration but does not necessarily have to be connected to the student's academic studies. Average grant: £500.

RGS–IBG Hong Kong Research Grant
Grants Officer, Royal Geographical Society (with IBG), 1 Kensington Gore, London SW7 2AR. Tel: +44 20 7591 3073, fax: +44 20 7591 3031, email: grants@rgs.org, website: www.rgs.org
Available for postgraduate students of any nationality who intend to undertake geographical research in the Greater China region (People's Republic of China, Taiwan, Macau SAR and Hong Kong SAR). The research must be of relevance to their overall studies. The proposed research must be of more than 4 weeks' duration and should be of an applied nature. Average grant: £2500.

RGS–IBG Gilchrist Fieldwork Award
Grants Officer, Royal Geographical Society (with The Institute of British Geographers), 1 Kensington Gore, London SW7 2AR. Tel: +44 20 7591 3073, fax: +44 20 7591 3031, email: grants@rgs.org, website: www.rgs.org/grants
Overseas research by small teams of senior university academics or researchers, most of British nationality, for a field season of at least 6 weeks. Average grant: £12,000 (even numbered years only).

RGS–IBG Individual Travel Awards
Grants Officer, Royal Geographical Society (with IBG), 1 Kensington Gore, London SW7 2AR. Tel: +44 20 7591 3073, fax: +44 20 7591 3031, email: grants@rgs.org, website: www.rgs.org
Two awards: one for geographical research in the field, likely to last at least 6 months (Violet Cressey-Marcks Fisher Travelling Fellowship); one for overseas travel in connection with biological study, teaching or research (Dax Copp Fellowship). Grants: £350–500.

RGS–IBG John Radford Award for Geographical Photography
Grants Officer, Royal Geographical Society (with IBG), 1 Kensington Gore, London SW7 2AR.
Tel: +44 20 7591 3073, fax: +44 20 7591 3031, email: grants@rgs.org, website: www.rgs.org
This is an award for young photographers wishing to undertake a geographically oriented photographic assignment. The award is open to nationals of all member states of the European Union but applicants must be aged between 18 and 30. The project focus should not be well documented but highlight a subject that might otherwise go unrecorded in the mainstream broadcasting/press. Average grant: £700.

RGS–IBG Journey of a Lifetime Award
Grants Officer, Royal Geographical Society (with The Institute of British Geographers), 1 Kensington Gore, London SW7 2AR. Tel: +44 20 7591 3073, fax: +44 20 7591 3031, email: grants@rgs.org, website: www.rgs.org
A bursary is on offer to someone who is undertaking a journey that will inspire an interest in peoples and places and who has the ability to communicate his or her experiences through the medium of radio broadcasting. The winner will receive training in sound-recording techniques from the BBC. A resulting programme or series will be produced for BBC Radio 4. Average grant: £4000.

RGS–IBG Monica Cole Research Grants
Grants Officer, Royal Geographical Society (with IBG), 1 Kensington Gore, London SW7 2AR.
Tel: +44 20 7591 3073, fax: +44 20 7591 3031, email: grants@rgs.org, website: www.rgs.org
This grant is made once every 3 years and is available to a female physical geographer undertaking original fieldwork overseas. There is no age qualification and both undergraduates and postgraduates can apply. Applicants should be registered at a UK institute of higher education. Average grant: £1000.

RGS–IBG Neville Shulman Challenge Award
Grants Officer, Royal Geographical Society (with IBG), 1 Kensington Gore, London SW7 2AR.
Tel: +44 20 7591 3073, fax: +44 20 7591 3031, email: grants@rgs.org, website: www.rgs.org
To further the understanding and exploration of the planet while promoting personal development through the intellectual or physical challenges involved in undertaking research and/or expeditions. Grants: up to £5000.

RGS–IBG Ralph Brown Expedition Award
Grants Officer, Royal Geographical Society (with The Institute of British Geographers), 1 Kensington Gore, London SW7 2AR. Tel: +44 20 7591 3073, fax: +44 20 7591 3031, email: grants@rgs.org, website: www.rgs.org
A major annual award for the leader of a research expedition associated with the study of inland or coastal wetlands, rivers or the shallow (< 200 m) marine environment. Applicants must be aged over 25 and a member of the RGS–IBG. Close involvement with host country institutions essential. Average grant: £15,000.

RGS–IBG Slawson Awards
Grants Officer, Royal Geographical Society (with The Institute of British Geographers), 1 Kensington Gore, London SW7 2AR. Tel: +44 20 7591 3073, fax: +44 20 7591 3031, email: grants@rgs.org, website: www.rgs.org
Grants for PhD students undertaking geographical field research involving key development issues with strong social values. Applicants must be registered with a UK institute of higher education and must hold Fellowship of the Society. Grants: up to £3000.

APPENDIX 8: GRANT-GIVING ORGANISATIONS FOR EXPEDITIONS

RGS-IBG Small Research Grants Scheme

Grants Co-ordinator, Royal Geographical Society (with IBG), 1 Kensington Gore, London SW7 2AR.
Tel: +44 20 7591 3073, fax: +44 20 7591 3031, email: grants@rgs.org, website: www.rgs.org
This scheme provides research grants for Fellows of the RGS–IBG. Some preference will be given to grants that are intended to be pump-priming preparatory work for larger grant applications. Applications are particularly welcomed from those in the early stages of their research careers. A PhD is normally expected. Individual grants: up to a maximum of £3000.

Rolex Awards for Enterprise

PO Box 1311, 1211 Geneva 26, Switzerland. Website: www.rolexawards.com
The Rolex Awards for Enterprise aim to encourage a spirit of enterprise in individuals around the world by supporting outstanding efforts in areas that advance human knowledge and well-being. Applications are invited in the areas of science, technology, exploration, environment and culture. Projects must expand knowledge of our world, improve the quality of life on the planet or contribute to the betterment of humankind. In judging applications, the Selection Committee determines whether the candidates show exceptional spirit of enterprise. Projects are also judged on the basis of their feasibility, originality and potential impact on the world and on society. Winners receive $US100,000 and a gold Rolex chronometer.

Royal Anthropological Institute Emslie Horniman Fund for Anthropological Research

50 Fitzroy Street, London W1P 5HS. Website: www.therai.org.uk
Anthropological fieldwork outside the UK by individual university graduates below doctoral level. Citizens of the UK, the Commonwealth and Ireland only eligible. Expeditions as such are *not* considered. Grants: £1000–7000.

Royal Anthropological Institute Ruggles-Gates Fund for Biological Anthropology

50 Fitzroy Street, London W1P 5HS. Website: www.therai.org.uk
Research in biological anthropology by individuals (only within the fields of human population biology, human genetics, human ethnology, palaeoanthropology or evolutionary anthropology). No restrictions on candidate's nationality. Expeditions as such are *not* considered. Grants: up to £600 per annum. Deadline: 31 March.

Royal Archaeological Institute

c/o Society of Antiquaries, Burlington House, Piccadilly, London W1V OHS.
Website: www.britac.ac.uk//cba/rai/
Annual grants awarded for archaeological and historical research. Applications considered for archaeological fieldwork, survey and aspects of excavation and post-excavation research, architectural recording and analysis, artefact and art historical research. Grants: between £200 and £1000.

Royal Geographical Society (with the Institute of British Geographers) Grants

Grants Officer, Royal Geographical Society (with The Institute of British Geographers), 1 Kensington Gore, London SW7 2AR. Tel: +44 20 7591 3073, fax: +44 20 7591 3031, email: grants@rgs.org, website: www.rgs.org/grants

Royal Institution of Chartered Surveyors Education Trust

12 Great George Street, London SW1P 3AD. Tel: +44 20 7334 3873, fax: +44 20 7334 3795, website: www.ricseducationtrust.org; www.rics-foundation.org/education/applicationform.pdf
Grants for field surveying and mapping and its application to environmental research. Grants: £500–7500.

Royal Scottish Geographical Society

Graham Hills Building, 40 George Street, Glasgow G1 1QE. Tel: +44 141 552 3330, fax: +44 141 552 3331, website: www.geo.ed.ac.uk/~rsgs

RSGS Expedition Grants

For qualification or training in geography of at least one member. Expedition must have Scottish base or Scottish membership. Objectives of expedition must be inherently geographical. Recipients are required to submit a report of findings and financial statement. Average grant: £500.

RSGS Travel and Small Research Grants

The RSGS Travel Grants scheme is designed to provide individual members of the RSGS in higher education institutes with grants up to a maximum of £500 in support of geographical research. Awards are specifically given to help with the cost of travel and subsistence associated with attendance at conferences and symposia where the applicant intends to present a research paper.
The RSGS Small Research Grants scheme is designed to provide individual members of the RSGS in higher education institutions with grants not exceeding £1000 in support of geographical research. The award is specifically given to help with the cost of data collection and fieldwork. Average grant: £500.

Royal Society Research Grants Scheme

6 Carlton House Terrace, London SW1Y 5AG. Website: www.royalsoc.ac.uk/funding
For research in any scientific or technological discipline within the natural sciences by academic workers in UK institutions of higher education. No stipends payable. Grants: up to £10,000.

Rufford Small Grants for Nature Conservation

Website: www.rufford.org
For individuals and groups running small conservation projects of approximately one year's duration. Applications from non-first world (less developed) areas are strongly encouraged. It is hoped that many recipients of Rufford Small Grants will progress in their field and go on to apply for a main Whitley Award. Average grant: £5000.

Shell Personal Development Awards

Website: www.shell.com/careers
For first or second year undergraduates who want to stretch themselves and reach new goals, e.g. by contributing to community projects, learning a new language or skill, planning an expedition or developing an existing skill, e.g. in sport, performance arts or music. Open to students studying in the UK or the Netherlands. Average grant: £500.

Shipton/Tilman Grant

W.L. Gore & Associates, Inc., 105 Vieve's Way, Elkton, MD 21922, USA. Website: www.gore-tex.com
For endeavours that demonstrate the exploration philosophy of Shipton and Tilman – small, lightweight and innovative.

Society for Underwater Technology

Paula Clatworthy, Administrative Secretary, Society for Underwater Technology, 80 Coleman Street, London EC2R 5BJ. Tel: +44 20 7382 2601, fax: +44 20 7382 2684, email: paula@sutadmin.demon.co.uk, website: www.sut.org.uk
The society, through its Educational Support Fund, is making efforts to foster the interest of suitable graduates in marine science and technology, with an active campaign to attract the highest calibre students into the scheme. The SUT sponsors UK and overseas students at undergraduate and MSC level. Grants: £2000–5500 for an undergraduate, £4000–8000 for a postgraduate.

Sports Council for Wales

Sophia Gardens, Cardiff CF11 9SW. Tel: +44 29 2030 0500, fax: +44 29 2030 0600
The Sports Council for Wales supports the principle of grant aid for overseas expeditions, which will enhance the development of a sport. Consideration will be given to Welsh-based expeditions where the majority of members qualify by birth, parentage or residence in Wales for at least 12 months in the past 2 years. Grants: between £500 and £1500.

Trans-Antarctic Association

c/o Scott Polar Research Institute, Lensfield Road, Cambridge CB2 1ER. Website: www.transantarctic.org.uk
For fieldwork in Antarctica and associated research by nationals of the UK, South Africa, Australia and New Zealand. Note: not for Arctic work or for work by other nationals. Grants: £100–1500.

Unesco Man and the Biosphere Programme, Young Scientists Research Grant Scheme

1 rue Miollis, 75732 Paris, Cedex 15, France. Tel: +33 145 68 10 00, website: www.unesco.org/mab
To facilitate research work of young scientists in Unesco Man and the Biosphere Programme field projects, international comparative studies and biosphere reserves. Applicants should not be older than 40 years. Requests must be supported by the MAB committees in the researcher's country of origin and in the host country. Grants: up to $US5000.

Whitley Laing Foundation

50 Queensdale Road, London W11 4SA. Tel: +44 20 7602 3443, fax: +44 20 7603 3935, email: info@whitleyaward.org, website: www.whitleyaward.org
An annual award scheme for field projects that will make a pragmatic, substantial and lasting contribution to nature conservation in developing countries. Awards to cover use of renewable energy, technology, human rights issues and general conservation work. Applicants must work with the host country. Grants: £5000–50,000.

Wilderness Award

Inglewood, New Road, High Littleton, Somerset BS39 6JH. Website: www.wildernesslectures.com
For an individual undertaking an unusual and exciting project in a wilderness area. He or she will give a lecture in the following year's Wilderness Lectures Series. This is an explorer's award and consequently semi-commercial and/or trips with charitable objectives are not supported: please read the guidelines on the website before applying. Average grant: £500.

Wingate Scholarships

The Harold Hyam Wingate Foundation, 2nd Floor, 20–22 Stukeley Street, London WC2B 5LR.
Tel: +44 20 7438 9513, fax: +44 20 7242 3568, email: clark@wingate.org.uk, website: www.wingate.org.uk
Scholarships are awarded to individuals of great potential or proven excellence who need financial support to undertake creative or original work of intellectual, scientific, artistic, social or environmental value, or to outstandingly talented musicians for advanced training. They are designed to help with the costs of a specific project that may last for up to 3 years. Wingate Scholarships do not fund electives or adventure-type trips, or trips in connection with taught courses. Average grant: £6500.

Winston Churchill Memorial Trust

15 Queen's Gate Terrace, London SW7 5PR. Tel: +44 20 7584 9315, website: www.wcmt.org.uk
Various categories each year including the Mike Jones Award for canoeing, and often one on exploration and adventure. No educational or professional qualifications are needed. British citizens only. Average grant: £3000+. Deadline: 31 October.

World Pheasant Association

PO Box 5, Lower Basildon, Reading RG8 9PF. Tel: +44 118 984 5140, fax: +44 118 984 3369,
email: office@pheasant.org.uk, website: www.gn.apc.org/worldpheasant/Default.htm
Projects must deal with the conservation of *Galliformes* (including their habitat). Wherever possible, it is desirable to link projects to IUCN/WPA Action Plan targets, but other activities will be considered. Grants: £150–250.

Wyndham Deedes Travel Scholarship to Israel

The Director, Anglo-Israel Association, 9 Bentinck Street, London W1M 5RP.
Website: www.shamash.org/ejin/brijnet/aia/deedes.htm
To enable British graduates and others with special qualifications or interests, who are normally resident in the UK, to make an intensive study of some aspect of Israel. Grants: up to £2000.

Young Explorers' Trust

Ted Grey, Stretton Cottage, Wellow Road, Ollerton, Newark, Notts NG22 9AX. Tel/fax: +44 1623 861027,
email: ted@yet2.demon.co.uk, website: www.theyet.org
Expeditions with most members aged below 20 years old (not usually undergraduate expeditions) involved in discovery and exploration in remote areas. Aims can include community projects, fieldwork and/or physical adventure (climbing/sailing, etc.). New groups and ones with disadvantaged members are encouraged to apply. Grants: £100–500 or equipment bursaries.

Young Scientists for Rain Forests Award

Conservation Foundation, 1 Kensington Gore, London SW7 2AR. Tel: +44 20 7591 3111,
website: www.conservationfoundation.co.uk
For ethnomedical and ethnobotanical studies in rain forests. Grants: up to £10,000.

For further information on grants for expeditions and fieldwork, see www.rgs.org/grants

ABOUT THE ROYAL GEOGRAPHICAL SOCIETY (WITH THE IBG)

The Royal Geographical Society (with the Institute of British Geographers) is dedicated to the advancement of geographical knowledge and understanding, and its application to help solve the challenges of our world. A learned society and the professional body for geography and geographers, the RGS–IBG is the largest and most active society of this kind in the world.

Founded in 1830, the Society was given a Royal Charter in 1859 for "the advancement of geographical science". It has been engaged ever since in furthering geographical knowledge through research, including exploration, and in publishing and disseminating new information for the benefit of scholarship, education, policy and wider public understanding.

SUPPORTING RESEARCH

The Society was pivotal in establishing geography as a research and teaching discipline in British universities in the nineteenth century, and continues actively to support geography as an academic discipline. It does this by providing grants, convening over 20 specialist Research Groups, organising academic conferences and seminars including a 3-day annual international conference, and publishing three scholarly journals: *Area*, *Transactions* and the *Geographical Journal*. The Society's Postgraduate Forum provides a supportive network for young researchers in geography.

PROMOTING AND ENHANCING EDUCATION

The Society believes that all young people should learn about the world in which they live. The study of geography helps to create socially, culturally and environmentally responsible citizens who are aware of the need for sustainable development. School geography, through the stimulating use of contemporary issues and real world examples, develops a particularly wide range of skills relevant to employment, as well as an appreciation of contemporary issues. Geographical knowledge is essential for the decision-makers of tomorrow.

To support those involved in geography education the Society provides continuing professional development training through its Education Forum, as well as online learning resources for the use of teachers, briefings, career advice and university course guides.

The Society, with the expertise and knowledge of its membership, presents a strong voice for geography in both higher and secondary education. It also works hard to promote the importance of geography to government by monitoring and influencing government legislation.

ADVICE AND TRAINING FOR FIELDWORK AND EXPEDITIONS

Much geographical endeavour has traditionally been field based. The Society supports such studies, as well as scientific expeditions in more remote and challenging environments, in a number of different ways. This includes the provision of advice, information and training through the Expedition Advisory Centre, and the publication of manuals, notably the *Expedition Handbook* and *Expedition Medicine*.

The Society organises its own major scientific research projects in collaboration with researchers around the world. These have been traditionally field based, but the concept has recently been expanded to include desk research and programmes nearer to home within Europe.

GRANTS PROGRAMME

The Society empowers others through its grants programme which funds and facilitates all aspects of geographical endeavour by individuals and teams. The funded projects contribute to furthering geographical knowledge of the world, and to individual and group learning.

Grants are available for research and exploration. These include awards for undergraduate research expeditions, postgraduates and established researchers. More substantial field research and exploration awards include the Gilchrist and Ralph Brown awards. Grants range in value from £350 to £15,000.

Grants are also available for teachers to support new initiatives and, in addition, there are awards for photography and radio broadcasting. An average of £120,000 is currently awarded each year within the grants programme.

EVENTS PROGRAMME

The Society has a thriving programme of events, with over 150 events per year aimed at different audiences. These include:

- entertaining lectures, films and celebrity interviews to enthuse and inform people about the world
- education events to update and inform teachers, sixth form students and advisers on curriculum, careers, and health and safety issues
- professional and academic conferences and seminars to discuss the latest research findings in geographical, social and environmental sciences
- training events for expeditions and fieldwork.

The Monday night lecture programme is delivered by well-known and expert speakers in the Ondaatje Theatre – London's premier lecture venue – at the Society's premises in London. The RGS–IBG also has eight regional branches in the UK outside the south-east, and one overseas branch in Hong Kong.

INFORMATION SERVICES AND RESOURCES

The Society holds over two million maps, books, photographs, journals, reports, manuscripts and artefacts, which together form one of the most important geographical information resources in the world. Many of the items derive from the Society's support for scientific exploration and research from the nineteenth century onwards.

The holdings reflect the development of the study of geography and western knowledge of the world from the mediaeval period to the present day. They provide great insight into the activities and publications of geographers, travellers, explorers and other scientists, and they are a tremendous educational resource. The photographic materials in particular have great relevance to a wide range of ethnic groups in modern multicultural Britain.

One of the Society's most ambitious projects in its recent history has been to widen access to its world-renowned information resources. The "Unlocking the Archives" Project, supported by the Heritage Lottery Fund, will provide public access to substantial areas of the archival materials for the first time, vastly improved storage to safeguard the items for the future, and educational enhancement to reach a wide audience.

MEMBERSHIP – OPEN TO ALL

The Society welcomes all those interested in geography to join as members. The membership contributes to a thriving Society through giving expertise, financial and practical support, and enthusiasm.

The membership encompasses academics and other scholars, those using geography in professional work, teachers and students, business people, and geography enthusiasts including those with a keen interest in learning through travel. They all share a fascination with people, places and environments, and all of them recognise and support what geographical understanding offers individuals and society as a whole.

There are three categories of individual membership:

• Fellow: for those who can demonstrate a sufficient involvement in geography or an allied subject through training, profession, research, publications or other work of a similar nature.
• Ordinary Member: for those who have an interest in geography, such as through travel.
• Young Geographer: for those between the ages of 14 and 24.

Fellows may apply for the additional benefit of the professional status of Chartered Geographer if they are using and keeping up to date their geographical knowledge and/or skills in their employment.

In addition the Society offers membership categories for organisations:

• Education Membership for schools
• Higher Education Membership for individual departments and for exploration societies in universities and in colleges of higher education
• Corporate Membership for business and non-educational public sector bodies.

For further information on the work of the RGS–IBG or for details of membership please contact:

The Information Officer, Royal Geographical Society (with IBG), 1 Kensington Gore, London SW7 2AR. Tel: + 44 20 7591 3000, fax: + 44 20 7591 3001, email: info@rgs.org, website www.rgs.org

ABOUT THE RGS-IBG EXPEDITION ADVISORY CENTRE

The RGS–IBG Expedition Advisory Centre offers a wide variety of resources for anyone planning an expedition or field research overseas.

The Expedition Advisory Centre's primary focus is field research at undergraduate level and schools expeditions, but whatever type of expedition you are planning there will probably be something the Expedition Advisory Centre can do to help. The Centre publishes books on many aspects of expedition planning and arranges seminars and workshops throughout the year.

Each November the Centre organises *Explore*, the annual Expedition Planners' Seminar in London. Whether your objective is research, conservation, education or adventure, this weekend seminar is the place to find the inspiration, contacts and practical advice you need.

The RGS–IBG also houses a unique reference collection of expedition reports. Over 5000 reports contain details of the achievements and research results of expeditions to almost every country of the world.

The catalogue of Reports, and details of over 8000 planned and past expeditions, are held on a database of sporting, scientific and youth expeditions from 1965 to the present day, to enable expedition leaders to share their experience and expertise. You can search the Expeditions Database from the EAC website.

The Expedition Advisory Centre receives core support from Shell International Ltd. Shell has provided sponsorship for over a decade so that the Centre can maintain and improve its services to schools, universities and the scientific and academic communities in general, and so promote interest in, and research into, geographical and environmental concerns worldwide.

RGS–IBG EXPEDITION ADVISORY CENTRE

1 Kensington Gore, London SW7 2AR. Tel: + 44 20 7591 3000, fax: + 44 20 7591 3001, email: eac@rgs.org, website: www.rgs.org/eac

AUTHOR BIOGRAPHIES

Corrin Adshead

Corrin served in the Brigade of Gurkhas in Hong Kong, Nepal and Brunei. He has since worked for several expedition companies and was the Expedition Coordinator for Trekforce Expeditions, London. He has led jungle expeditions to Indonesia, Malaysia and Belize, taking teams into remote areas of rain forest and living and working with indigenous forest communities. He runs jungle-training courses for expedition leaders and volunteers and spends long periods living and working in the African continent.

Tom Ang

Tom is a photographer, writer and traveller. A specialist in travel and digital photography, his main interest is in Central Asia. He won a Thomas Cook Award for Best Illustrated Travel Book for the Marco Polo Expedition. He is Senior Lecturer in Photographic Practice at the University of Westminster and is the author of ten books on photography (*Picture Editing*; *Digital Photography*; *Silver Pixels*; *Tao of Photography*; *Dictionary of Photography and Digital Imaging*; *Digital Photographers' Handbook*; *Advanced Digital Photography*; *Photoshop for Photography*; *KISS Digital Photography*; and *Private Album*).

Michael Asher

Michael Asher served in both the Parachute Regiment and the SAS. During his desert explorations he has covered more than 20,000 miles on foot and with camels. A fluent Arabic speaker, he spent three years living with a traditional Bedouin tribe in the Sudan, and in 1987 he and his wife, Mariantonietta Peru, crossed the Sahara desert from west to east by camel and on foot – a distance of 4500 miles – in 9 months. The author of 15 books, almost all concerned with the desert and its people, he has presented three TV documentaries for Channel 4, including the controversial *The Real Bravo Two Zero*.

Adrian Barnett

Adrian has done biological and conservation fieldwork in the Andes, Amazon and West Africa. A research associate at the California Academy of Sciences, he is also a founding partner of Akodon Ecological Consulting. His fieldwork has been concerned with high-altitude and tropical habitats, small mammals and endangered monkeys. Formerly Senior Research Officer with Friends of the Earth's Rainforest Campaign, he is author of *Expedition Field Techniques:*

Small Mammals (1995) and *Primates* (1995) published by the RGS–IBG Expedition Advisory Centre.

Clive Barrow
Clive has been involved in the planning and leadership of expeditions for the past 18 years. His expedition "career" started in the Army with projects to the Canadian Rockies and Chile, and he went on to become a full-time expedition leader for Raleigh International. For 7 years he was Operations Director for World Challenge Expeditions Ltd where he set up systems to handle up to 100 expeditions to the less developed world each year. He is now an expeditions consultant to World Challenge, and he regularly conducts work on a consultancy basis for Raleigh International.

Juliet Burnett
Juliet was the Director of the RGS–IBG marine field research project, the Shoals of Capricorn Programme, and had worked on Shoals since its beginnings in 1995. A marine biologist by training, Juliet has also worked as a political lobbyist and campaigns officer for ACTIONAID. Juliet has been involved in reef-related marine research expeditions in Belize, Mauritius, the US Virgin Islands and Tonga (South Pacific). She has a particular interest in marine education and in the involvement of local communities, and is currently working to extend her diving experience to include training for divers with disabilities.

Phil Coates
With over 15 years of expedition and adventure travel experience, from the high Himalayas and hot deserts to tropical rain forests and the frozen Arctic, Phil has worked on film and video productions and assignments in temperatures ranging from -40°C to +40°C. In the spring of 1998 Phil successfully walked to the Magnetic North Pole, filming the expedition for Australian television in the process. During his career, Phil has been responsible for introducing new digital technology to programme makers while at Sony Broadcast and was Managing Director of Location Logistics, a specialist media consultancy, offering management and creative services to wilderness-based productions.

Malcolm Coe
Formerly Lecturer in Animal Ecology and now an Emeritus Fellow of St Peter's College, Oxford, Malcolm specialises in tropical ecology, and has spent 45 years working in Africa on equatorial mountains, large mammals, dung beetles and carrion. From 1970 to 1980 he directed studies of the giant tortoises of Aldabra Atoll in the Indian Ocean. He was co-leader of the RGS/National Museum of Kenya: Kora Research Project in Kenya (1983–5) and Leader/Scientific Director of the RGS/Department of Wildlife Mkomazi Ecological Research Programme (1990–97) in northern Tanzania. He is currently studying herbivores and the dispersal of acacia seeds in Africa

Richard Crane
Richard has spent 20 years in overseas development, initially on the fund-raising side with the charity Intermediate Technology. He did his undergraduate years at Durham and his PhD at Reading University. He has been co-organiser of successful fund-raising expeditions to benefit Intermediate Technology: "Running the Himalayas", "Bicycles up Kilimanjaro" and "Journey to the Centre of the Earth".

489

Karen Darke
Karen was a keen runner and mountaineer before becoming paralysed in a rock-climbing accident, and has since pursued alternative ways to access the outdoors – canoeing, skiing and
hand-cycling. She has hand-cycled in various corners of the globe, from Central Asia over the
Himalayas, the length of the Japanese archipelago, New Zealand, Turkey and Iceland. Originally a geologist, she joined Shell, where she now works in a learning and development role,
and has recently completed a part-time secondment to the Royal Geographical Society (with
IBG) to support the development of scientific fieldwork and geographical research opportunities promoting diversity and inclusion.

Paul Deegan
Paul's mountaineering and exploratory expeditions have taken him to the Andes, the Alps, the
Himalaya, the Pamirs, Alaska and East Africa. His articles have appeared in several publications including *Geographical, Global Adventure, High Mountain Sports*, the *Sunday Times* and
the *Daily Telegraph*. Paul is the author of the British Mountaineering Council's award-winning
book, *The Mountain Traveller's Handbook*.

Ian Douglas
Ian was a member of the Oxford University Expedition to Cyrenaica in 1960, and helped to
lead the first Brathay Exploration Group Expedition to Tunisia in 1962. He began research on
hydrology and geomorphology in Australian rain forests in 1963 and since 1966 has had a long
experience of Malaysian rain forests. He is currently Emeritus Professor at Manchester University and coordinates the UK NERC Lowland Permeable Catchment Research Programme
(LOCAR), as well as directing the Manchester hydrology programme at the Danum Valley
Field Studies Centre in Ulu Segama, Sabah, Malaysian Borneo.

Rita Gardner
Rita is Director of the Royal Geographical Society (with IBG). Research interests include
Quaternary environmental change, aeolian geomorphology and soil erosion. She has carried
out most of her research in South India and Kashmir, Oman (as a member of the RGS–IBG
Wahiba Sands Project) and in Nepal, the latter as a Director of the RGS–IBG Institute of
Hydrology Nepal Research Project, 1990–98. Formerly a lecturer at King's College London, and
then Director of the Environmental Science Unit at Queen Mary and Westfield College
London, she has been teaching undergraduates and tutoring field dissertations for over 18
years.

Nigel Gifford
Nigel is the founder of "High & Wild", the adventure travel planning, expedition and logistics
company based in Wells, Somerset. He is a conservationist, mountaineer and skydiver. Author
of *Expeditions and Exploration* and *The* Daily Telegraph *Adventurous Traveller* (with Richard
Madden), Nigel is an adviser to Green Globe 21 on resources and sustainable tourism and
founder of the "Ellie Poo Paper Co.", assisting in the protection in the wild of *Elephas maximus
maximus.*

Andrew Goudie
Andrew is a former Professor and Head of Department at the School of Geography and the
Environment, University of Oxford. He is now the Master of St Cross College, Oxford. He was

deputy leader of the RGS projects in Karakoram and Kora (Kenya), and leader of the 1988 Kimberley Research Project (Australia). He is a geomorphologist with interests in deserts, the tropics, climatic change and the human impact. In 1991 he received the Founder's Medal of the RGS–IBG.

James Greenwood
Inspired by Tschiffeley's solo expedition on horseback – 10,000 miles from Buenos Aires to Washington DC – James decided at 24 years of age to embark on his own "Global Ride" – 20,000 miles around the world – on horseback. Mostly alone, James was the first Briton to circle the world by horse. Ten years later he came home and settled in the Forest of Dean. His book *No Guns, Big Smiles* was published and contains a humorous and insightful account of the first stage of his journey – 12 months riding two criollo horses through Argentina, Bolivia and Peru.

Rupert Grey
Rupert Grey is a partner in Farrer & Co. He is solicitor to and/or on the Board of a number of charities involved with young people, including the Liverpool Institute for Performing Arts and YouthNet. He advises the Royal Geographical Society on copyright issues. He has been on many expeditions, both as a photographer and leader, and his articles and photographs have been widely published.

Miranda Haines
A freelance journalist, Miranda was formerly Editor of *Geographical*, the official magazine of the Royal Geographical Society (with IBG). She started journalistic life in the Parisian newsroom of the *International Herald Tribune*. After 3 years of real training with the Americans in Paris she was prepared for anything and returned to London to freelance for the *Trib* and other publications. For an easier life Miranda took a permanent job as editor of *Traveller* – a travel magazine for independent travellers – which is where she first became a Fellow of the Society.

Clive Jermy
A graduate of the University of London and with postgraduate research at the University of Leicester, Clive joined the Natural History Museum, London, in 1958 as Head of the Fern Section, retiring in 1992 as Head of Curation in the Department of Botany. A Fellow of the RGS since 1957, he soon became involved in Society expedition affairs, and at various times was a member of the Expeditions and Research Committee, and Council, and Honorary Head of the EAC in its first 3 years. He has travelled widely in the tropics, especially Malaysia, Indonesia and Papua New Guinea, and was the coordinator of the scientific programme on the RGS Gunung Mulu Expedition 1977–8. He was a Trustee and Chairman of the Young Explorers' Trust.

Peter Knowles
Peter is an ex-Chairman of the British Canoe Union (BCU) Expeditions Committee. He first kayaked the Grand Canyon in 1973 and was run over by a huge 35-foot motorised raft in one of the rapids. He has had a love–hate relationship with rafts ever since. In the last 30 years he been exploring and running rivers in many different countries and continents and was selected as one of 20 "modern explorers" in an exhibition at the RGS. Recent expeditions include first descents and river exploration in Bhutan and Iran. He is now a publisher specialising in international river guidebooks, including the popular *White Water Nepal*.

Vanessa Lee
Vanessa is a senior solicitor in the general insurance department of Barlow Lyde & Gilbert, specialising in employer and public liability claims. Vanessa has experience of advising on health and safety matters and defending prosecutions against schools arising from incidents whilst on school trips.

Nick Lewis
Nick works for Poles Apart, an environmental and location services company that specialises in extreme environments. A geologist and environmental scientist by training, he is also an experienced mountaineer and field guide. His environmental expertise was developed in the oil industry, where he has conducted numerous environmental assessments and contaminated land investigations. He has climbed all over the world with expeditions to Alaska, the Yukon, Kyrgyzstan, Pakistan and Patagonia. He has guided parties in the Arctic and Antarctic.

Chris Loynes
Chris began expeditioning as a member of undergraduate research trips to Iceland and Norway with Sheffield University. He has since led many British and European youth expeditions for Colchester and Brathay Exploration Groups. Having studied Outdoor Leader Training in Australia, USA and Canada on a Churchill Fellowship, he is currently working as an educational consultant and lectures at St Martin's College on outdoor studies and development training.

Nicholas McWilliam
Nicholas McWilliam is a geography lecturer at Anglia Polytechnic University, Cambridge, and consultant with OryxMapping, developing low-cost geographical information systems for protected areas in Tanzania and researching emergency mapping facilities for humanitarian relief agencies in Lesotho. He previously worked for the British Antarctic Survey's GIS Centre, for the RGS–IBG on their Mkomazi Ecological Research Programme, and for the Expedition Advisory Centre. Other expeditions have taken him to Sumatra, Siberia and the Alps, with other trips to the mountains of Ladakh, Tien Shan, Norway, Altai and various parts of East Africa.

Edward Maltby
Edward is Professor of Environmental and Physical Geography, Head of the Wetland Ecosystem Research Group and Director of the Royal Holloway Institute for Environmental Research (RHIER). His first degree was in geography at the University of Sheffield, graduating in 1971, and he gained his PhD from the University of Bristol in 1977. His research has had a strong international flavour extending to the Falklands, the USA, Jamaica, Canada, Australia, New Zealand, Romania, Vietnam, China, Thailand, Kuwait and throughout western Europe. He is former Chair of the IUCN Committee on Ecosystem Management and is currently coordinator of a major EC Wetland Ecosystem Research Project called EVALUWET.

John Matthews
John is Professor of Physical Geography in the Department of Geography at the University of Wales, Swansea, Director of the Swansea Radiocarbon Dating Laboratory, and editor of *The Holocene*, an interdisciplinary journal focusing on recent environmental change. He has led 30 research expeditions to the Arctic–Alpine regions of southern Norway, and his main research

interests are in the geoecology of recently deglaciated terrain (glacier forelands) and the effects of Holocene climatic changes on the landscape. In 1988 he received the RGS Ness Award for leadership of glaciological expeditions.

Paul Munton
Originally trained as a zoologist, Paul has since then been involved in wildlife conservation. After postgraduate work he studied the Arabian Tahr in the desert mountains of the Sultanate of Oman, and subsequently worked in Saudi Arabia introducing city-dwelling Saudis to the Bedu of the interior; for WWF International in Switzerland; and for the NCC in the UK. He was Director of the Biological Research Programme of the RGS–IBG's Oman Wahiba Sands Project, and went on to work for Oman's Ministry of Environment on their National Conservation Strategy Protected Area System Plan.

Hallam Murray
Hallam spent 15 years in book publishing. He has cycled from California to Tierra del Fuego and across the Falklands (two journeys of a total of 17,000 miles). Other bicycle expeditions have included India and "en famille" in Mauritius. He is now a freelance lecturer, writer and photographer. Hallam has travelled in over 40 countries. Family expeditions of approximately 3 months each include Ecuador, Chile, Mauritius and Bangladesh.

Suresh Paul
Suresh is currently based at Glenmore Lodge, the National Mountain Centre, where he runs Equal Adventure Developments (EAD). EAD designs and develops equipment for disabled outdoor athletes, as well as managing a range of other research and development projects aimed at removing barriers and promoting access and adventure for all. Suresh has been interested in the outdoors and expeditions since his youth. He graduated from climbing bus shelters and brick walls in north London to participating in a range of youth expedition projects which form the foundation for his present involvement in a wide range of expeditions and research projects.

Roger Payne
Roger is Sports and Development Director of UIAA, the International Mountaineering and Climbing Federation. He is a qualified mountain guide and as well as numerous Alpine seasons he has climbed in Alaska, Peru, India, Nepal, Pakistan, New Zealand, Khazakstan, Kirghizstan and China. As Executive Officer of the British Mountaineering Council he answered many enquiries about expeditions to the greater ranges and was involved in the BMC/Mount Everest Foundation grant screening process for many years.

David Rootes
David has worked in the polar regions for more than 25 years. He has spent three winters on a research station in Antarctica and numerous summer seasons in both the Arctic and Antarctic. He is a partner at Poles Apart, a company specialising in remote area logistics and environmental surveys, and a co-owner of Antarctic Logistics and Expeditions/Adventure Network International, which provides flights to Antarctica and the Arctic.

Tom Sheppard
Tom has accumulated over 105,000 miles of desert/overland experience, much of it away from

previous routes or, for reasons of logistics, solo. In 1975 he led the first-ever coast-to-coast crossing of the Sahara from the Atlantic to the Red Sea for which he gained the RGS's Ness Award. He is the author of *Vehicle-dependent Expedition Guide* and *Off-roader Driving*, published by Desert Winds.

Peter Simmonds
Peter is a geodetic land surveyor by trade, serving with the British Army. His work has taken him all around the globe, carrying out surveys for many tasks. His current post is as the Navigation Instructor at the Royal School of Military Survey, where he instructs service personnel in the arts of land navigation.

Richard Snailham
Richard was a senior lecturer at the Royal Military Academy, Sandhurst for 25 years and has subsequently been author and feature writer at Raleigh International. He lectures widely, and has written 6 travel books and numerous articles. Expeditions have so far taken him to the Middle East, Africa, Latin America and Asia. He was Honorary Foreign Secretary of the Royal Geographical Society and is Honorary Vice-President of the Scientific Exploration Society.

Neil Walker
Neil Walker has been a journalist for 30 years. He is the managing director of Revolution Recordings, a production company with 10 years' experience of making independent feature programmes for the BBC and others. Neil is not able to give individual advice on how to record, but he does run tailor-made workshops for those who wish to improve their understanding of the medium. Contact by email: Neil@Walkerone.freeserve.co.uk.

David Warrell
Currently Professor of Tropical Medicine and Infectious Diseases and Head, Nuffield Department of Clinical Medicine, University of Oxford, David is Chairman of the RGS–IBG Medical Cell and past President of the Royal Society of Tropical Medicine and Hygiene and International Federation of Tropical Medicine. David has long-term clinical and research interests in tropical medicine. He lived in Nigeria for 5 years and in Thailand for 7 years and has worked in these countries and also in Ethiopia, Kenya, Tanzania, Sierra Leone, South Africa, Burma, Sri Lanka, Bangladesh, Papua New Guinea, Brazil, Ecuador, Colombia and Peru. His main research interests are malaria, rabies, louse-borne relapsing fever, and venomous bites and stings. He has written/edited books, chapters and papers on these subjects and is co-editor of four editions of *The Oxford Textbook of Medicine* and two editions of *Expedition Medicine* and *Essential Malariology*.

Andrew Warren
Andrew is a desert geomorphologist, and recently retired from University College London. His desert speciality is sand dunes. He has been on expeditions to southern Algeria, Tunisia, the Ténéré desert in Niger, the Namib Desert, central Sudan, southern Libya, south-west Egypt and Oman, and has worked in Nebraska, Pakistan and Central Asia. He is joint author of *Desert Geomorphology* (1992) published by University College Press.

Mark Whittingham
Mark has worked as an underwriter for the Zurich Insurance Co. for 5 years and is presently

employed as a Senior Account Executive with Aon Risk Services, where he has worked for 15 years. He is a Chartered Insurance Practitioner and specialises in custom-made insurance programmes for expeditions and companies undertaking adventurous travel.

Justine Williams
Justine is Head of Public Fundraising, ITDG. Challenging environments need the right technology and skills. Through training and the transference of technical knowledge ITDG has helped thousands of women and men in Africa, South Asia and Latin America: food security, renewable energy, smoke reduction – small is still beautiful!

Dick Willis
Dick has been a member or leader of caving expeditions to France, Spain, Italy, Greece, USSR, Java, New Guinea, Irian Jaya, Borneo, Thailand and China. He was the recipient of the 1990 RGS Ness Award for contributions to speleology and is a Winston Churchill Travelling Fellow. He is a consultant in the internet industry but also works as a freelance travel writer and lecturer.

Nigel Winser
A life scientist by training, Nigel was involved in the organisation of undergraduate expeditions to the Sahara, Ethiopia and the Tana River in Kenya. He joined the RGS in 1976 and helped establish the Expedition Advisory Centre. He is now Deputy Director and head of the Expeditions and Fieldwork Division. He has been responsible for the RGS's international multi-disciplinary research programmes in East Africa, the Middle East, the Karakoram Mountains in Pakistan and forest research in Sarawak and Brunei Darussalam since 1977. The most recent survey, the Shoals of Capricorn Programme, organised jointly with the Royal Society, focused on the undersea Mascarene Plateau in the Indian Ocean.

Shane Winser
Shane heads the RGS–IBG's Expedition Advisory Centre which provides advice, information and training to some 500 plus scientific and adventurous expeditions each year. A zoology graduate with a postgraduate diploma in Information Science, she has helped organise the RGS–IBG's own research projects to the tropical forests of Sarawak and Brunei, the mountains of the Karakoram, and the drylands of western Australia, Kenya and Oman.

ACKNOWLEDGEMENTS

This edition of the *Expedition Handbook,* like the earlier editions, is entirely dependent on the willingness of the authors to share their expertise. Many are members of the RGS–IBG who also help us run our seminars and training workshops; without them the Expedition Advisory Centre (EAC) would not function. We are therefore deeply indebted to all the contributors, both authors and photographers, for giving of their time freely and willingly to help others.

The continued financial support of Shell International Ltd, who have provided sponsorship for the RGS–IBG EAC for over 15 years, has enabled us to maintain and improve our services to schools, universities, and the research and academic communities. This handbook is an important contribution to these services, and towards promoting interest in and research into geographical and environmental concerns worldwide.

Without the support of the RGS–IBG staff, especially Nigel Winser, RGS–IBG Deputy Director and Head of the Expeditions and Fieldwork Division, and Nicholas McWilliam, the co-editor of the first edition in 1993–94, this second edition of the *Expedition Handbook* would never have come to publication.

Finally, enormous thanks are due to my tireless colleagues in the RGS–IBG Expedition Advisory Centre who have all played their part: Georgina Hebblethwaite, Tim Labrum, Anna McCormack, Louise Every, Tom Martin and Tom Bourne.

S.W.

INDEX

G
Gateway towns 182–3
Geographical magazine 401, 405
Geography Discipline Network 30
Geomorphology studies 291–5, 302–303,
    309–10
GPS (Global positioning system) 62–3,
    76, 200, 218, 228, 366
Grants vi, 6, 87–9, 190, 187, 343, 417,
    470–82, 484
Group development 13–14
    See also Teamwork
Guides and porters 188, 215, 360–61

H
Hammocks 211
*Health and Safety of Pupils on
    Educational Visits (HASPEV)* 125–27,
    134
Health *see* Medical matters
Health and safety *see* Safety
    management
Heat illnesses 168
Hepatitis 164–65
Home agent 115
Horse, travel by 369–72
Host country liaison 32–7, 81, 214–15,
    221, 450, 456
HIV/AIDS 108, 164
Hydrology studies 279–81
Hygiene 227, 365
Hypothermia 168, 225–6

I
Immunisations *see* Vaccinations
Inclusive expeditions 18–31, 238,
    469
Indigenous people 214–15, 221, 300
Insurance 111, 137–48, 150, 354, 452
Intellectual property rights 34, 214
Interview technique 97–8, 411

Invertebrate studies 269, 271–2, 289–90,
    365

J
Japanese encephalitis
Jet lag 167–8
Jungle boots 207
Jungle expeditions *see* Tropical forest
    expeditions

K
Kayak expeditions *see* Canoeing
    expeditions
Kidnap 144–5

L
Land use studies 301
Language skills 34
Leader's logbook 16
Leadership 8–17, 112, 131–2
Leeches 208
Lecturing, 91, 406–408
Legal matters 27–8, 117–21, 124–36, 111,
    137–9, 141–2, 399–400
    For school visits 122–36
Liability *see* Legal matters
Local Education Authority Outdoor
    Education Advisers 127–9
Local people *see* Indigenous peoples

M
Machetes 212
Magazines, writing for 400,
    401–404
Malaria 156–9, 208
Maps 53–5, 76, 198–200, 339–40, 444
Marine expeditions *see* Underwater
    expeditions
Marketing and the media 84–6, 95–8,
    104
Media 92, 95–8, 455